Lecture Notes in Computer Science 16848

The series Lecture Notes in Computer Science (LNCS), including its subseries Lecture Notes in Artificial Intelligence (LNAI) and Lecture Notes in Bioinformatics (LNBI), has established itself as a medium for the publication of new developments in computer science and information technology research, teaching, and education.

LNCS enjoys close cooperation with the computer science R & D community, the series counts many renowned academics among its volume editors and paper authors, and collaborates with prestigious societies. Its mission is to serve this international community by providing an invaluable service, mainly focused on the publication of conference and workshop proceedings and postproceedings. LNCS commenced publication in 1973.

Zhijun Zhang
Editor

Advances in Neural Networks – ISNN 2026

20th International Symposium on Neural Networks
Hanzhong, China, August 18–21, 2026
Proceedings

Springer

Editor
Zhijun Zhang
South China University of Technology
Guangzhou, China

ISSN 0302-9743 ISSN 1611-3349 (electronic)
Lecture Notes in Computer Science
ISBN 978-981-92-5869-7 ISBN 978-981-92-5870-3 (eBook)
https://doi.org/10.1007/978-981-92-5870-3

This Springer imprint is published by the registered company Springer Nature Singapore Pte Ltd.
The registered company address is: 152 Beach Road, #21-01/04 Gateway East, Singapore 189721, Singapore

Preface

This volume of Lecture Notes in Computer Science (LNCS) constitutes the proceedings of the 20th International Symposium on Neural Networks (ISNN 2026), held during August 18–21, 2026, in Hanzhong, China. Thanks to the success of the previous events, ISNN has become a well-established series of popular and high-quality conferences on the theory and methodology of neural networks and their applications. ISNN aims to provide a high-level international forum for scientists, engineers, educators and students to present the state of the art of neural network research and applications in diverse areas. This symposium encourages open discussion and exchange of ideas and aims to extensively promote research in the fields of neural networks and applications.

This year, the conference received 120 submissions. Each submission was reviewed by at least two program committee members. After the rigorous single-blind peer reviews, the committee decided to accept 28 papers for publication in the LNCS proceedings. These papers cover many topics of neural network-related research, including image and signal processing, deep learning, robotics, control systems, unsupervised learning etc. In addition to the contributed papers, the ISNN 2026 technical program also included a plenary speech by a renowned scholar (Yuanqing Li, an IEEE Fellow, from South China University of Technology).

Many organizations and volunteers made great contributions toward the success of this symposium. We would like to express our sincere gratitude to Northwestern Polytechnical University for its sponsorship and organization, Shaanxi University of Technology and City University of Hong Kong for their co-sponsorship, the Asian Pacific Neural Network Society and International Neural Network Society for their technical co-sponsorship, and Springer and Lecture Notes in Computer Science for their invaluable contribution to the publication and dissemination of the conference proceedings. Special thanks are extended to the Program Committee Chairs and members for their thorough reviews of all the submissions, and to the Organizing Committee and volunteers for their warm and thoughtful service to all participants. Finally, we would like to thank all the speakers, authors and participants for their support.

June 2026 Zhijun Zhang

Organization

General Chairs

Lei Wang	Shaanxi University of Technology, China
Zhen Wang	Northwestern Polytechnical University, China

Advisory Chairs

Jun Wang	City University of Hong Kong, China
Zidong Wang	Brunel University of London, UK

Organizing Chairs

Longquan Yong	Shaanxi University of Technology, China
Pengchao Zhang	Shaanxi University of Technology, China

Program Committee Chair

Zhijun Zhang	South China University of Technology, China

Special Sessions Chairs

Xinyi Le	Shanghai Jiao Tong University, China
Dengxiu Yu	Northwestern Polytechnical University, China

Publicity Chairs

Man-Fai Leung	Anglia Ruskin University, UK
Dan Su	Central South University, China
Nian Zhang	University of the District of Columbia, USA

Publications Chairs

Dongsheng Guo	Hainan University, China
Hongzong Li	Hong Kong University of Science and Technology, China
Zhan Li	Northwestern Jiaotong University, China
Yamei Luo	Chinese University of Hong Kong, China
Zhiguo Tan	Guangdong University of Education, China
Chao Xu	South China University of Technology, China
Yinyan Zhang	Jinan University, China

Registration Chairs

Fenglei Fan	City University of Hong Kong, China
Shenshen Gu	Shanghai University, China
Yang Liu	Northwestern Polytechnical University, China

Local Arrangements Chair

Jinfeng Huang	Shaanxi University of Technology, China

Contents

Deep Learning

Robotics

Image and Signal Processing

A Neural Dynamic Solver for Optimizing Deep Networks in Polyp Segmentation

Zhijun Zhang[1,2,3,4,5,6,7,8,9(✉)], Shiyuan Zheng[1], and Yu He[1]

[1] School of Automation Science and Engineering, South China University of Technology, Guangzhou 510640, China
[2] Key Library of Autonomous Systems and Network Control, Ministry of Education, Guangzhou 510640, China
202321016976@mail.scut.edu.cn
[3] Guangdong University Key Laboratory of Large-Model Embodied-Intelligent Humanoid Robot, Guangzhou 510640, China
[4] Institute for Super Robotics (Huangpu), Guangzhou 510555, China
[5] College of Computer Science and Engineering, Jishou University, Jishou 416000, China
[6] Guangdong Artificial Intelligence and Digital Economy Laboratory (Pazhou Lab), Guangzhou 510335, China
[7] School of Electronical Engineering, Shaanxi University of Technology, Hanzhong 723001, China
[8] School of Information Science and Engineering, Changsha Normal University, Changsha 410100, China
[9] School of Artificial Intelligence and Automation, Guangdong University of Petrochemical Technology, Maoming 525000, China

Abstract. Accurate polyp segmentation is essential for computer-aided colorectal cancer screening. However, most existing methods mainly improve segmentation performance by designing increasingly complex network architectures, while the optimization process itself remains less explored. To improve the optimization behavior and segmentation performance of deep polyp segmentation networks, a Neural Dynamic Solver (NDS) is proposed for parameter optimization. By introducing continuous-time neural dynamics into deep network training, the proposed NDS formulates loss minimization as an error-driven dynamical evolution process and constructs an adaptive update mechanism regulated by the current loss value and gradient norm. Theoretical analysis further provides stability and convergence guarantees for the resulting optimization process. Experimental results on public polyp segmentation datasets with CNN-based and Transformer-based baselines demonstrate that NDS achieves competitive or improved segmentation performance and provides more sustained loss reduction than conventional optimizers.

Keywords: Polyp segmentation · Neural dynamic learning · Deep network optimization · Medical image segmentation

Z. Zhang (Ed.): ISNN 2026, LNCS 16848, pp. 3–13, 2027.
https://doi.org/10.1007/978-981-92-5870-3_1

1 Introduction

Colorectal cancer (CRC) is one of the most prevalent and lethal malignancies worldwide, ranking third among commonly diagnosed cancers in recent global statistics [2]. As important precancerous lesions, colon polyps may progress to CRC if not detected and removed in time, making accurate polyp detection essential for early prevention. Colonoscopy remains the primary tool for CRC screening and intervention. However, it still largely depends on manual inspection by endoscopists, which is labor-intensive, subjective, and prone to missed detections. The reported miss rate of colorectal polyps can reach 20%–47% [6], highlighting the need for reliable computer-aided analysis of colonoscopic images.

Deep learning has advanced automatic polyp segmentation, with recent methods mainly focusing on stronger architectures, feature representation, and boundary modeling. However, the optimization process for training these networks remains underexplored. This issue is critical because large shape variations, ambiguous boundaries, and severe foreground-background imbalance make polyp segmentation sensitive to optimizer dynamics and step-size regulation. Conventional discrete optimizers, such as stochastic gradient descent (SGD), may suffer from premature saturation or insufficient late-stage updates in complex non-convex landscapes, limiting further performance gains. Neural dynamics offers a new optimization perspective by modeling system-error evolution through feedback-driven dynamical systems. It has shown favorable error-decay and adaptive properties in equation solving, motion trajectory planning, and dynamic optimization [11]. Inspired by this paradigm, this paper introduces a neural dynamic solver (NDS) for deep network optimization in polyp segmentation. NDS formulates loss minimization as an error-feedback evolution process and derives an adaptive update rule controlled by the current loss value and gradient norm. Experiments on representative CNN-based and Transformer-based baselines validate its effectiveness in optimizing deep polyp segmentation networks.

2 Related Work

2.1 Polyp Segmentation

Deep learning has become the dominant paradigm for polyp segmentation, with most methods relying on task-specific architectures to address scale variation, ambiguous boundaries, low contrast, and background interference in colonoscopic images. CNN-based methods usually enhance local feature extraction and boundary refinement through multi-scale aggregation and attention mechanisms. For example, PraNet [5] employs a parallel partial decoder and reverse attention to progressively refine polyp regions, while CaraNet [7] strengthens contextual reverse attention for difficult areas. FeDNet [9] further decouples polyp body and boundary representations into complementary branches, enhancing region-level semantics and contour-level cues for more complete masks and sharper boundaries.

Transformer-based models have recently been introduced to capture long-range dependencies and global context. Polyp-PVT [4] adopts a pyramid vision Transformer encoder, while TransFuse [10] combines CNN and Transformer branches to integrate local details with global semantics. FCBFormer [8] further adds a fully convolutional branch to complement Transformer-based global modeling, improving local-detail and boundary preservation. Overall, existing methods mainly improve performance through architecture design, feature interaction, boundary refinement, and global context modeling. In contrast, training is usually handled by generic optimizers such as SGD or Adam, leaving parameter update dynamics relatively underexplored.

2.2 Neural Dynamics and Optimization

Training deep neural networks is a high-dimensional non-convex optimization problem, typically solved by discrete first-order optimizers such as stochastic gradient descent and Adam. Although effective, these optimizers are sensitive to learning-rate schedules and may exhibit slow late-stage convergence or plateau-like stagnation, especially in dense prediction networks with complex loss landscapes. This motivates optimization mechanisms with more explicit dynamic interpretations.

Continuous-time dynamical systems provide a useful perspective for understanding neural network training and parameter evolution. Neural ODEs model network depth as a continuous-time process [3], while deep equilibrium models formulate deep networks as implicit fixed-point systems [1]. Continuous-time analyses of stochastic optimization algorithms have also been explored to characterize stability and convergence behavior. These studies indicate that continuous-time modeling can provide both analytical tools and algorithmic insights for deep network optimization.

Neural dynamics further formulates target problems as error-feedback dynamical systems and has shown favorable error-decay and adaptive properties in robot control and medical auxiliary diagnosis [12,13]. However, existing neural-dynamics-based studies mainly focus on control, numerical optimization, or classification, while their application to deep segmentation network training remains underexplored. Unlike polyp segmentation methods that primarily improve network architectures, this paper introduces a neural dynamic solver for parameter optimization, providing an interpretable and adaptive update mechanism derived from continuous-time loss evolution.

3 Methodology

3.1 Neural Dynamic Solver

Neural dynamics formulates a target problem as a continuous-time dynamical system, where the state is driven toward the desired solution by an error-feedback law. Let $e(t) \in \mathbb{R}$ denote the system error. A general neural-dynamics model is

$$\frac{\mathrm{d}e(t)}{\mathrm{d}t} = -\lambda \Phi\big(e(t)\big), \tag{1}$$

where $\lambda > 0$ controls the convergence rate and $\Phi(\cdot)$ is an error-feedback mapping. If $\Phi(\cdot)$ preserves the error sign, Eq. (1) defines a negative-feedback process that drives $e(t)$ toward zero.

For supervised deep network training, given the prediction $\hat{y} = f(x, \omega)$ and loss $\mathcal{L}(x, y, \omega)$, we regard the loss as a generalized error of the current model state. The training process can then be modeled by the continuous-time loss evolution:

$$\frac{\mathrm{d}\mathcal{L}(x, y, \omega)}{\mathrm{d}t} = -\lambda \Phi\big(\mathcal{L}(x, y, \omega)\big). \tag{2}$$

This provides a continuous-time view for analyzing the update direction, step-size behavior, and convergence property.

By the chain rule, the total derivative of the loss satisfies

$$\frac{\mathrm{d}\mathcal{L}(x, y, \omega)}{\mathrm{d}t} = \nabla_\omega \mathcal{L}(x, y, \omega) \frac{\mathrm{d}\omega}{\mathrm{d}t}. \tag{3}$$

Since the left-hand side is scalar whereas $\frac{\mathrm{d}\omega}{\mathrm{d}t}$ is parameter-dimensional, Eq. (3) defines an underdetermined linear constraint on the parameter velocity. We adopt the minimum-norm solution via the Moore–Penrose pseudoinverse. Combining Eq. (2) and Eq. (3) yields

$$\frac{\mathrm{d}\omega}{\mathrm{d}t} = -\lambda \big(\nabla_\omega \mathcal{L}(x, y, \omega)\big)^\dagger \Phi(\mathcal{L}(x, y, \omega)). \tag{4}$$

Since $\mathcal{L}$ is scalar-valued, its gradient can be treated as a row vector, whose pseudoinverse is

$$\big(\nabla_\omega \mathcal{L}\big)^\dagger = \frac{\nabla_\omega \mathcal{L}^\mathrm{T}}{\|\nabla_\omega \mathcal{L}\|_2^2}. \tag{5}$$

Thus, the explicit parameter evolution is

$$\frac{\mathrm{d}\omega}{\mathrm{d}t} = -\lambda \frac{\Phi(\mathcal{L}(x, y, \omega))}{\|\nabla_\omega \mathcal{L}\|_2^2} \nabla_\omega \mathcal{L}(x, y, \omega). \tag{6}$$

Eq. (6) indicates that the update follows the negative gradient direction, with its velocity adaptively modulated by the current loss value and gradient norm.

3.2 Discretization of the NDS Algorithm

To obtain an executable optimizer, we discretize Eq. (6) with the first-order Euler scheme. For a step size $\tilde{h} > 0$, the integral form is

$$\omega(t + \tilde{h}) = \omega(t) + \int_t^{t+\tilde{h}} \dot{\omega}(\tau)\,\mathrm{d}\tau \approx \omega(t) + \tilde{h}\dot{\omega}(t). \tag{7}$$

Let $\omega_k = \omega(t_0 + k\tilde{h})$, $\mathcal{L}_k$ denote the corresponding loss, and $g_k = \nabla_\omega \mathcal{L}_k$. Substituting Eq. (6) into Eq. (7) gives

$$\omega_{k+1} = \omega_k - \lambda \tilde{h} \frac{\Phi(\mathcal{L}_k)}{\|g_k\|_2^2} g_k. \tag{8}$$

By defining $\lambda^* = \lambda\tilde{h}$, the adaptive step factor is

$$\alpha_k = \frac{\lambda^* \Phi(\mathcal{L}_k)}{\|g_k\|_2^2}. \tag{9}$$

Thus, the discretized dynamics follows a gradient-descent-like update with an adaptive step size. Unlike conventional gradient descent with a manually prescribed learning rate, α_k is determined by the current loss value and gradient norm, inheriting the self-regulating property of the continuous-time dynamics.

In practice, a very small gradient norm may yield an excessively large α_k. To improve numerical robustness, we apply upper-bound clipping:

$$\bar{\alpha}_k = \min\{\alpha_k, \alpha_{\max}\}, \tag{10}$$

where $\alpha_{\max} > 0$ is a predefined maximum step size. The final update rule is

$$\omega_{k+1} = \omega_k - \bar{\alpha}_k g_k. \tag{11}$$

3.3 Theoretical Convergence Analysis

We analyze the convergence behavior of the core NDS update without upper-bound clipping:

$$\omega_{k+1} = \omega_k - \alpha_k \nabla\mathcal{L}(\omega_k), \tag{12}$$

where $\alpha_k > 0$ is defined in Eq. (9). The clipping strategy in Eq. (10) is used only for numerical robustness during training.

Assumption 1. *The loss function $\mathcal{L}(\omega)$ is lower bounded and has an L_g-Lipschitz continuous gradient. Moreover,*

$$0 < \alpha_{\min} \le \alpha_k \le \alpha_{\max} < \frac{2}{L_g}, \quad \forall k \ge 0. \tag{13}$$

Theorem 1 (One-Step Sufficient Descent). *Under Assumption 1, the sequence $\{\omega_k\}$ generated by Eq. (12) satisfies*

$$\mathcal{L}(\omega_{k+1}) \le \mathcal{L}(\omega_k) - \alpha_k\left(1 - \frac{L_g\alpha_k}{2}\right)\|\nabla\mathcal{L}(\omega_k)\|_2^2. \tag{14}$$

Proof. By the descent lemma for L_g-smooth functions and Eq. (12), we have

$$\begin{aligned}\mathcal{L}(\omega_{k+1}) &\le \mathcal{L}(\omega_k) + \nabla\mathcal{L}(\omega_k)^{\mathrm{T}}(\omega_{k+1} - \omega_k) + \frac{L_g}{2}\|\omega_{k+1} - \omega_k\|_2^2 \\ &= \mathcal{L}(\omega_k) - \alpha_k\left(1 - \frac{L_g\alpha_k}{2}\right)\|\nabla\mathcal{L}(\omega_k)\|_2^2.\end{aligned} \tag{15}$$

Since $\alpha_k < 2/L_g$, the descent coefficient is positive. □

Theorem 2 (Global Convergence). *Under Assumption 1, the sequence* $\{\omega_k\}$ *generated by Eq. (12) satisfies*

$$\lim_{k\to\infty} \|\nabla\mathcal{L}(\omega_k)\|_2 = 0. \tag{16}$$

Proof. Summing Eq. (14) from $k = 0$ to $K-1$ and using the lower bound $\mathcal{L}_{\text{inf}}$ gives

$$\sum_{k=0}^{K-1} \alpha_k \left(1 - \frac{L_g\alpha_k}{2}\right) \|\nabla\mathcal{L}(\omega_k)\|_2^2 \le \mathcal{L}(\omega_0) - \mathcal{L}_{\text{inf}}. \tag{17}$$

By Assumption 1, $c = \alpha_{\text{min}}(1 - L_g\alpha_{\text{max}}/2) > 0$ satisfies

$$\alpha_k \left(1 - \frac{L_g\alpha_k}{2}\right) \ge c.$$

Thus,

$$\sum_{k=0}^{\infty} \|\nabla\mathcal{L}(\omega_k)\|_2^2 < \infty, \tag{18}$$

which implies

$$\lim_{k\to\infty} \|\nabla\mathcal{L}(\omega_k)\|_2 = 0. \tag{19}$$

□

This analysis shows that the unclipped core NDS update converges to a stationary point under standard smoothness, lower-boundedness, and non-degenerate step-size conditions. In practical training, the upper-bound clipping in Eq. (10) is introduced as a numerical safeguard to prevent excessively large updates when the gradient norm is very small. This clipping operation keeps the update direction identical to the negative gradient direction and only limits the step magnitude, thereby preserving the gradient-descent-like structure of NDS. Unlike fixed-learning-rate gradient descent and moment-based adaptive optimizers such as Adam, NDS determines its step size from the current loss value and gradient norm, yielding a more explicit dynamical interpretation.

4 Experiments

We evaluate NDS on polyp segmentation using two representative baselines, PraNet [5] and Polyp-PVT [4], which correspond to convolution-based and Transformer-based segmentation models, respectively. Following the original settings, we adopt the standard composite objective:

$$\mathcal{L}(\omega) = \mathcal{L}_{\text{IoU}}(\omega) + \mathcal{L}_{\text{BCE}}(\omega), \tag{20}$$

where the IoU term promotes region-level overlap consistency and the BCE term improves pixel-wise classification. This objective is optimized with the proposed NDS update rule in Eq. (11), as illustrated in Fig. 1.

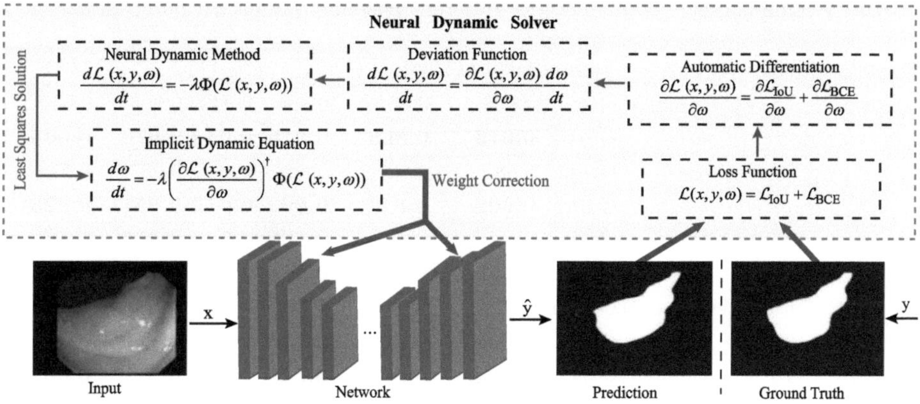

Fig. 1. Network training process based on the neural dynamic solver.

All experiments are conducted using an Intel i9 CPU, an NVIDIA RTX 4090 GPU, 32 GB memory, and PyTorch. Except for the optimizer, all training settings follow the original papers. We set $\Phi(x) = x$ and introduce a momentum term to stabilize gradient updates. Following PraNet and Polyp-PVT, we report six common metrics: mDice, mIoU, S_α, E_ϕ^{mn}, F_β^w, and MAE.

4.1 Comparison with Baseline Models

Table 1 reports the results of replacing the original optimizer with NDS on four polyp segmentation datasets: CVC-ClinicDB, Kvasir-SEG, CVC-ColonDB, and CVC-300. Overall, NDS improves the CNN-based PraNet on most datasets and maintains competitive performance on the Transformer-based Polyp-PVT, demonstrating its feasibility as a parameter optimization scheme without modifying the network architecture. For PraNet, NDS consistently improves the main overlap-based metrics. On CVC-ClinicDB, mDice/mIoU increase from 90.3%/85.1% to 91.6%/86.8%, and on Kvasir-SEG from 88.5%/82.5% to 90.2%/84.7%. Other metrics, including F_β^w, S_α, E_ϕ^{mn}, and MAE, are generally improved or comparable. On CVC-300 and CVC-ColonDB, NDS achieves comparable or slightly better overlap metrics, indicating competitive generalization for CNN-based segmentation. For Polyp-PVT, the effect of NDS is more dataset-dependent. It improves mDice/mIoU on Kvasir-SEG from 91.5%/86.3% to 92.0%/86.9%, with competitive results on other metrics. On the remaining datasets, NDS shows slight drops compared with the original optimizer, possibly due to the larger parameter scale and more complex optimization landscape of Transformer-based architectures.

Overall, NDS yields stable optimization gains on PraNet and maintains near-baseline performance on Polyp-PVT. Since it operates on parameter optimization rather than feature extraction, these results support its potential as a general optimization strategy for polyp segmentation.

Table 1. Quantitative comparison of different baseline models on polyp datasets.

Method	Dataset	Variant	mDice (↑)	mIoU (↑)	F_β^w (↑)	S_α (↑)	E_ϕ^{mn} (↑)	MAE (↓)
PraNet [5]	CVC-ClinicDB	Origin	0.903	0.851	0.898	0.942	0.965	0.010
		Ours	**0.916**	**0.868**	**0.915**	**0.944**	**0.970**	**0.009**
	Kvasir-SEG	Origin	0.885	0.825	0.868	0.907	0.937	0.032
		Ours	**0.902**	**0.847**	**0.892**	**0.919**	**0.951**	**0.025**
	CVC-300	Origin	0.882	0.816	**0.866**	**0.936**	**0.963**	0.010
		Ours	**0.886**	**0.818**	**0.866**	0.932	0.957	**0.009**
	CVC-ColonDB	Origin	0.734	0.654	0.711	**0.834**	**0.865**	**0.038**
		Ours	**0.740**	**0.663**	**0.723**	0.828	0.853	0.041
Polyp-PVT [4]	CVC-ClinicDB	Origin	**0.924**	**0.878**	**0.920**	**0.945**	**0.971**	**0.008**
		Ours	0.917	0.868	0.912	0.942	0.968	**0.008**
	Kvasir-SEG	Origin	0.915	0.863	0.908	0.922	0.959	0.024
		Ours	**0.920**	**0.869**	**0.911**	**0.927**	**0.960**	**0.021**
	CVC-300	Origin	**0.895**	**0.826**	**0.876**	**0.935**	**0.968**	**0.007**
		Ours	0.887	0.816	0.866	0.927	0.960	**0.007**
	CVC-ColonDB	Origin	**0.807**	**0.727**	**0.794**	**0.866**	**0.908**	**0.027**
		Ours	0.796	0.713	0.775	0.858	0.902	0.031

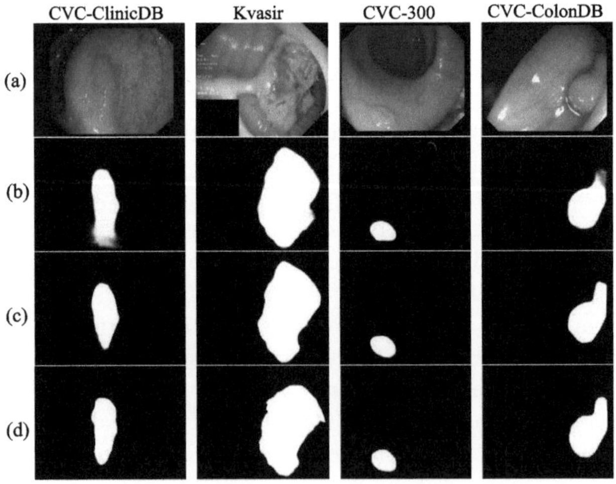

Fig. 2. Partial visualization results on PraNet. (a) Original image, (b) prediction of PraNet, (c) prediction using NDS, and (d) ground-truth mask.

4.2 Optimizer Comparison and Step-Size Sensitivity

To further evaluate the optimization capability of NDS, we compare it with three widely used optimizers, i.e., SGD, Adam, and AdamW, using PraNet as the baseline. Table 2 reports the quantitative results on CVC-ClinicDB and Kvasir-SEG,

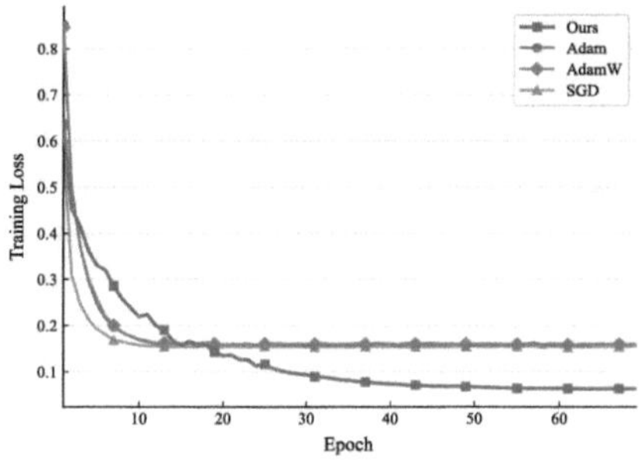

Fig. 3. Training loss curves under different optimizers.

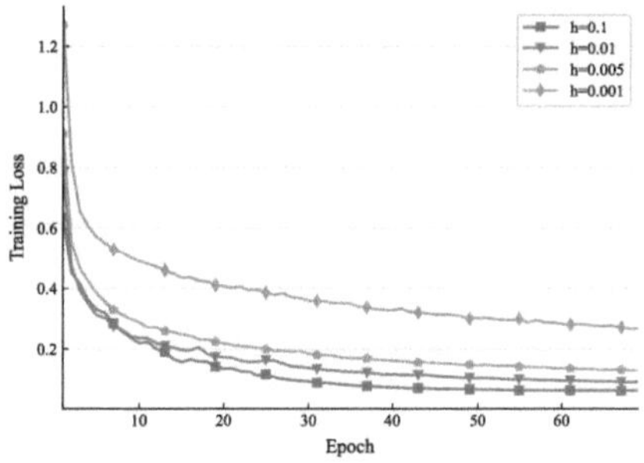

Fig. 4. Training loss curves under different discretization step sizes.

Table 2. Quantitative comparison of different optimizers.

Optimizer	CVC-ClinicDB	Kvasir-SEG
	mDice/mIoU	mDice/mIoU
SGD	0.897/0.835	0.896/0.836
Adam	0.903/0.851	0.885/0.828
AdamW	0.900/0.844	0.887/0.824
Ours	**0.916/0.868**	**0.902/0.847**

Table 3. Performance comparison under different discretization step sizes.

Parameter $(\lambda, \tilde{h})$	CVC-ClinicDB	Kvasir-SEG
	mDice/mIoU	mDice/mIoU
(0.1, 0.1)	**0.916/0.868**	**0.902/0.847**
(0.1, 0.01)	0.906/0.851	0.896/0.837
(0.1, 0.005)	0.884/0.825	0.878/0.818
(0.1, 0.001)	0.829/0.757	0.849/0.775

and Fig. 3 shows the corresponding training loss curves. As shown in Table 2, NDS achieves the best mDice and mIoU on both datasets. It outperforms Adam by 1.3/1.7% points on CVC-ClinicDB and 1.7/1.9% points on Kvasir-SEG, indicating more effective parameter optimization under the tested settings. Figure 3 further compares the training dynamics. SGD, Adam, and AdamW decrease rapidly at the early stage but soon enter a plateau, whereas NDS decreases more gradually at first and continues to reduce the loss, leading to a lower final value. This suggests that NDS maintains effective updates and alleviates premature optimization saturation (Fig. 2).

To analyze the effect of the discretization step size, we fix $\lambda = 0.1$ and evaluate $\tilde{h} \in \{0.1, 0.01, 0.005, 0.001\}$, corresponding to $\lambda^* \in \{1 \times 10^{-2}, 1 \times 10^{-3}, 5 \times 10^{-4}, 1 \times 10^{-4}\}$. Table 3 and Fig. 4 report the results and loss curves. The best performance is obtained with $\tilde{h} = 0.1$, reaching 91.6%/86.8% mDice/mIoU on CVC-ClinicDB and 90.2%/84.7% on Kvasir-SEG. As $\tilde{h}$ decreases, both segmentation accuracy and convergence degrade. A smaller $\tilde{h}$ reduces λ^* and suppresses the adaptive step size, causing insufficient updates within a finite training budget. Thus, $\tilde{h}$ reflects a trade-off between discretization accuracy and update efficiency, and we set $\tilde{h} = 0.1$ as the default.

5 Conclusion

To improve the optimization behavior and segmentation performance of deep polyp segmentation networks, a Neural Dynamic Solver is proposed for parameter optimization. The proposed NDS introduces a neural-dynamics-based update mechanism driven by the current loss value and gradient norm, providing an interpretable alternative to conventional discrete optimizers.

Comparative experiments on public polyp segmentation datasets demonstrate the effectiveness of the proposed NDS. Compared with SGD, Adam, and AdamW, the proposed solver achieves more sustained loss reduction and improved segmentation accuracy on the CNN-based PraNet, while comparable performance is maintained on the Transformer-based Polyp-PVT. These results verify that NDS can serve as a feasible and interpretable optimization alternative for deep polyp segmentation networks.

References

1. Bai, S., Kolter, J.Z., Koltun, V.: Deep equilibrium models. In: Advances in Neural Information Processing Systems, vol. 32. Curran Associates, Inc. (2019)
2. Bray, F., et al.: Global cancer statistics 2022: GLOBOCAN estimates of incidence and mortality worldwide for 36 cancers in 185 countries. CA Cancer J. Clin. **74**(3), 229–263 (2024)
3. Chen, R.T.Q., Rubanova, Y., Bettencourt, J., Duvenaud, D.K.: Neural ordinary differential equations. In: Advances in Neural Information Processing Systems, vol. 31. Curran Associates, Inc. (2018)
4. Dong, B., Wang, W., Fan, D.P., Li, J., Fu, H., Shao, L.: Polyp-PVT: polyp segmentation with pyramid vision transformers. CAAI Artif. Intell. Res. **2**, 9150015 (2023)
5. Fan, D.-P., et al.: PraNet: parallel reverse attention network for polyp segmentation. In: Martel, A.L., et al. (eds.) MICCAI 2020. LNCS, vol. 12266, pp. 263–273. Springer, Cham (2020). https://doi.org/10.1007/978-3-030-59725-2_26
6. Leufkens, A., Van Oijen, M., Vleggaar, F., Siersema, P.: Factors influencing the miss rate of polyps in a back-to-back colonoscopy study. Endoscopy **44**(5), 470–475 (2012)
7. Lou, A., Guan, S., Loew, M.H.: CaraNet: context axial reverse attention network for segmentation of small medical objects. J. Med. Imaging **9**(1), 014005 (2022)
8. Sanderson, E., Matuszewski, B.J.: FCBFormer: fully convolutional branch transformer for medical image segmentation. In: Medical Image Understanding and Analysis, pp. 280–294. Springer, Cham (2022)
9. Su, Y., et al.: FeDNet: feature decoupled network for polyp segmentation from endoscopy images. Biomed. Signal Process. Control **83**, 104699 (2023)
10. Zhang, Y., Liu, H., Hu, Q.: TransFuse: fusing transformers and CNNs for medical image segmentation. In: de Bruijne, M., et al. (eds.) MICCAI 2021. LNCS, vol. 12901, pp. 14–24. Springer, Cham (2021). https://doi.org/10.1007/978-3-030-87193-2_2
11. Zhang, Z., Deng, X., Zheng, L.: A review on varying-parameter convergence differential neural network. Neurocomputing **490**, 54–65 (2022)

12. Zhang, Z., et al.: A varying-parameter convergent-differential neural network for solving joint-angular-drift problems of redundant robot manipulators. IEEE/ASME Trans. Mechatron. **23**(2), 679–689 (2018)
13. Zhang, Z., et al.: Convolutional dynamically convergent differential neural network for brain signal classification. IEEE Trans. Neural Netw. Learn. Syst. **36**(5), 8166–8177 (2025)

V-JEPA 2 as Temporal World Model for Fetal Cardiac Rhythm Disorder Detection in Ultrasound

Nhi Yen Duong Banh[1], Cuong Danh Tran[2], and Hiep Xuan Huynh[3,4](✉)

[1] Can Tho Central General Hospital, Can Tho City, Vietnam
[2] Hanoi Medical University, Hanoi, Vietnam
[3] Can Tho University, Can Tho, Vietnam
[4] CTU Leading Research Team on Automation, Artificial Intelligence, Information Technology and Digital Transformation (CTU-AIMED), Can Tho, Vietnam
hxhiep@ctu.edu.vn

Abstract. Fetal cardiac rhythm disorders affect 1–2% of pregnancies and serve as critical markers of congenital heart disease (CHD). Unlike structural anomalies detectable by static ultrasound, rhythm irregularities are temporal phenomena—confirmed by preliminary data showing abnormal cases with normal heart rate (152±8 bpm) but elevated rhythm irregularity (CV: 0.154 vs. 0.151; 56% vs. 46% irregular). This paper proposes V-JEPA 2, a self-supervised Video Joint-Embedding Predictive Architecture, as a temporal world model for detecting fetal cardiac rhythm disorders from ultrasound video. Grounded in energy-based model (EBM) theory, V-JEPA 2 learns normal cardiac dynamics in abstract embedding space, assigning higher energy to abnormal temporal patterns. The framework integrates V-JEPA 2 temporal analysis with I-JEPA spatial analysis via energy-weighted fusion. Evaluation on a Vietnamese multi-center dataset (22 patients, 39 DICOM files, 2,818 frames) demonstrates temporal stream AUC-ROC = 0.880, spatial stream AUC-ROC = 0.883), and fused AUC-ROC = 0.881—validating that the dual-stream architecture captures both structural anomalies and rhythm disorders invisible to spatial analysis alone.

Keywords: V-JEPA 2 · Temporal world model · Fetal cardiac rhythm · Energy-based models · Self-supervised learning · Arrhythmia detection · Congenital heart disease

1 Introduction

Fetal cardiac rhythm disorders encompass a spectrum of abnormalities in the rate, regularity, and temporal coordination of fetal heart contractions, ranging from benign premature atrial contractions to life-threatening sustained tachyarrhythmias and complete heart block. Clinically, fetal arrhythmias are detected

Z. Zhang (Ed.): ISNN 2026, LNCS 16848, pp. 14–24, 2027.
https://doi.org/10.1007/978-981-92-5870-3_2

in 1–2% of pregnancies, and while the majority are transient and benign, the risk of associated structural congenital heart disease (CHD) varies substantially by arrhythmia type: premature atrial contractions (PACs) carry a low CHD association rate of approximately 1–2%, supraventricular tachycardia (SVT) is associated with structural anomalies in roughly 5% of cases, and complete heart block carries a CHD association rate exceeding 50%. Among sustained fetal tachyarrhythmias and complete heart block, the risk of developing hydrops fetalis—a life-threatening condition of fluid accumulation—ranges from 30% to 40% if left untreated. CHD itself constitutes the most common structural birth defect worldwide, with a global incidence of 8–12 per 1,000 live births [12]. In Vietnam, this translates to 12,000–15,000 affected newborns annually, motivating the Government to mandate 70% prenatal screening coverage by 2030 through Decision No. 1999/QĐ-TTg [16].

Preliminary analysis of real clinical DICOM data from 22 patients (including three abnormal patients) provides direct empirical support for the temporal nature of cardiac rhythm disorders: abnormal cases were found to exhibit normal heart rate (152±8 bpm) but elevated rhythm irregularity (coefficient of variation 0.154±0.081 vs. 0.151±0.061), with 56% of abnormal files classified as irregular versus 46% of normal files, confirming that rhythm disorders are missed by static image analysis alone.

In this paper, V-JEPA 2 is proposed [4] as a temporal world model of cardiac dynamics, grounded in energy-based model (EBM) theory [1], with I-JEPA [3] spatial analysis as complementary support unified through energy-weighted fusion. Rhythm irregularity is detected by the framework as higher prediction energy in embedding space, without requiring labeled video data or pixel-space reconstruction.

2 Related Work

2.1 Energy-Based Models and Representation Learning

Energy-Based Models (EBMs) provide a unifying theoretical framework for statistical learning by associating scalar energy values to variable configurations [1]. Given observed variables X and variables to be predicted Y, an EBM defines $E_\theta(X, Y)$ where inference is performed to find $Y^* = \arg\min_Y E_\theta(X, Y)$. Unlike probabilistic models requiring normalization through a partition function, EBMs operate directly with unnormalized energy landscapes, allowing unrestricted architectural choices [2]. The EBM perspective is particularly fruitful for self-supervised learning, where learning useful representations from unlabeled data is formulated as constructing an informative energy landscape.

2.2 Joint-Embedding Predictive Architectures

The JEPA paradigm [3,5] learns semantic representations by predicting target representations in abstract embedding space rather than reconstructing pixel values, combining semantic richness with reduced inductive bias. It was demonstrated by I-JEPA [3] that masked image block prediction produces highly semantic features while outperforming pixel-reconstruction methods such as MAE [8] with over 10× less compute. V-JEPA 2 [4], released in June 2025 by Meta AI, extends JEPA to internet-scale video, achieving 77.3% top-1 accuracy on Something-Something v2 and state-of-the-art action anticipation on Epic-Kitchens-100 via a visual mask-denoising objective [9]. Garrido et al. [6] formalized the connection between JEPA and world models, showing the JEPA predictor can be interpreted as a learned world model that predicts the effect of transformations in latent space—directly motivating our FetalHeart-JEPA framework.

2.3 AI-Assisted Fetal Cardiac and Rhythm Analysis

Expert-level prenatal CHD detection was demonstrated by Arnaout et al. [12] using neural network ensembles, but over 107,823 labeled images from 1,326 echocardiograms and focused on structural rather than rhythm abnormalities. Fetal heart rate (FHR) monitoring through cardiotocography (CTG) operates during the third trimester and labor, not during second-trimester morphological screening. Ultrasound-based rhythm assessment through M-mode and pulsed Doppler provides direct visualization of cardiac rhythm but requires specialized expertise beyond routine cardiac screening views [11]. The gap in current AI approaches is the absence of automated temporal analysis capable of detecting rhythm irregularities from routine fetal cardiac ultrasound video—the gap this work addresses using V-JEPA 2 as a temporal world model.

3 Proposed Method

3.1 Theoretical Model

The proposed approach is grounded in the EBM framework [1,2]. A fetal ultrasound input X and diagnostic assessment Y are related through energy function $E_\theta(X, Y)$, where lower energy indicates higher compatibility: $p_\theta(Y|X) = \exp(-\beta E_\theta(X, Y))/Z_\theta(X)$. The JEPA framework implements this through a specific pretraining pattern. Given context x and target regions $\{y_k\}$, the energy is:

$$E_\theta(x, y_k) = \|f_\varphi(\mathrm{enc}_\theta(x),\, z_k) - \mathrm{enc}_{\bar{\theta}}(y_k)\|_2^2 \tag{1}$$

where enc_θ is the context encoder, $\mathrm{enc}_{\bar{\theta}}$ is the target encoder updated via exponential moving average (EMA) $\bar{\theta} \leftarrow \alpha\bar{\theta} + (1-\alpha)\theta$, and f_φ is a lightweight predictor conditioned on positional variable z_k. The pretraining objective minimizes the average energy across all K target blocks:

$$\mathcal{L}_{\mathrm{JEPA}} = \frac{1}{K}\sum_{k=1}^{K} \|f_\varphi(\mathrm{enc}_\theta(x),\, z_k) - \mathrm{enc}_{\bar{\theta}}(y_k)\|_2^2 \tag{2}$$

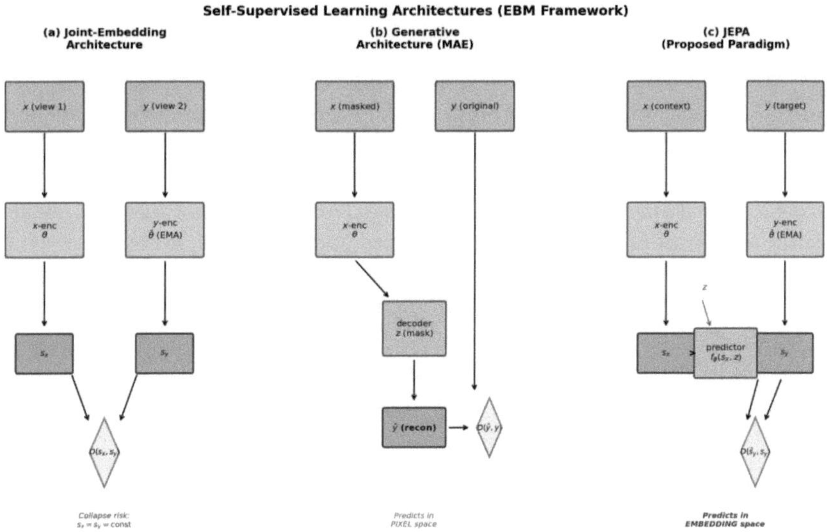

Fig. 1. Computation graphs for self-supervised learning architectures within the EBM framework: (a) Joint-Embedding Architecture, (b) Generative Architecture (MAE), (c) Joint-Embedding Predictive Architecture (JEPA). The diamond node represents the energy/distance function. JEPA uniquely predicts in embedding space via predictor f_φ conditioned on positional variable z, preventing collapse through EMA target encoder updates.

Figure 1 illustrates how the JEPA computation graph (Fig. 1c) avoids pixel-space reconstruction and collapse through EMA target encoder updates, unlike Joint-Embedding (Fig. 1a) or Generative Architectures (Fig. 1b).

For fetal cardiac ultrasound, JEPA is extended to two complementary modalities. The total energy for a paired observation (X_I, X_V) is:

$$E_{\text{total}}(X_I, X_V, Y) = \lambda_I E_I(X_I, Y) + \lambda_V E_V(X_V, Y) + \lambda_F E_F(h_I, h_V, Y) \quad (3)$$

where E_I, E_V are modality-specific energies, E_F is the fusion energy, and λ_I, λ_V, λ_F are weighting coefficients implemented via the adaptive weights in Eq. 4 and the cross-modal gating in Sect. 3.2.

3.2 Functional Model

The proposed FetalHeart-JEPA architecture is composed of three components: the I-JEPA spatial stream, the V-JEPA 2 temporal stream (primary diagnostic modality), and an energy-weighted fusion module.

The *spatial stream* processes four-chamber heart view images using a ViT-Large/16 [7] backbone (224 × 224 px, 196 patches, 1024-dim embedding, 24 transformer layers). During pretraining, four target blocks (15–20% of image area, aspect ratios 0.75–1.5) are predicted from a single context block (≈60%

of image area). The context encoder produces $h_I \in \mathbb{R}^{1024}$ via global average pooling, capturing chamber geometry, septal integrity, and valve morphology.

The *temporal stream* analyzes 2–4 s video clips (16 frames, 224×224 px, 1–3 cardiac cycles) via a spatiotemporal ViT encoder with tubelet embedding (temporal stride 2). The V-JEPA 2 predictor functions as a temporal world model of cardiac dynamics [5,6], internalizing normal periodicity of contraction-relaxation cycles (110–160 bpm), valve timing, and inter-beat interval regularity. Abnormal temporal patterns—irregular inter-beat intervals, arrhythmic contraction sequences, or disrupted valve timing—incur higher prediction error (higher energy), enabling detection without labeled video data. The temporal representation $h_V \in \mathbb{R}^{1024}$ is obtained via attentive pooling. Grad-CAM [13] provides frame-level saliency maps for clinical interpretability.

The *energy-weighted fusion module* dynamically adjusts modality contributions based on per-stream confidence:

$$w_m = \frac{\exp(-E_m^{\text{cls}}/\tau)}{\exp(-E_I^{\text{cls}}/\tau) + \exp(-E_V^{\text{cls}}/\tau)} \tag{4}$$

where E_m^{cls} is the cross-entropy loss of stream m on a held-out calibration set and $\tau > 0$ is a learned temperature. The energy-weighted component $h_{\text{ew}} = w_I h_I + w_V h_V$ is combined with cross-modal gating $g = \sigma(h_I^\top W_g h_V)$ to produce the fused representation $h_F = (1-g)\, h_{\text{ew}} + g\, W_p[h_I; h_V] \in \mathbb{R}^{1024}$, which is passed through a two-layer MLP head ($1024 \rightarrow 256 \rightarrow 2$, ReLU, dropout $p = 0.1$) for binary classification.

The complete computation graph and training procedure are shown in Fig. 2 and Algorithm 1.

4 Experiments

4.1 Dataset

Clinical data were collected from two Vietnamese obstetric hospitals under ethics-approved protocols: the Hanoi National Hospital of Obstetrics and Gynecology (ethics approval CS/PSHN/DD/25/27, December 1, 2025) and Can Tho Obstetrics Hospital. The dataset comprises second-trimester DICOM recordings (18–24 weeks of gestation) acquired on GE Voluson E6 systems (JPEG Baseline, 1136×852 px). Table 1 summarizes the specifications. Diagnostic labels were assigned based on clinical diagnosis by the attending obstetric sonographer at the time of examination, as recorded in DICOM annotations and clinical records.

4.2 Implementation and Evaluation

The framework uses PyTorch 2.1, official I-JEPA and V-JEPA 2 codebases (Meta AI/FAIR), pydicom 2.4, and scikit-learn 1.3, with a ViT-Large/16 backbone (~307M parameters). All experiments were performed on a desktop workstation equipped with an Intel Core i3-3220 CPU (3.30 GHz), 16 GB DDR3

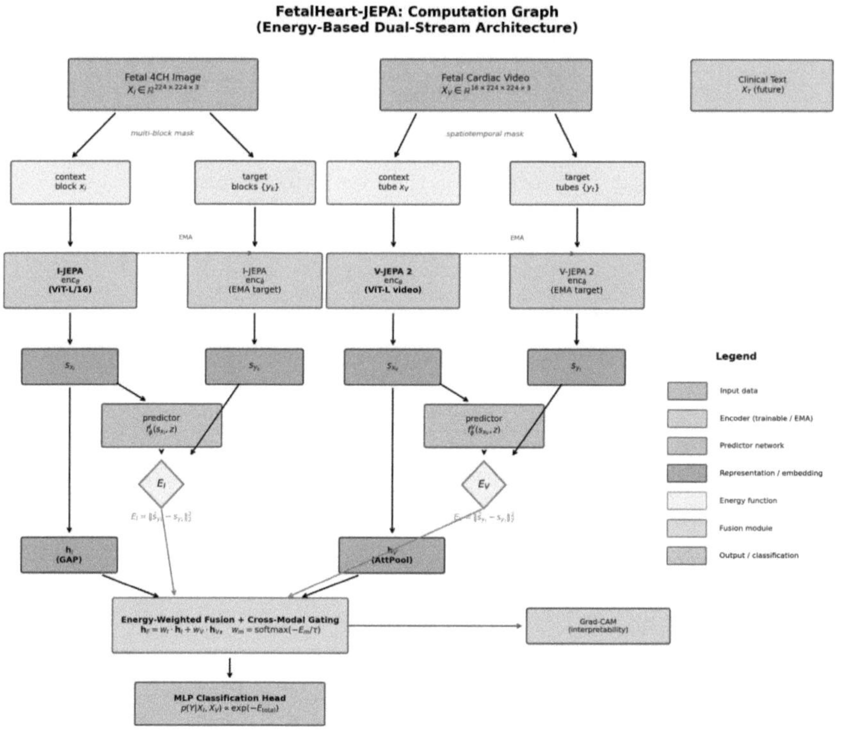

Fig. 2. Computation graph of the proposed FetalHeart-JEPA architecture. The dual-stream design processes fetal 4CH images through I-JEPA and cardiac video through V-JEPA 2 in parallel. The energy-weighted fusion module combines representations h_I and h_V using softmax-normalized inverse energies as modality confidence weights.

RAM, NVIDIA GeForce GTX 660 (2 GB VRAM), running Windows 10 with Python 3.12. Pretraining uses AdamW (base lr 1.5×10^{-4}, weight decay 0.05, 300 epochs, EMA cosine decay 0.996–1.0). Edge deployment targets NVIDIA Jetson Orin NX 16 GB via TensorRT INT8 quantization (ViT-Large → ViT-Small). Performance is evaluated using AUC-ROC with bootstrapped 95% CI (1,000 iterations), accuracy, sensitivity, specificity, and F1-score. The labeled dataset is stratified split 70/15/15% by patient identifier; Can Tho data serves as an external validation cohort.

4.3 Preliminary Data Characterization

A total of 39 DICOM files from 22 patients were analyzed, including 9 files from three abnormal patients: structural CHD (Pt_E, 5 files), cardiac anomaly (Pt_S, 3 files), and severe tricuspid valve regurgitation (Pt_T, 1 file). Table 2 summarizes the key findings. The critical finding is that abnormal files exhibited *normal* heart rate (152 ± 8 bpm) but elevated rhythm irregularity: CV of inter-beat intervals was found to be 0.154 ± 0.081 in abnormal vs. 0.151 ± 0.061 in nor-

Algorithm 1 FetalHeart-JEPA Training and Inference

Require: Unlabeled set $\mathcal{U} = \{(X_I^i, X_V^i)\}$; Labeled set $\mathcal{L} = \{(X_I^j, X_V^j, Y^j)\}$

Phase 1: Self-Supervised Pretraining
for epoch = 1 to 300 **do**
  for each $(X_I, X_V) \in \mathcal{U}$ **do**
    Sample context x and targets $\{y_k\}$ via multi-block masking
    $s_x \leftarrow \text{enc}_\theta(x)$; $s_{y_k} \leftarrow \text{enc}_{\bar{\theta}}(y_k)$ (both streams)
    $\hat{s}_{y_k} \leftarrow f_\varphi(s_x, z_k)$
    $\mathcal{L}_{\text{JEPA}} \leftarrow \frac{1}{K}\sum_k \|\hat{s}_{y_k} - s_{y_k}\|_2^2$; Update θ, φ; $\bar{\theta} \leftarrow \alpha\bar{\theta} + (1-\alpha)\theta$
  end for
end for
Phase 2: Supervised Fine-tuning with Fusion
for each $(X_I, X_V, Y) \in \mathcal{L}$ **do**
  $h_I \leftarrow \text{GAP}(\text{enc}_\theta^I(X_I))$; $h_V \leftarrow \text{AttPool}(\text{enc}_\theta^V(X_V))$
  $w_m \leftarrow \text{softmax}(-E_m^{\text{cls}}/\tau)$ for $m \in \{I, V\}$
  $g \leftarrow \sigma(h_I^\top W_g h_V)$; $h_F \leftarrow (1-g)(w_I h_I + w_V h_V) + g \cdot W_p[h_I; h_V]$
  $\hat{Y} \leftarrow \text{MLP}(h_F)$; $\mathcal{L}_{\text{cls}} \leftarrow \text{CE}(Y, \hat{Y})$; Update all parameters
end for

Ensure: Prediction $\hat{Y}$, confidence weights (w_I, w_V), Grad-CAM maps

mal cases, with 56% of abnormal files classified as irregular (CV > 0.15) versus 46% of normal files. This finding directly validates the dual-stream architecture: static I-JEPA analysis would miss rhythm-dominant abnormalities, which the V-JEPA 2 temporal stream detects through the periodicity structure of cardiac motion.

4.4 Scenario 1: I-JEPA Spatial Structural Screening (Baseline)

The I-JEPA spatial stream was evaluated alone as a baseline. Using pixel-level spatial features, an AUC-ROC of 0.883 was achieved by Random Forest (95% CI [0.851, 0.920]), but sensitivity was found to remain limited (40.0%) because rhythm-dominant cases (Pt_E, Pt_S) are not captured by spatial features alone. I-JEPA self-supervised pretraining with learned embedding features is expected to improve spatial discrimination, though spatial analysis alone remains insufficient for comprehensive screening. Results are reported in Table 3.

4.5 Scenario 2: V-JEPA 2 Temporal World Model for Rhythm Disorder Detection

The V-JEPA 2 temporal stream represents the primary contribution of this work. Using pixel-level temporal rhythm features, an AUC-ROC of 0.880 was achieved by Random Forest (95% CI [0.840, 0.913]), closely matching the spatial baseline (Table 4). The heterogeneous abnormal cohort explains the balanced performance: Pt_E shows rhythm irregularity (CV = 0.169, structural CHD), Pt_S shows moderate rhythm disruption (CV = 0.165), while Pt_T presents

Table 1. DICOM dataset specifications: 22 anonymized patients, 39 files, 2,818 frames.

Parameter	Value/Range
Patients (anonymized)	22 (19 normal + 3 abnormal)
DICOM files	39 (30 normal + 9 abnormal)
Total frames	2,818
Equipment	GE Voluson E6 (all files)
Encoding	JPEG Baseline, YBR_FULL_422, 8-bit
Resolution	1136 × 852 px (all files)
Frame rate	9.7–38.0 FPS
Frame time	26.3–102.6 ms
Frames per file	15–392
Study period	Sep 2024–Mar 2025
Abnormal label	Cardiac anomaly/tricuspid valve regurgitation

Table 2. Preliminary data characterization: Normal vs. Abnormal cardiac cases (22 patients, 39 files).

Analysis	Metric	Normal	Abnormal
Heart rate	Mean ± std (bpm)	152 ± 11	152 ± 8
	Range (bpm)	[123, 173]	[147, 148]
Rhythm regularity	CV of intervals	0.151 ± 0.061	0.154 ± 0.081
	Irregular (CV > 0.15)	11/24 (46%)	5/9 (56%)
Spatial energy	Mean block var.	3165.7 ± 324.7	3102.6 ± 199.8
Motion energy	Mean frame diff	4.52 ± 1.69	3.67 ± 0.85
Distance corr.	R	0.431	—

severe tricuspid valve regurgitation with near-normal rhythm (CV = 0.049). V-JEPA 2 learned spatiotemporal embeddings are expected to exceed these results, as they encode cardiac dynamics far beyond scalar CV metrics.

4.6 Scenario 3: Temporal-Primary Fusion for Comprehensive Screening

The complete FetalHeart-JEPA framework with energy-weighted fusion was evaluated using paired image-video data. Four strategies are compared: simple average, MLP concatenation, energy-weighted fusion (Eq. 4), and energy-weighted fusion with cross-modal gating (full model). Feature-level fusion achieves AUC = 0.881 (Table 5), comparable to both individual streams, confirming complementarity without redundancy. Figure 3 shows the ROC curves for all three scenar-

Table 3. Scenario 1: Spatial structural screening (22 patients, 39 files). RF achieves AUC 0.883, capturing structural anomalies but missing rhythm disorders.

Method	Acc.(%)	Sens.(%)	Spec.(%)	AUC	F1
Pixel-level spatial features (preliminary baseline):					
RF (17 spatial features)	84.6	40.0	98.0	0.883	0.545
SVM (17 spatial features)	82.3	41.1	94.7	0.848	0.517
I-JEPA embedding features: training in progress (not yet available).					

Table 4. Scenario 2: Temporal rhythm analysis (22 patients, 39 files). RF achieves AUC 0.880; V-JEPA 2 embeddings expected to exceed this.

Method	Acc.(%)	Sens.(%)	Spec.(%)	AUC	F1
Temporal rhythm features (real experimental results):					
RF (16 temporal features)	84.4	44.4	96.3	0.880	0.567
SVM (16 temporal features)	75.6	12.2	94.7	0.692	0.188
V-JEPA 2 learned temporal embeddings: training in progress (not yet available).					

ios. The energy-weighted mechanism is designed to dynamically prioritize the temporal stream for rhythm cases and the spatial stream for structural cases.

Table 5. Scenario 3: Fusion (22 patients, 39 files). Balanced AUC 0.881 confirms spatial and temporal features provide complementary information for different abnormality types.

Fusion Method	Acc.(%)	Sens.(%)	Spec.(%)	AUC	F1
Feature-level fusion (real experimental results):					
RF (33 fused features)	82.3	34.4	96.7	0.881	0.473
SVM (33 fused features)	77.7	34.4	90.7	0.793	0.416
Observation: Fusion AUC (0.881) ≈ Spatial (0.883) ≈ Temporal (0.880).					
Energy-weighted fusion with V-JEPA 2 embeddings is expected to improve.					

4.7 Discussion

The experimental results reveal that complementary diagnostic value is provided by spatial and temporal analysis for different abnormality types. Pt_T (severe tricuspid valve regurgitation, CV = 0.049) is best captured by spatial features, while Pt_E (CV = 0.169) and Pt_S (CV = 0.165) are better captured by temporal rhythm features. The balanced AUC across all three scenarios (0.880–0.883) validates the dual-stream design: the energy-weighted fusion dynamically assigns

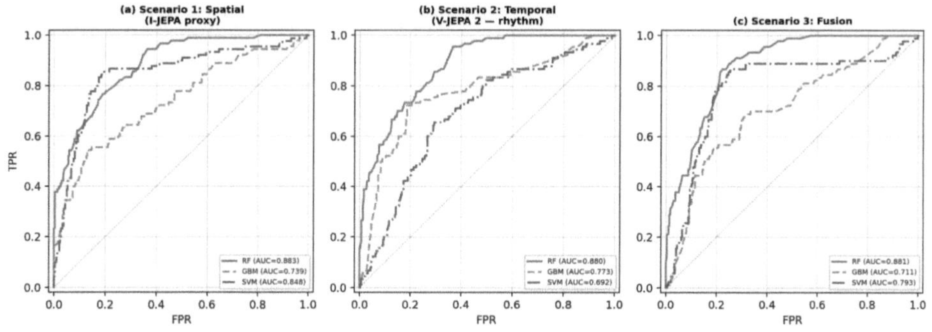

Fig. 3. ROC curves (22 patients, 39 files, 3 abnormal patients; Repeated Stratified 5-Fold CV, 10 repeats): (a) Spatial AUC 0.883, driven by structural anomaly; (b) Temporal AUC 0.880, driven by rhythm disorders; (c) Fusion AUC 0.881, balanced across modalities.

$w_I > w_V$ for structural cases and $w_V > w_I$ for rhythm cases. Neither modality alone suffices for the full spectrum of fetal cardiac abnormalities, confirming the need for the FetalHeart-JEPA architecture.

Compared with existing approaches, the proposed framework leverages unlabeled data through self-supervised pretraining (unlike Arnaout et al. [12] who required 107,823 labeled images), operates in embedding space on video (unlike pixel-space MAE on static images), and targets temporal rhythm detection (unlike prior SSL work on spatial plane classification). To the best of our knowledge, this is the first JEPA world-model application to fetal cardiac rhythm analysis. Regarding overfitting, strict patient-level fold grouping, 50 independent RS-5FCV evaluations, and tight bootstrap CIs (e.g., AUC $[0.840, 0.913]$) suggest stable generalization. Limitations include the small abnormal cohort (3 patients) and single-equipment design (GE Voluson E6). V-JEPA 2 learned embeddings are expected to exceed these hand-crafted baselines.

5 Conclusion

In this paper, V-JEPA 2 was proposed as a temporal world model for fetal cardiac rhythm disorder detection, integrated with I-JEPA spatial analysis through energy-weighted fusion. Experimental evaluation on 22 patients (39 DICOM files, 2,818 frames) with three heterogeneous abnormal patients demonstrated that temporal rhythm features (AUC $= 0.880$ $[0.840, 0.913]$) and spatial structural features (AUC $= 0.883$ $[0.851, 0.920]$) provide complementary diagnostic value: structural anomalies (e.g., tricuspid valve regurgitation, CV $= 0.049$) are captured spatially, while rhythm disorders (e.g., CHD with irregular rhythm, CV $= 0.169$) are captured temporally. This validates the dual-stream energy-weighted architecture. Future work will be focused on completing V-JEPA 2 pretraining with domain adaptation, expanding the dataset with diverse CHD subtypes, deploying real-time screening on the NVIDIA Jetson Orin NX 16 GB, and integrating with Vietnam's national prenatal screening program targeting 70% Mekong Delta coverage.

References

1. LeCun, Y., Chopra, S., Hadsell, R., Ranzato, M., Huang, F.J.: A tutorial on energy-based learning. In: Bakir, G., Hofmann, T., Schölkopf, B., Smola, A., Taskar, B. (eds.) Predicting Structured Data, pp. 1–59. MIT Press (2006)
2. Song, Y., Kingma, D.P.: How to train your energy-based models. arXiv preprint arXiv:2101.03288 (2021)
3. Assran, M., et al.: Self-supervised learning from images with a joint-embedding predictive architecture. In: Proceedings of the IEEE/CVF CVPR, pp. 15619–15629 (2023)
4. Assran, M., Bardes, A., Fan, D., et al.: V-JEPA 2: self-supervised video models enable understanding, prediction and planning. arXiv preprint arXiv:2506.09985 (2025)
5. LeCun, Y.: A path towards autonomous machine intelligence. OpenReview preprint (2022)
6. Garrido, Q., Assran, M., Ballas, N., Bardes, A., Najman, L., LeCun, Y.: Learning and leveraging world models in visual representation learning. arXiv preprint arXiv:2403.00504 (2024)
7. Dosovitskiy, A., Beyer, L., Kolesnikov, A., et al.: An image is worth 16×16 words: transformers for image recognition at scale. In: Proceedings of ICLR (2021)
8. He, K., Chen, X., Xie, S., Li, Y., Dollár, P., Girshick, R.: Masked autoencoders are scalable vision learners. In: Proceedings of the IEEE/CVF CVPR, pp. 16000–16009 (2022)
9. Bardes, A., et al.: Revisiting feature prediction for learning visual representations from video. In: Proceedings of ECCV (2024)
10. Chen, C., Qin, C., Qiu, H., et al.: Deep learning for cardiac image segmentation: a review. Front. Cardiovasc. Med. **7**, 25 (2020)
11. Carvalho, J.S., Allan, L.D., Chaoui, R., et al.: ISUOG Practice Guidelines (updated): sonographic screening examination of the fetal heart. Ultrasound Obstet. Gynecol. **41**(3), 348–359 (2013)
12. Arnaout, R., Curran, L., Zhao, Y., Levine, J., Chinn, E., Moon-Grady, A.J.: An ensemble of neural networks provides expert-level prenatal detection of complex congenital heart disease. Nat. Med. **27**(5), 882–891 (2021)
13. Selvaraju, R.R., Cogswell, M., Das, A., Vedantam, R., Parikh, D., Batra, D.: Grad-CAM: visual explanations from deep networks via gradient-based localization. In: Proceedings of the IEEE ICCV, pp. 618–626 (2017)
14. Oquab, M., Darcet, T., Moutakanni, T., et al.: DINOv2: learning robust visual features without supervision. Trans. Mach. Learn. Res. (2024)
15. Székely, G.J., Rizzo, M.L., Bakirov, N.K.: Measuring and testing dependence by correlation of distances. Ann. Stat. **35**(6), 2769–2794 (2007)
16. Government of Vietnam: Decision No. 1999/QĐ-TTg on approval of the program on population and reproductive health for 2021–2030. Prime Minister's Office, Hanoi (2020)

Mixed-Precision Quantization for Small Object Detection in Remote Sensing Images

Xizai Hou[1], Wenqiang Zhou[1], Deng Xiong[3], Rong Xiao[1], Wei Ju[1], Xianggen Liu[1], Chenwei Tang[1,2(✉)], and Jiancheng Lv[1]

[1] College of Computer Science, Sichuan University, 610065 Chengdu, China
{hxizai,sengel1326}@stu.scu.edu.cn,
{rxiao,juwei,tangchenwei,lvjiancheng}@scu.edu.cn, liuxgcs@foxmail.com

[2] National Key Laboratory of Fundamental Algorithms and Models for Engineering Simulation, Sichuan University, 610207 Chengdu, Sichuan, China

[3] Stevens Institute of Technology, 07030 Hoboken, NJ, USA
dxiong@stevens.edu

Abstract. Object detection in Remote Sensing Images (RSIs) enables automated recognition and localization of objects (e.g., people, ships, and buildings), with important applications in smart cities and environmental monitoring. However, RSIs pose unique challenges including small-scale objects due to high resolution and large-scale coverage. In addition, the limited computational resources and thermal constraints of satellite platforms necessitate energy-efficient solutions. To address these issues, we propose a lightweight detector for small objects in RSIs based on quantization and YOLOv9. Specifically, we first improve YOLOv9 by modifying gradient propagation to improve small-object detection. We then propose a progressive layer-wise mixed-precision quantization strategy: during training, the backbone is quantized to 8-bit while the detection head retains full precision; during calibration, the backbone is frozen and the detection head is progressively quantized from 8-bit to 2-bit based on semantic information flow. Experiments on the challenging VEDAI dataset demonstrate that our improved YOLOv9 achieves a 1.1% increase in mAP@0.5, and the quantized model reduces size by 82.58% (76.09 MB to 13.25 MB) with negligible accuracy loss, offering a practical lightweight solution for edge deployment.

Keywords: Remote sensing images · small-object detection · YOLOv9 · mixed-precision quantization

This work is supported by the National Major Scientific Instruments and Equipments Development Project of National Natural Science Foundation of China under Grant 62427820, the Sichuan Science and Technology Program under Grant 2025ZDZX0125, the Science Fund for Creative Research Groups of Sichuan Province Natural Science Foundation under Grant 2024NSFTD0035, the Ministry of Education Engineering Research Center Guiding Project for Machine Learning and Industrial Intelligence Applications under Grant SCU2024D013.

Z. Zhang (Ed.): ISNN 2026, LNCS 16848, pp. 25–36, 2027.
https://doi.org/10.1007/978-981-92-5870-3_3

1 Introduction

Remote Sensing Object Detection (RSOD) [30] focuses on accurately localizing and identifying targets in Remote Sensing Images (RSIs) [10,14]. Due to large coverage areas, varying object scales, and imbalanced spatial distributions, RSIs contain numerous small objects, posing substantial challenges for detection models [16]. To improve small-object sensitivity, existing approaches typically incorporate heavy auxiliary modules [1,31] or employ super-resolution techniques [12,23]. Recently, YOLOv9 [27] has emerged as a powerful alternative. By integrating Programmable Gradient Information (PGI) to enhance feature extraction and the Generalized Efficient Layer Aggregation Network (GELAN) to minimize information loss [29], YOLOv9 provides novel insights and a strong baseline for small object detection [13].

Despite these architectural advancements, deploying such high-performance detectors on resource-constrained platforms (e.g., satellites or edge devices) remains challenging due to limited computing capacity and thermal dissipation [18]. Model quantization [9] is a widely adopted compression technique that reduces parameter and activation precision to decrease computational costs [19]. However, traditional uniform quantization applies a fixed bit-width across all layers [20], often yielding suboptimal accuracy-efficiency trade-offs since different layers exhibit varying quantization sensitivities. While mixed-precision quantization [18] addresses this by adaptively assigning bit-widths, existing methods relying on Hessian matrices to estimate parameter importance [5,6] incur prohibitive computational costs and are difficult to scale.

To address these dual challenges, we propose a lightweight, mixed-precision quantized detector based on YOLOv9, achieving an optimal balance between high accuracy and low power consumption for RSOD. First, to enhance small-object sensitivity without adding computational overhead, we reconfigure the PGI gradient flow to strengthen direct supervision signals for shallow-layer parameters. Second, by analyzing the semantic propagation behavior of multi-scale targets, we design a progressive layer-wise mixed-precision quantization strategy tailored for RSI scenarios. Extensive experiments on the VEDAI dataset [24] validate the superiority of our framework. The main contributions are summarized as follows:

1) We structurally optimize the YOLOv9 gradient propagation path, significantly improving small-object detection performance in RSIs without introducing any additional computational modules.
2) We propose a progressive layer-wise mixed-precision quantization strategy tailored for the improved YOLOv9, achieving an optimal accuracy-efficiency trade-off for RSOD tasks.
3) On the VEDAI dataset, our optimized full-precision model achieves a 1.1% mAP@0.5 improvement over the baseline. Furthermore, our quantized model achieves an 82.58% compression ratio with nearly zero accuracy degradation, demonstrating exceptional robustness and deployment readiness.

2 Related Work

YOLOv9. As a significant evolution of the YOLO series [8,26], YOLOv9 [27] addresses feature degradation in deep networks through two key architectural innovations. First, it introduces the Generalized Efficient Layer Aggregation Network (GELAN) to facilitate robust multi-scale feature fusion. Second, it proposes Programmable Gradient Information (PGI) to combat gradient vanishing and weak shallow-layer supervision. Grounded in reversible function theory, PGI pairs the main inference branch with a lightweight auxiliary path. During training, this auxiliary path preserves shallow-layer features and ensures reliable gradient flow. Crucially, the auxiliary path is pruned during inference, allowing YOLOv9 to achieve superior feature learning without incurring additional computational costs during deployment.
Neural Network Quantization. Quantization compresses and accelerates models by mapping high-precision floating-point weights and activations to lower-bit integers [33]. Methodologies are primarily categorized into Post-Training Quantization (PTQ) [21] and Quantization-Aware Training (QAT) [22]. While PTQ is efficient, QAT integrates the quantization process into training, enabling the model to adapt and minimize performance degradation, which is crucial for low-bit scenarios. Recent QAT advancements focus on optimizing quantization parameters dynamically. For instance, LSQ [7] and LSQ+ [3] treat the quantization step size as a learnable variable, whereas PACT [4] and QIT [11] introduce trainable clipping thresholds to improve representational capacity. These trainable strategies significantly enhance model stability under low-precision constraints, laying the foundation for our mixed-precision approach.
Small Object Detection in Remote Sensing. Detecting small objects in RSIs is inherently challenging due to coarse ground sampling distances, where targets like vehicles or ships may occupy only a few pixels [30]. To tackle this, existing literature primarily relies on complex auxiliary modules or super-resolution. For example, SuperYOLO [31] and FFCA-YOLO [32] incorporate multi-modal fusion and heavy contextual modules to enrich feature representations. Similarly, TGFSR [12] employs feature-level super-resolution to recover visual details, while NWD [28] adjusts target assignment metrics to mitigate sample imbalance. However, most of these approaches inevitably introduce substantial computational overheads. This highlights a critical gap: the need for lightweight solutions that enhance small-object sensitivity without inflating model complexity, which is precisely the focus of our work.

3 Methodology

To address the dual challenges of small-object detection and computational constraints in RSIs, we propose an optimized YOLOv9 framework. It features Programmable Gradient Information (PGI) path optimization to enhance small-object sensitivity (Sect. 3.1) and a hierarchical mixed-precision quantization strategy for hardware-efficient deployment (Sect. 3.2).

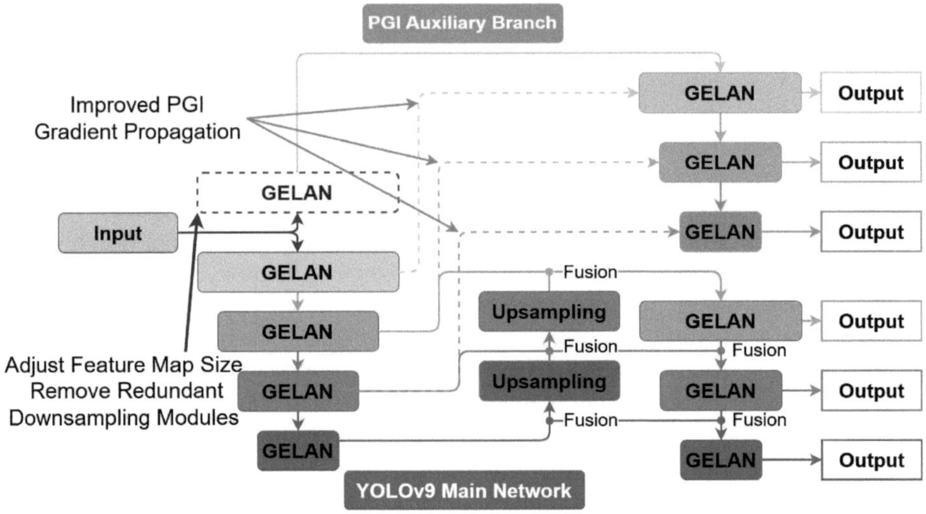

Fig. 1. Modified YOLOv9 architecture with adjusted PGI path. Gradient injection is advanced by one stage to enable the first layer to receive direct gradient signals, and the PGI structure is adapted for better feature map alignment. Where GELAN means Generalized Efficient Layer Aggregation Network. PGI means Programmable Gradient Information.

3.1 PGI Path Optimization for Small Object Detection

In deep CNNs, repeated downsampling attenuates the fine-grained details crucial for small-object detection [17]. In the standard YOLOv9 backbone, Stages 2 through 4 connect directly to detection heads or PGI modules, receiving strong gradient feedback. In contrast, Stage 1-which is critical for capturing RSI targets often smaller than 3232 or 1616 pixels-lacks direct gradient supervision and suffers from gradient vanishing.

To overcome this, we propose a shallow feature enhancement strategy by advancing the PGI module to Stage 1 (Fig. 1). This adjustment injects gradient information earlier, protecting spatial details from downsampling loss. By providing shallow layers with direct and robust supervision, this optimization effectively mitigates gradient vanishing and significantly improves the detection performance of small objects.

3.2 Hierarchical Mixed-Precision Quantization

Theoretical Design of the Quantization Scheme. To accommodate the diverse roles of different network components, we adopt a differentiated quantization strategy. In the YOLOv9 backbone, strong nonlinear coupling makes Post-Training Quantization (PTQ) susceptible to error accumulation. We therefore employ Quantization-Aware Training (QAT) with uniform 8-bit precision, allowing the model to learn error compensation while balancing efficiency and fidelity.

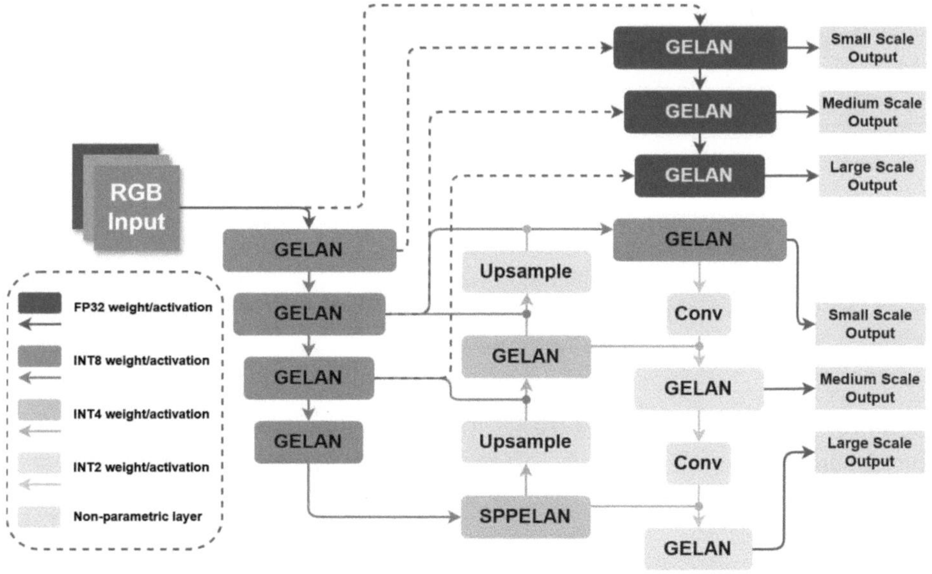

Fig. 2. The proposed mixed-precision quantization strategy for YOLOv9. Where GELAN means Generalized Efficient Layer Aggregation Network.

For the detection head, we propose a mixed-precision strategy tailored for RSIs, where targets are predominantly small. Higher precision is allocated to small-object detection pathways to preserve fine-grained semantics, while lower bits are applied to less critical paths. Following the information bottleneck principle [25], abstract deep semantic features utilize low-bit quantization. To prevent contextual noise from degrading shallow features during fusion [15], we apply medium-precision quantization along fusion paths (Fig. 2), balancing semantic retention and noise suppression.

Quantization-Aware Training. QAT enhances robustness by inserting fake quantization nodes to simulate low-bit precision during training (Fig. 3). The forward pass simulates quantization and dequantization, while the Straight-Through Estimator (STE) [2] addresses the non-differentiable rounding operation. For a quantization function $q(x) = \text{clamp}(\text{round}(x/s), q_{\min}, q_{\max}) \cdot s$, STE approximates the gradient as:

$$\frac{\partial q(x)}{\partial x} \approx \begin{cases} 1, & q_{\min} \leq \frac{x}{s} \leq q_{\max}, \\ 0, & \text{otherwise.} \end{cases} \tag{1}$$

This allows the loss gradient to propagate as $\frac{\partial L}{\partial x} \approx \frac{\partial L}{\partial q(x)}$, enabling the network to adapt to quantization noise.

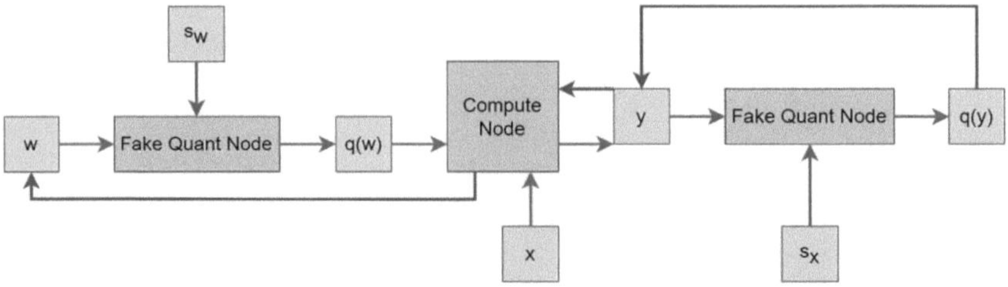

Fig. 3. Data flow (blue) and gradient flow (red) in QAT. The Straight-Through Estimator (STE) bypasses non-differentiable rounding during backpropagation. (Color figure online)

Learned Step Size Quantization. To further enhance the expressiveness of the low-precision model, we integrate LSQ [7], which treats the quantization step size s as a trainable parameter. For an input v, the quantization-dequantization process is defined as:

$$\hat{v} = \left\lfloor \text{clamp}\left(\frac{v}{s}, Q_{\text{neg}}, Q_{\text{pos}}\right) \right\rceil \cdot s, \tag{2}$$

where Q_{neg} and Q_{pos} define the integer range based on bit width. The gradient of $\hat{v}$ with respect to s is approximated via STE:

$$\frac{\partial \hat{v}}{\partial s} \approx \begin{cases} Q_{\text{neg}}, & \frac{v}{s} \leq Q_{\text{neg}} \\ \left\lfloor \frac{v}{s} \right\rceil - \frac{v}{s}, & Q_{\text{neg}} < \frac{v}{s} < Q_{\text{pos}} \\ Q_{\text{pos}}, & \frac{v}{s} \geq Q_{\text{pos}} \end{cases} \tag{3}$$

To accelerate early-stage convergence, s is initialized based on the data distribution as $s = 2|v|/\sqrt{Q_{\text{pos}}}$ [7].

4 Experiment

4.1 Experimental Setup

Dataset. We conduct experiments on the publicly available VEDAI dataset [24], designed for vehicle detection in aerial imagery. The dataset contains 1,246 images that originate from aerial photographs provided by Utah AGRC, covering diverse terrains such as grasslands, mountains, and highways. All images are captured at a fixed orbital altitude with a resolution of $16,000 \times 16,000$ pixels, available in both RGB and infrared formats, and preprocessed into $1,024 \times 1,024$ or 512×512 patches. Our experiments focus on 8 vehicle categories (e.g., cars, pickups), with emphasis on small object detection performance.

Evaluation Metrics. We use three standard metrics: Precision, Recall, and Mean Average Precision (mAP), with mAP as the primary indicator, reflecting both classification and localization quality. These metrics are based on True Positive (TP), False Positive (FP), and False Negative (FN), using an IoU threshold of 0.5 by default. IoU, Precision, and Recall are defined as follows:

$$IoU = \frac{|A \cap B|}{|A \cup B|}, \quad Precision = \frac{TP}{TP + FP}, \quad Recall = \frac{TP}{TP + FN}. \tag{4}$$

By varying confidence thresholds, we obtain Precision-Recall (P-R) curves. The area under each curve gives the AP for a class, and mAP is the average over all classes. In addition, we report mAP@0.5:0.95, which averages AP over IoU thresholds from 0.5 to 0.95 (step 0.05), offering a more comprehensive evaluation than the single-threshold mAP@0.5.

Implementation Details. As the original dataset lacks predefined training and testing splits, we randomly divide the RGB images into training and testing sets at a ratio of 8:2. Each annotation includes the bounding box center coordinates, rotation angle (relative to the x-axis), four corner points, category ID, and binary flags indicating occlusion and truncation. To address class imbalance, categories with fewer than 50 instances (e.g., motorcycles, buses, airplanes) are excluded, and the remaining categories are relabeled (0-7). The annotations are subsequently converted to the YOLO format. During the training phase, all convolutional layers in the YOLOv9 backbone are replaced with QAT-compatible layers. Additionally, the PGI connections are modified to direct gradients toward shallower layers, thereby enhancing small-object detection. The detection head is retained in full precision (FP32) to prevent premature quantization noise. The model is trained using stochastic gradient descent (SGD) with an initial learning rate of 0.01, momentum of 0.937, weight decay of 5×10^{-4}, and a batch size of 12 for 300 epochs, resulting in an intermediate model referred to as ***YOLOv9-RSQ-Raw***. During PTQ, the intermediate model is reloaded, and the detection heads convolutional layers are replaced with LSQ blocks, initialized using parameters from the pre-trained FP32 layers. The PGI auxiliary head, which is not used during inference, remains in full precision (FP32). To ensure correct LSQ behavior, YOLOs fused acceleration modules are disabled. The backbone is frozen, and the detection head undergoes 15 epochs of calibration training with a learning rate of 0.02.

4.2 Comparative Experimental Results

To verify the effectiveness of the hierarchical-aware quantization scheme, we first apply QAT on the backbone network to obtain an intermediate model, ***YOLOv9-RSQ-Raw***. Based on the semantic propagation characteristics of small object detection tasks, we then apply a fine-grained hierarchical quantization strategy: the paths responsible for small-scale feature outputs were quantized to 8 bits, those participating in small-scale semantic fusion were quantized to 4 bits, and the remaining paths were quantized to 2 bits. After 15 calibration epochs, the final model, ***YOLOv9-RSQ***, was obtained. Table 1 presents the comparison of core detection metrics between ***YOLOv9-RSQ***, the unquantized

(a) YOLOv9-RSQ-Raw (b) Original Image (c) YOLOv9-RS

Fig. 4. Comparison between shallow-layer outputs of the backbone network and the original image. The target region in the upper right and the noise region in the lower right are highlighted.

baseline ***YOLOv9-RS***, the intermediate model ***YOLOv9-RSQ-Raw*** (backbone quantized, detection head unquantized), and a representative lightweight model ***SuperYOLO***. As shown in Table 1, the fully quantized ***YOLOv9-RSQ*** maintains the same mAP@0.5 score as the full-precision ***YOLOv9-RS***, demonstrating that detection performance is well preserved despite aggressive quantization. It is worth noting that both precision and recall exhibit non-monotonic trends during quantization: precision initially decreases then recovers, while recall consistently improves.

To investigate this, we visualized shallow feature maps in Fig. 4 and analyzed the underlying causes as follows. Comparing feature maps of both models on the same input, the full-precision model activates not only the target regions but also noise with similar textures. In contrast, the QAT-processed backbone suppresses these irrelevant activations, indicating that low-precision computation serves as an implicit regularizer, reducing overfitting and enhancing sensitivity to hard samples. This improves recall but may reduce precision due to localization noise introduced by quantization. During calibration, PTQ perturbs detection head parameters significantly. Subsequent fine-tuning under a noisier distribution enables the model to learn a more robust decision boundary, yielding consistent improvements in both precision and recall. Ultimately, ***YOLOv9-RSQ*** matches the full-precision baselines mAP@0.5, confirming the effectiveness of the proposed hierarchical quantization strategy in compressing the model while maintaining detection performance.

4.3 Ablation Studies

We conducted a series of ablation studies to validate two key contributions: 1) The effectiveness of PGI path optimization in improving gradient propagation; 2) The ability of the mixed-precision quantization strategy to compress the model without compromising detection performance.

Table 1. Comparison Between Full-Precision and Quantized Models.

Models	Precision ↑	Recall ↑	mAP@0.5 ↑	mAP@0.5:0.95 ↑
YOLOv9-RSQ	0.812	**0.769**	**0.818**	0.541
YOLOv9-RS	**0.867**	0.694	**0.818**	**0.564**
YOLOv9-RSQ-Raw	0.783	0.758	0.799	0.557
SuperYOLO	-	-	0.725	-

Table 2. Impact of PGI Gradient Path Adjustment on Full-Precision Model Performance.

Models	Precision ↑	Recall ↑	mAP@0.5 ↑	mAP@0.5:0.95 ↑
YOLOv9-C-NoPGI	0.796	**0.750**	0.809	0.534
YOLOv9-C	0.766	**0.750**	0.807	0.535
YOLOv9-RS	**0.867**	0.694	**0.818**	**0.564**

Validation of PGI Path Optimization. To assess the impact of PGI path optimization, we compared the following three models: 1) ***YOLOv9-C-NoPGI***: without the PGI auxiliary path; 2) ***YOLOv9-C***: with PGI but no path optimization; 3) ***YOLOv9-RS***: with PGI path optimization on top of YOLOv9-C. As shown in Table 2, YOLOv9-RS outperforms the other two models in terms of Precision and mAP@0.5, despite a slight drop in Recall. Structural analysis reveals that in YOLOv9-C and its variant, PGI gradients mainly influence the 2nd to 4th layers of the backbone, missing the shallowest (1st) layer. In contrast, the optimized PGI path in YOLOv9-RS allows even the 1st layer to receive backward signals from PGI, effectively guiding shallow layer parameter updates and improving small object detection-consistent with our theoretical expectations and validating the optimizations efficacy.

Table 3. Impact of Different Methods on Model Detection Performance, where *Opt.*, *Back.*, and *Head* denote structure optimization, backbone QAT, and various detection head quantization schemes, respectively.

Optimization Strategy			**Evaluation Metrics**			
Opt.	*Back.*	*Head*	**Precision ↑**	**Recall ↑**	**mAP@0.5 ↑**	**mAP@0.5:0.95 ↑**
✓	✓	MP	**0.812**	**0.769**	**0.818**	0.541
✗	✓	MP	0.730	0.666	0.634	0.463
✓	✗	MP	0.613	0.639	0.634	0.402
✓	✓	INT2	0.697	0.675	0.719	0.428
✓	✓	INT4	0.762	0.725	0.800	0.538
✓	✓	INT8	0.783	0.734	0.809	**0.543**

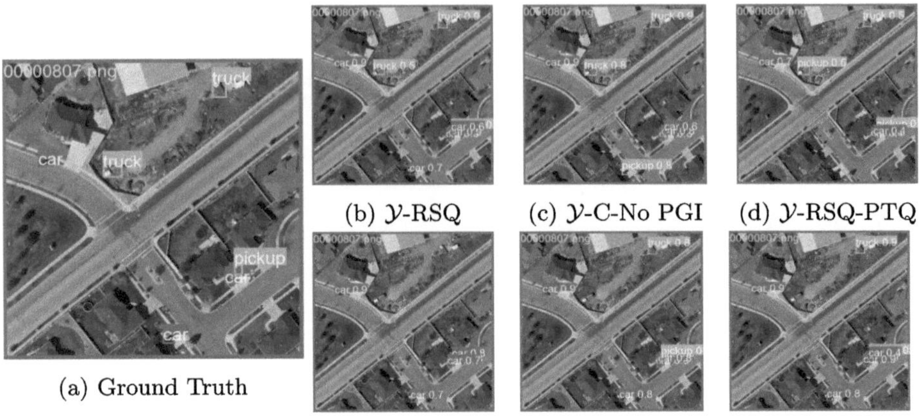

(a) Ground Truth (b) $\mathcal{Y}$-RSQ (c) $\mathcal{Y}$-C-No PGI (d) $\mathcal{Y}$-RSQ-PTQ (e) $\mathcal{Y}$-RSQ-INT2 (f) $\mathcal{Y}$-RSQ-INT4 (g) $\mathcal{Y}$-RSQ-INT8

Fig. 5. Visualization of ablation results for quantization methods, where $\mathcal{Y}$ means YOLOv9.

Validation of the Quantization Strategy. To evaluate the staged and progressive mixed-precision quantization strategy, we designed controlled experiments to isolate the effects of structure optimization, backbone QAT, and various detection head quantization schemes. From the Table 3, we observe that without structural optimization, even with QAT and mixed-precision quantization, model performance significantly degrades. This suggests that sensitivity to small objects is inherently insufficient in the original structure, and quantization noise amplifies this weakness.

Finally, we compared the detection head quantized using INT2, INT4, INT8, and the proposed mixed-precision (MP) strategy. Visualization shows in Fig. 5, where $\mathcal{Y}$ denotes the YOLOv9 baseline. Results show that increasing bit-width consistently improves performance, and the mixed-precision strategy achieves the best or second-best results across all metrics. These findings confirm the effectiveness of our proposed strategy, which applies higher precision to semantic-critical paths and lower precision to less relevant ones, successfully balancing model compression and detection performance.

5 Conclusion

This paper proposes an optimized YOLOv9 model that balances high accuracy for small-object detection with low computational cost. First, we redesign the PGI propagation path to enhance shallow-layer gradient supervision, significantly improving small-object sensitivity. Second, we introduce a semantic-driven mixed-precision quantization strategy, allocating high, medium, and low bit-widths to critical, deep, and irrelevant paths, respectively. This achieves substantial model compression with minimal performance degradation. Experiments on the VEDAI dataset demonstrate that our lightweight model performs comparably to the full-precision baseline. Future work will explore integrating multimodal data (e.g., infrared or microwave imagery) to enhance robustness in complex scenarios.

Acknowledgments. This work was supported by the Science and Technology Major Project of Sichuan Province under Grant 2024ZDZX0003.

References

1. Bai, Y., Li, R., Gou, S., Zhang, C., Chen, Y., Zheng, Z.: Cross-connected bidirectional pyramid network for infrared small-dim target detection. IEEE Geosci. Remote Sens. Lett. **19**, 1–5 (2022)
2. Bengio, Y., Léonard, N., Courville, A.: Estimating or propagating gradients through stochastic neurons for conditional computation. arXiv preprint arXiv:1308.3432 (2013)
3. Bhalgat, Y., Lee, J., Nagel, M., Blankevoort, T., Kwak, N.: Lsq+: improving low-bit quantization through learnable offsets and better initialization. In: Proceedings of the IEEE/CVF Conference on Computer Vision and Pattern Recognition Workshops. pp. 696–697 (2020)
4. Choi, J., Wang, Z., Venkataramani, S., Chuang, P.I.J., Srinivasan, V., Gopalakrishnan, K.: Pact: parameterized clipping activation for quantized neural networks. arXiv preprint arXiv:1805.06085 (2018)
5. Dong, Z., Yao, Z., Arfeen, D., Gholami, A., Mahoney, M.W., Keutzer, K.: Hawq-V2: hessian aware trace-weighted quantization of neural networks. Adv. Neural. Inf. Process. Syst. **33**, 18518–18529 (2020)
6. Dong, Z., Yao, Z., Gholami, A., Mahoney, M., Keutzer, K.: Hawq: hessian aware quantization of neural networks with mixed-precision. In: IEEE (2019)
7. Esser, S.K., McKinstry, J.L., Bablani, D., Appuswamy, R., Modha, D.S.: Learned step size quantization. arXiv preprint arXiv:1902.08153 (2019)
8. Ge, Z., Liu, S., Wang, F., Li, Z., Sun, J.: Yolox: exceeding yolo series in 2021. arXiv preprint arXiv:2107.08430 (2021)
9. Gholami, A., Kim, S., Dong, Z., Yao, Z., Mahoney, M.W., Keutzer, K.: A survey of quantization methods for efficient neural network inference. In: Low-Power Computer Vision, pp. 291–326. Chapman and Hall/CRC (2022)
10. Han, B., He, L., Ke, J., Tang, C., Gao, X.: Weighted parallel decoupled feature pyramid network for object detection. Neurocomputing **593**, 127809 (2024)
11. Jung, S., et al.: Learning to quantize deep networks by optimizing quantization intervals with task loss. In: Proceedings of the IEEE/CVF Conference on Computer Vision and Pattern Recognition. pp. 4350–4359 (2019)
12. Li, J., Zhang, Z., Tian, Y., Xu, Y., Wen, Y., Wang, S.: Target-guided feature super-resolution for vehicle detection in remote sensing images. IEEE Geosci. Remote Sens. Lett. **19**, 1–5 (2021)
13. Li, J., Feng, Y., Shao, Y., Liu, F.: Idp-yolov9: improvement of object detection model in severe weather scenarios from drone perspective. Appl. Sci. **14**(12), 5277 (2024)
14. Li, Z., et al.: Deep learning-based object detection techniques for remote sensing images: a survey. Remote Sensing **14**(10), 2385 (2022)
15. Lin, T.Y., Dollár, P., Girshick, R., He, K., Hariharan, B., Belongie, S.: Feature pyramid networks for object detection. In: Proceedings of the IEEE Conference on Computer Vision and Pattern Recognition. pp. 2117–2125 (2017)

16. Liu, J., Zhang, J., Ni, Y., Chi, W., Qi, Z.: Small-object detection in remote sensing images with super resolution perception. IEEE Journal of Selected Topics in Applied Earth Observations and Remote Sensing (2024)
17. Liu, L., et al.: Deep learning for generic object detection: a survey. Int. J. Comput. Vision **128**, 261–318 (2020)
18. Liu, X., Wang, T., Yang, J., Tang, C., Lv, J.: Mpq-yolo: ultra low mixed-precision quantization of yolo for edge devices deployment. Neurocomputing **574**, 127210 (2024)
19. Liu, X., et al.: BOB-YOLO: balancing optimization binarized yolo via module-wise latency (2024)
20. Liu, Z., Cheng, K.T., Huang, D., Xing, E.P., Shen, Z.: Nonuniform-to-uniform quantization: towards accurate quantization via generalized straight-through estimation. In: Proceedings of the IEEE/CVF Conference on Computer Vision and Pattern Recognition. pp. 4942–4952 (2022)
21. Liu, Z., Wang, Y., Han, K., Zhang, W., Ma, S., Gao, W.: Post-training quantization for vision transformer. Adv. Neural. Inf. Process. Syst. **34**, 28092–28103 (2021)
22. Nagel, M., Fournarakis, M., Bondarenko, Y., Blankevoort, T.: Overcoming oscillations in quantization-aware training. In: International Conference on Machine Learning. pp. 16318–16330. PMLR (2022)
23. Rabbi, J., Ray, N., Schubert, M., Chowdhury, S., Chao, D.: Small-object detection in remote sensing images with end-to-end edge-enhanced gan and object detector network. Remote Sensing **12**(9), 1432 (2020)
24. Razakarivony, S., Jurie, F.: Vehicle detection in aerial imagery: a small target detection benchmark. J. Vis. Commun. Image Represent. **34**, 187–203 (2016)
25. Tishby, N., Zaslavsky, N.: Deep learning and the information bottleneck principle. In: 2015 IEEE Information Theory Workshop (itw). pp. 1–5. IEEE (2015)
26. Wang, C.Y., Bochkovskiy, A., Liao, H.Y.M.: YOLOv7: trainable bag-of-freebies sets new state-of-the-art for real-time object detectors. In: Proceedings of the IEEE/CVF Conference on Computer Vision and Pattern Recognition. pp. 7464–7475 (2023)
27. Wang, C.Y., Yeh, I.H., Mark Liao, H.Y.: YOLOv9: learning what you want to learn using programmable gradient information. In: European conference on computer vision. pp. 1–21. Springer, Cham (2024). https://doi.org/10.1007/978-3-031-72751-1_1
28. Xu, C., Wang, J., Yang, W., Yu, H., Yu, L., Xia, G.S.: Detecting tiny objects in aerial images: a normalized wasserstein distance and a new benchmark. ISPRS J. Photogramm. Remote. Sens. **190**, 79–93 (2022)
29. Youwai, S., Chaiyaphat, A., Chaipetch, P.: Yolo9tr: a lightweight model for pavement damage detection utilizing a generalized efficient layer aggregation network and attention mechanism. J. Real-Time Image Proc. **21**(5), 163 (2024)
30. Zang, Z., Lin, C., Tang, C., Wang, T., Lv, J.: Zero-shot aerial object detection with visual description regularization. In: Proceedings of the AAAI Conference on Artificial Intelligence. vol. 38, pp. 6926–6934 (2024)
31. Zhang, J., Lei, J., Xie, W., Fang, Z., Li, Y., Du, Q.: SuperYOLO: super resolution assisted object detection in multimodal remote sensing imagery. IEEE Trans. Geosci. Remote Sens. **61**, 1–15 (2023)
32. Zhang, Y., Ye, M., Zhu, G., Liu, Y., Guo, P., Yan, J.: FFCA-YOLO for small object detection in remote sensing images. IEEE Trans. Geosci. Remote Sens. **62**, 1–15 (2024)
33. Zhou, W., et al.: Precision neural network quantization via learnable adaptive modules. arXiv preprint arXiv:2504.17263 (2025)

CuPL-ME: CuPL-Based Zero-Shot Learning via Expert-Guided Markdown Prompt Optimization

Lin Long[1], Ying Wang[1], Rong Xiao[1], Wangyang Ying[3], Wei Ju[1], Chenwei Tang[1,2(✉)], and Jiancheng Lv[1]

[1] College of Computer Science, Sichuan University, Chengdu 610065, China
{longlin1,wangying2yvonne}@stu.scu.edu.cn,
{rxiao,juwei,tangchenwei,lvjiancheng}@scu.edu.cn

[2] National Key Laboratory of Fundamental Algorithms and Models for Engineering Simulation, Sichuan University, Chengdu 610065, Sichuan, China

[3] School of Computing and AI, Arizona State University, 85281 Tempe, AZ, USA
yingwangyang@gmail.com

Abstract. Contrastive Language-Image Pre-Training (CLIP) has advanced Zero-Shot Learning (ZSL), while LLM-based prompt optimization improves CLIP by generating category-specific semantic descriptions. However, existing methods often produce generic descriptions and fail to highlight fine-grained discriminative features, limiting visual-semantic alignment in fine-grained classification. To address this, we propose CuPL-ME, an enhanced CuPL framework for fine-grained zero-shot classification. CuPL-ME introduces domain-specific expert roles into LLM prompting, such as botanists, ornithologists, and food visual specialists, to guide the generation of more precise visual descriptions. It also adopts *Markdown* formatting to structure and emphasize key visual features. Experiments show CuPL-ME improves accuracy by +1.27% on Food101, +1.95% on Flowers 102, and +6.94% on CUB-200-2011, demonstrating its effectiveness in fine-grained zero-shot classification.

Keywords: Fine-grained zero-shot learning · prompt optimization · large language model · domain expertise

This work is supported by the National Major Scientific Instruments and Equipments Development Project of National Natural Science Foundation of China under Grant 62427820, the Sichuan Science and Technology Program under Grant 2025ZDZX0125, the Science Fund for Creative Research Groups of Sichuan Province Natural Science Foundation under Grant 2024NSFTD0035, the Ministry of Education Engineering Research Center Guiding Project for Machine Learning and Industrial Intelligence Applications under Grant SCU2024D013.

Z. Zhang (Ed.): ISNN 2026, LNCS 16848, pp. 37–49, 2027.
https://doi.org/10.1007/978-981-92-5870-3_4

1 Introduction

Zero-Shot Learning (ZSL) [1,2] enables models to recognize unseen visual categories by leveraging auxiliary semantic knowledge (e.g., attributes or descriptions), overcoming the absence of labeled training data and enhancing generalization to novel classes. The advent of Contrastive Language-Image Pre-training (CLIP) [3] has established new paradigms for vision tasks, including zero-shot image classification [4,5] and open-vocabulary object detection [6,7]. To further enhance CLIP's performance and applicability, researchers have proposed various improvements. For example, continuous prompt optimization approaches like Context Optimization (CoOp) [8] and Conditional Context Optimization (CoCoOp) [9] model text inputs as learnable vectors. In addition, LLM-based textual augmentation methods such as Customized Prompts via Language Models (CuPL) [10] employ powerful LLMs to generate discriminative, category-specific textual descriptions that are structured into classification prompts for CLIP without model fine-tuning.

However, CuPL fails to fully exploit the domain expertise embedded in LLMs when generating textual descriptions. This limitation results in two critical shortcomings: 1) the lack of expert-guided prompting leads to generic feature descriptions that miss fine-grained, discriminative details crucial for distinguishing visually similar categories; and 2) the unstructured output format obscures important visual characteristics, increasing the risk of omitting critical features and degrading classification accuracy. These shortcomings are especially detrimental in fine-grained image classification tasks, where subtle distinctions often define class boundaries. As shown in Fig. 1, for the visually similar **"Black-eyed Susan"** and **"Sunflower"** categories in the fine-grained Flowers 102 dataset [11], the semantic descriptions generated by LLaMA3.3[1] demonstrate significant

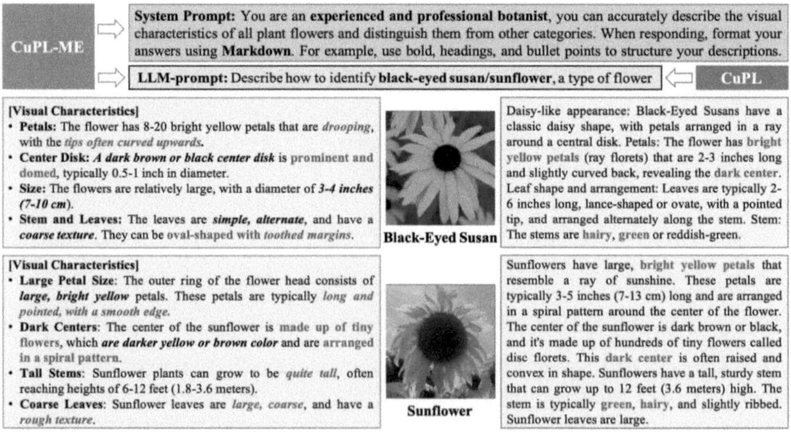

Fig. 1. Comparison between CuPL and our CuPL-ME (using LLaMA3.3).

[1] https://ollama.com/library/llama3.3.

overlap in phrasing, e.g., *"bright yellow petals"*, *"dark center"*, and *"hairy and green"*, when guided by CuPL.

To address these limitations, we propose **CuPL-ME**, an enhanced version of CuPL that improves the quality of generated feature descriptions by incorporating expert-driven text generation and structured representation. As shown in Fig. 1, our proposed CuPL-ME employs a systematic prompt for LLaMA3.3: *"You are an experienced and professional* ***botanist****, you can accurately describe the visual characteristics of all plant flowers and distinguish them from other categories. When responding, format your answers using* ***Markdown****. For example, use bold, headings, and bullet points to structure your descriptions."*, and guides LLM to generate more fine-grained semantic descriptions. Then, the generated semantic descriptions capture more precise, expert-level attributes, such as *"petals that are drooping, with the tips often curved upwards"* for **"Black-eyed Susan"** and *"petals are typically long and pointed, with a smooth edge"* for **"Sunflower"**. The generated tailored and fine-grained semantic descriptions can be directly fed into pre-trained CLIP models, enabling precise alignment between visual and semantic feature subspaces. Critically, this approach requires no fine-tuning of CLIP, yet significantly enhances recognition accuracy for unseen classes in fine-grained zero-shot classification tasks. Our contributions are summarized as follows:

- CuPL-ME integrates domain expertise through specialized prompting, guiding LLMs to generate detailed semantic descriptions, overcoming the granularity limitations of conventional prompt optimization and significantly enhancing performance in fine-grained image classification tasks.
- CuPL-ME leverages Markdown-structured representation to emphasize discriminative visual characteristics, allowing CLIP to construct robust visual-semantic embedding spaces without fine-tuning and direct attention to critical characteristics.
- CuPL-ME outperforms CuPL baseline by 1.27% and 1.95% on Food101 and Flowers102, respectively, as well as achieves a 6.94% improvement over CLIP on CUB-200-2011, demonstrating superior performance on fine-grained zero-shot classification tasks.

2 Related Work

The core of ZSL [12] is to transfer knowledge from known categories to unseen ones by modeling semantic or visual feature spaces [13]. Early methods formulated classification as attribute prediction [14,15], while later studies combined CNN-based visual features with Skip-gram semantic embeddings [16] to improve visual-semantic alignment [17]. With generative models [18], unseen-class semantic or visual features can be synthesized, converting ZSL into supervised learning [19]. Recently, VLMs such as CLIP [3], ALIGN [20], and Flamingo [21] have further advanced cross-modal ZSL. Among them, CLIP is widely used for its strong image-text representation. It encodes text and images separately, projects them into a shared space, and aligns them through contrastive learning. Since CLIP is sensitive to prompts, manual prompt design is difficult. CoOp [8] and CoCoOp [9] learn continuous prompt vectors, but these prompts are less interpretable and lack explicit semantics. CuPL [10] instead uses LLMs to generate category descriptions for richer textual guidance. However, it still underuses LLM expert knowledge and lacks structured emphasis on fine-grained discriminative features, limiting fine-grained zero-shot classification.

3 Method

3.1 Motivation Analysis

In fine-grained zero-shot image classification [22], inter-category distinctions often manifest through subtle variations. For instance, species in the Flowers 102 [11] dataset may differ minimally in petal morphology, chromatic properties, or textural features, while bird categories in the CUB-200-2011 [23] exhibit marginal variations in beak curvature, wing configuration, or tail shape. CuPL employs 3 straightforward and general LLM-prompts, including: *1) "Describe what a(n) {} looks like: , 2) "Describe a(n) {}: , 3) "What are the identifying characteristics of a(n) {}?* , or LLM-prompts specially designed for each dataset, to query LLMs for semantic descriptions of relevant categories. However, this simplistic approach may lead the model to emphasize general features, resulting in low-quality textual descriptions. As shown in Fig. 1, particularly when describing visually similar categories, LLMs frequently generate features common to both classes, while the distinctive features critical for accurate classification are often neglected. The text generated by CuPL using LLMs fails to emphasize these key discriminative features, which are frequently obscured by numerous generic attributes, hindering the classification model's ability to capture them. This clearly indicates that CuPL does not fully exploit the potential of LLMs. By refining prompt engineering strategies, it is possible to more effectively leverage the knowledge embedded in LLMs, thereby enhancing the quality of the generated descriptive text. Such prompt optimization for LLMs can substantially enhance CLIP's zero-shot fine-grained classification accuracy without model fine-tuning.

3.2 Methodology

As shown in Fig. 2, we propose **Cu**stomized **P**rompts via **L**anguage models with **M**arkdown and **E**xpert knowledge (**CuPL-ME**), which optimizes system prompt design for LLMs from two perspectives: the incorporation of expert knowledge and the structured storage of textual information. This approach aims to generate descriptive texts that contain richer, fine-grained key features and emphasize critical discriminative attributes, thereby improving the classification accuracy of the CLIP model on zero-shot fine-grained classification tasks.

Expert knowledge-driven descriptive text generation. In the field of natural language processing, incorporating expert knowledge has been shown to

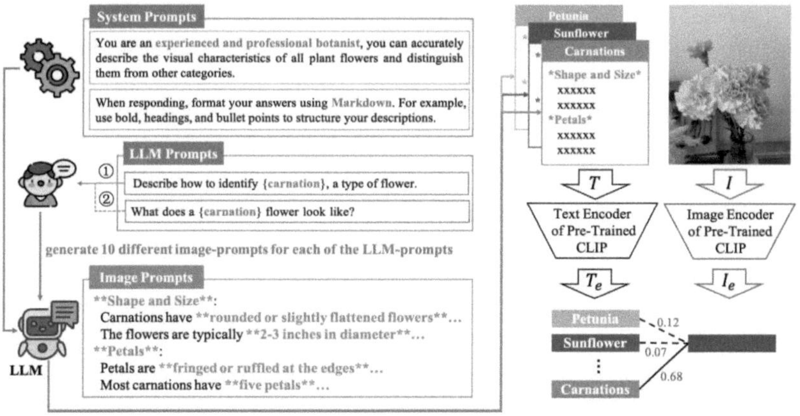

Fig. 2. Overall framework of the proposed **Cu**stomized **P**rompts via **L**anguage models with **M**arkdown and **E**xpert knowledge (**CuPL-ME**). We first generate semantic descriptions (**Image Prompts**) for each category using custom-engineered **System Prompts** and specialized **LLM prompts** utilizing LLMs like LLaMA3.3. Next, the constructed descriptive texts (**Image Prompts**) and images are modally aligned and categorized using the pre-trained CLIP model with a backbone of ViT-L/14 and the GPT-3 DaVinci-002 model.

significantly enhance model performance. In tasks such as medical diagnosis and legal analysis, complex concepts can be interpreted more accurately by modeling the perspectives of domain experts [24,25]. Motivated by this, we extend this concept to fine-grained image classification. Given that LLMs are trained on vast amounts of human knowledge, they exhibit strong capabilities in knowledge integration and contain expert-level information across various domains. We attempt to transform the expert domain knowledge of the LLMs into feature descriptions in image classification tasks by introducing expert roles, to solve the problem of the lack of relevance and discriminability of the text descriptions generated by the CuPL using the LLM approach. Specifically, based on the characteristics of the datasets, we designed ***Botanist*** and ***Ornithologist*** for the Flowers 102 and CUB-200-2011 datasets, respectively. In designing expert roles for the Food101 dataset, we made several attempts. Initially, we defined the expert role of ***Food Connoisseur*** based on common knowledge. But the generated descriptions included only a few visual attributes, while the remaining content focused primarily on ingredients and taste preferences. We subsequently experimented with the role of ***Commercial Food Photographer*** to incorporate additional visual attributes. However, the visual features generated remained limited. Finally, we introduced the ***Food Visual Specialist***, whose responses included a greater number of detailed visual attributes. Notably, this expert role employed summarizing statements to emphasize key features. We hypothesize that repeated emphasis on visual feature descriptions enhances the CLIP model s ability to capture salient semantic information. The superior performance of the Food Visual Specialist in subsequent experiments supports this hypothesis. Unless otherwise specified, the expert role used for the Food101 dataset in this paper is ***Food Visual Specialist***.

Structured Representation and Storage of Text Descriptions. CuPL directly utilizes textual descriptions generated by LLMs in plain **.txt** format a format that lacks structural and emphatic mechanisms to highlight feature information critical for classification. In fine-grained image classification tasks, it is essential that the textual descriptions produced by LLMs emphasize key distinguishing features, enabling subsequent classification models to effectively capture them. To address this limitation, we propose enhancing the output format of LLMs used in CuPL by adopting the **Markdown** format. This format supports emphasis mechanisms such as headings and special symbols (e.g., bold, underline, italic), which significantly improve the readability and distinguishability of the textual descriptions. To replace the LLM descriptive text output with the Markdown format, while fully leveraging the LLM s knowledge integration capabilities and ensuring stylistic consistency before and after processing, we design the **Markdown**-formatted system prompt. This improvement ensures that the text descriptions generated by the LLM are directly produced in Markdown format, thereby eliminating the need for manual and time-consuming conversion from raw text.

Generating and Utilizing Text Descriptions. Building on the aforementioned improvements, we utilize the API to access the LLM, e.g., LLaMA3.3, DeepSeek-v2[2], and Qwen2.5:72B[3], and generate text descriptions by constructing category-specific text prompts. Expert roles and Markdown text format are specified in the message field of the API request, enabling the LLM to respond accordingly. In CuPL, LLM-prompts are categorized into two types: CuPL (base), which includes three general-purpose prompts, and CuPL (full), which consists of some specialized prompts tailored to the characteristics of specific datasets. As shown in Fig. 3a and Fig. 3b, we adopt the CuPL (full) for the Flowers 102 and Food101 datasets. Since CuPL does not include the CUB-200-2011 dataset, we design three LLM-prompts specifically adapted to the characteristics of this dataset, as shown in Fig. 3c. After selecting and designing the system prompt and LLM-prompts, we obtained feature descriptions by querying the LLM. To maintain consistency with CuPL, we followed its query protocol conducting 10 repetitions per LLM prompt. Accordingly, if three LLM prompts are used, each category yields a total of 30 descriptive responses. The image classification model employed in this paper remains the CLIP model, where the Text encoder is based on the Transformer architecture [27], and the Image encoder utilizes the ViT-L/14 [28] architecture pre-trained by CLIP. It is important to emphasize that CuPL-ME does not introduce any modifications to the CLIP model itself. Therefore, no additional training is required.

[2] https://ollama.com/library/deepseek-v2.
[3] https://ollama.com/library/qwen2.5:72b.

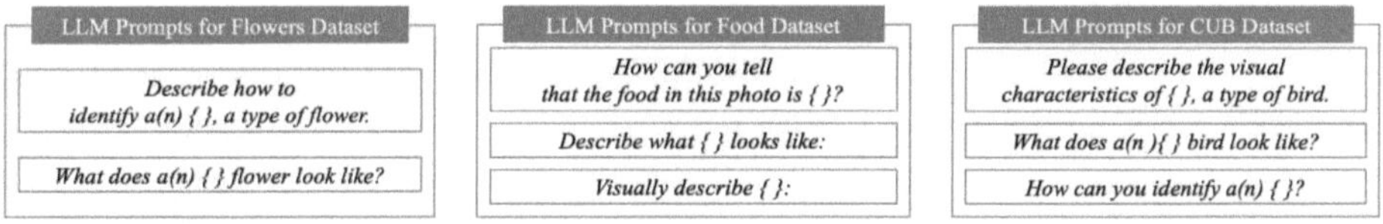

(a) (b) (c)

Fig. 3. Our LLM-Prompts templates, which are filled in with the category names for three fine-grained datasets, i.e., Flowers102 [11], Food101 [26], and CUB-200-2011 [23].

Table 1. Detail of three benchmark datasets, where N_C and N_I denote the number of classes and images, respectively.

Dataset	Category	N_C	N_I	Seen/Unseen
Food101	Food	101	101,000	/
Flowers102	Flowers	102	8,189	82/20
CUB-200-2011	Birds	200	11,788	150/50

As shown in Fig. 2, we utilize the text encoder of pre-trained CLIP model with a backbone of GPT-3 DaVinci-002 model[4] to extract features from a pre-constructed set of text descriptions T, yielding T_f, and the image encoder of pre-trained CLIP model with a backbone of ViT-L/14 [28] to extract features from the input image I, resulting in I_f. These feature vectors are then projected into a shared feature space to enable cross-modal alignment, as described in Eq. (1) and Eq. (2). Subsequently, we compute the cosine similarity between the image and text embeddings and compare the similarity scores for each image-text pair. The pair with the highest cosine similarity is considered the best match, and the category label associated with the corresponding text is assigned as the predicted label for the target image, as shown in Eq. (3).

$$T_f = \text{Encoder}_{\text{text}}(T), \quad I_f = \text{Encoder}_{\text{img}}(I), \tag{1}$$

$$T_e = \text{Norm}_t(T_f), \quad I_e = \text{Norm}_i(I_f), \tag{2}$$

$$\text{logits} = \text{Similarity}(I_e, T_e). \tag{3}$$

4 Experiment

4.1 Experiment Setting

Dataset. To ensure both fairness and difficulty, the chosen datasets include the benchmark datasets Food101 [26], Flowers102 [11], and CUB-200-2011 [23] datasets. Note that the CUB-200-2011 [23], a widely recognized dataset in the field of fine-grained zero-shot image classification, is not evaluated in the CuPL.

[4] https://platform.openai.com/docs/models.

As shown in Table 1, these datasets offer rich category diversity, providing a comprehensive basis for testing model generalization. Since the proposed CuPL-ME method does not undergo any additional training, all classes in each dataset are treated as unseen during evaluation. As shown in Fig. 4, each dataset presents fine-grained distinctions between categories (e.g., subtle differences in petal texture among flower species). In addition, variations in image conditions such as shooting angles, lighting, and occlusion further challenge the model s fine-grained classification capabilities.

(a) Flowers102 (b) Food101 (c) CUB-200-2011

Fig. 4. Sample images of three benchmark datasets.

Implementation Details. In this paper, image classification accuracy is used as the evaluation metric. A prediction is considered correct only when the predicted label exactly matches the ground-truth label; thus, higher accuracy indicates better performance. Unless otherwise specified, all reported results in this paper refer to this metric. Since the original LLM used in CuPL is not available, this paper first conducts a small-scale experiment to select an alternative LLM. Specifically, we evaluate DeepSeek-v2 (Model size: 8.9 GB), Qwen2.5:72B (Model size: 47 GB), and LLaMA3.3 (Model size: 42 GB) by reproducing the CuPL method on the Flowers102 dataset. The parameters used for LLMs and their corresponding definitions are provided in Table 2. The evaluation results for these three LLMs on the Flowers102 dataset are presented in Table 3. DeepSeek-v2 exhibits the weakest performance and is therefore excluded from further consideration. Although Qwen2.5:72B achieves the highest accuracy under the CuPL (base) setting, its performance significantly deteriorates under the CuPL (full) setting which is typically expected to yield higher accuracy and is thus also excluded. Consequently, this paper selects LLaMA3.3 as the LLM for the experiments. It is important to note that, to ensure experimental fairness, all CuPL results reported in this paper are reproduced.

Table 2. LLM parameter settings for text generation.

Parameter	Description	Value
temperature	Randomness	0.99
max_tokens	Token limit	50
stop	Stop condition	"."

4.2 Main Results

Comparison with CuPL. As shown in Table 4, the CuPL-ME method proposed in this paper achieves improvements of 1.27% and 1.95% over CuPL (full) on the Food101 and Flowers102 datasets, respectively, and demonstrates a 6.94% improvement over CLIP in zero-shot image classification on the CUB-200-2011 dataset. These results strongly validate the effectiveness of the proposed method. However, both CuPL (full) and the proposed approach fail to surpass the recognition accuracy of CuPL (base), or even that of CLIP s zero-shot classification on the Food101 dataset. In this paper, we retain the LLM-prompts from CuPL and continue to use those employed in CuPL (full). Upon analysis, we speculate that the observed performance limitation may stem from the LLM-prompts in CuPL (full) failing to adequately extract the key and distinguishing features of various food categories.

Table 3. CuPL zero-shot accuracy with various LLMs on Flowers102.

LLM	CuPL (base)	CuPL (full)
DeepSeek-v2	76.21%	77.76%
Qwen2.5:72B	84.29%	81.26%
LLaMA3.3	82.01%	83.03%

Table 4. Comparison of zero-shot accuracy on fine-grained datasets.

Dataset	CLIP	CuPL		Ours
		Base	Full	
Food101	93.09%	**93.62%**	91.68%	92.95%
Flowers102	78.24%	82.01%	83.03%	**84.98%**
CUB-200	56.54%	–		**63.48%**

Comparison of Different Experts in Food101. Table 5 presents the detailed results of the three expert roles Food Connoisseur, Commercial Food Photographer, and Food Visual Specialist evaluated on the Food101 dataset. As shown in Table 5, the Food Visual Specialist achieves the highest accuracy of 92.95%, outperforming the other two expert roles. This result suggests that this expert role more effectively activates the LLM and generates more informative feature text descriptions. Furthermore, it supports our hypothesis that repeatedly emphasizing key features increases the likelihood of the model capturing those features, thereby enhancing classification accuracy.

Classwise Accuracy. To thoroughly evaluate the performance of CuPL-ME in fine-grained recognition tasks, particularly in visualizing category-level confusion, we analyzed the experimental results using a confusion matrix. Figure 5 presents the confusion matrix visualization of the proposed CuPL-ME method on the fine-grained Flowers 102 dataset. As shown in Fig. 5, CuPL-ME achieves high classification accuracy across most categories, demonstrating its effectiveness in fine-grained recognition. However, the method exhibits notable confusion in distinguishing between the visually similar categories "mexican petunia" and "petunia". These categories share significant overlap in flower shape, size, and color, with the most critical distinguishing features residing in subtle variations in petal morphology (e.g., shallower lobes or more pronounced wrinkles at the petal tips of mexican petunia). Such subtle differences are easily obscured in images due to factors such as shooting angle, lighting conditions, background interference, or partial occlusion, thereby placing substantial demands on the model s feature extraction and classification capabilities. Furthermore, textual prompts may struggle to sufficiently capture and emphasize these subtle but highly discriminative visual features.

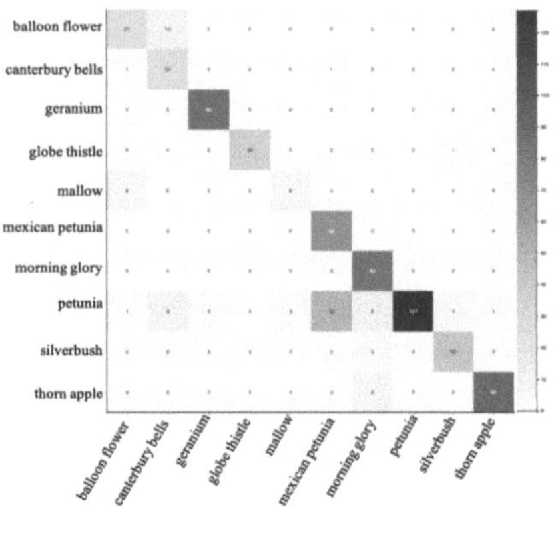

Fig. 5. Confusion matrix on Flowers102 dataset; darker diagonal squares denote higher identification frequency.

Exploration of Temperature. We further investigate the effect of the LLM Temperature parameter on text description generation. Temperature controls output randomness, typically within (0, 1]. Values close to 0 make the LLM select high-probability words, producing deterministic and consistent outputs, while values near 1 sample from a broader probability distribution, generating more diverse descriptions but potentially reducing coherence or grammatical quality. As shown in Fig. 6, both excessively high and low temperatures reduce classification accuracy. The best performance is achieved at 0.9, suggesting a balance between rich feature generation and textual validity.

Table 5. Comparison on Food101 dataset with different expert roles.

Expert Role	Accuracy
Food connoisseur	92.80%
Commercial food photographer	92.88%
Food visual specialist	**92.95%**

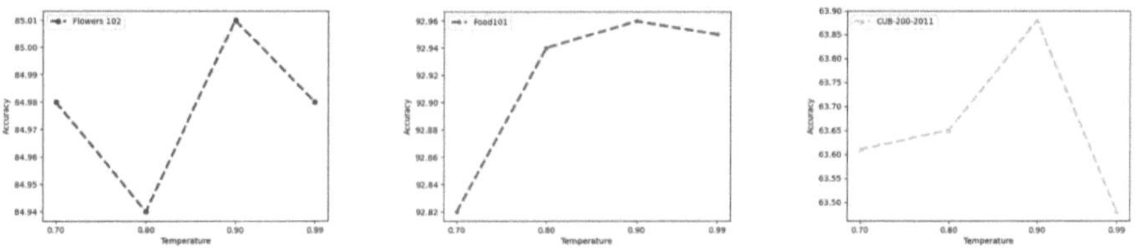

(a) Flowers102 (b) Food101 (c) CUB-200-2011

Fig. 6. Accuracy varies with Temperature on three benchmark datasets.

5 Conclusion

In this paper, we propose CuPL-ME for zero-shot fine-grained image classification by improving CLIP s semantic descriptions. CuPL-ME uses domain-specific expert roles and *Markdown*-structured outputs to generate and emphasize discriminative visual features without fine-tuning CLIP. Experiments on three datasets verify its effectiveness. Still, its partial reliance on CuPL s original prompts may limit category-specific feature extraction. Future work will optimize LLM prompts and explore CLIP encoder fine-tuning.

Acknowledgments. This work was supported by the Sichuan Province Science and Technology Program under Grants 2025ZDZX0125.

References

1. Socher, R., Ganjoo, M., Manning, C.D., Ng, A.: Zero-shot learning through cross-modal transfer. In: Advances in Neural Information Processing Systems, vol. 26, pp. 935–943 (2013)
2. Tang, C., He, Z., Li, Y., Lv, J.: Zero-shot learning via structure-aligned generative adversarial network. IEEE Transactions on Neural Networks and Learning Systems **33**(11), 6749–6762 (2021)
3. Radford, A., et al.: Learning transferable visual models from natural language supervision. In: International Conference on Machine Learning, pp. 8748–8763. PMLR (2021)
4. Karessli, N., Akata, Z., Schiele, B., Bulling, A.: Gaze embeddings for zero-shot image classification. In: Proceedings of the IEEE Conference on Computer Vision and Pattern Recognition (CVPR), pp. 4525–4534. IEEE (2017)
5. Esmaeilpour, S., Liu, B., Robertson, E., Shu, L.: Zero-shot out-of-distribution detection based on the pre-trained model CLIP. In: Proceedings of the AAAI Conference on Artificial Intelligence, vol. 36, no. 6, pp. 6568–6576 (2022)
6. Minderer, M., Gritsenko, A., Houlsby, N.: Scaling open-vocabulary object detection. In: Advances in Neural Information Processing Systems, vol. 36, pp. 72983–73007 (2023)
7. Zang, Z., Lin, C., Tang, C., Wang, T., Lv, J.: Zero-shot aerial object detection with visual description regularization. In: Proceedings of the AAAI Conference on Artificial Intelligence, vol. 38, no. 7, pp. 6926–6934 (2024)

8. Zhou, K., Yang, J., Loy, C.C., Liu, Z.: Learning to prompt for vision-language models. Int. J. Comput. Vision **130**(9), 2337–2348 (2022)
9. Zhou, K., Yang, J., Loy, C.C., Liu, Z.: Conditional prompt learning for vision-language models. In: Proceedings of the IEEE/CVF Conference on Computer Vision and Pattern Recognition, pp. 16816–16825 (2022)
10. Pratt, S., Covert, I., Liu, R., Farhadi, A.: What does a platypus look like? Generating customized prompts for zero-shot image classification. In: Proceedings of the IEEE/CVF International Conference on Computer Vision, pp. 15691–15701 (2023)
11. Nilsback, M.-E., Zisserman, A.: Automated flower classification over a large number of classes. In: 2008 Sixth Indian Conference on Computer Vision, Graphics & Image Processing, pp. 722–729. IEEE (2008)
12. Larochelle, H., Erhan, D., Bengio, Y.: Zero-data learning of new tasks. In: AAAI, pp. 646–651 (2008)
13. Farhadi, A., Endres, I., Hoiem, D., Forsyth, D.: Describing objects by their attributes. In: 2009 IEEE Conference on Computer Vision and Pattern Recognition, pp. 1778–1785. IEEE (2009)
14. Lampert, C.H., Nickisch, H., Harmeling, S.: Attribute-based classification for zero-shot visual object categorization. IEEE Trans. Pattern Anal. Mach. Intell. **36**(3), 453–465 (2013)
15. Palatucci, M., Pomerleau, D., Hinton, G.E., Mitchell, T.M.: Zero-shot learning with semantic output codes. In: Advances in Neural Information Processing Systems, vol. 22, pp. 1410–1418 (2009)
16. Mikolov, T., Chen, K., Corrado, G., Dean, J.: Efficient estimation of word representations in vector space. arXiv preprint arXiv:1301.3781 (2013)
17. Frome, A., et al.: DeViSE: a deep visual-semantic embedding model. In: Advances in Neural Information Processing Systems, vol. 26, pp. 2121–2129 (2013)
18. Goodfellow, I.J., et al.: Generative adversarial nets. In: Advances in Neural Information Processing Systems, vol. 27, pp. 2672–2680 (2014)
19. Xian, Y., Sharma, S., Schiele, B., Akata, Z.: f-VAEGAN-D2: a feature generating framework for any-shot learning. In: Proceedings of the IEEE/CVF Conference on Computer Vision and Pattern Recognition, pp. 10275–10284 (2019)
20. Jia, C., et al.: Scaling up visual and vision-language representation learning with noisy text supervision. In: International Conference on Machine Learning, pp. 4904–4916. PMLR (2021)
21. Alayrac, J.B., et al.: Flamingo: a visual language model for few-shot learning. In: Advances in Neural Information Processing Systems, vol. 35, pp. 23716–23736 (2022)
22. Gao, Y., et al.: Improving generalized zero-shot learning via cluster-based semantic disentangling representation. Pattern Recogn. **150**, 110320 (2024)
23. Wah, C., Branson, S., Welinder, P., Perona, P., Belongie, S.: The Caltech-UCSD Birds-200-2011 dataset. Technical report CNS-TR-2011-001, California Institute of Technology (2011)
24. Gu, Y., et al.: Domain-specific language model pretraining for biomedical natural language processing. ACM Transactions on Computing for Healthcare **3**(1), 1–23 (2021)
25. Lee, J., et al.: BioBERT: a pre-trained biomedical language representation model for biomedical text mining. Bioinformatics **36**(4), 1234–1240 (2020)
26. Bossard, L., Guillaumin, M., Van Gool, L.: Food-101 C Mining Discriminative Components with Random Forests. In: Fleet, D., Pajdla, T., Schiele, B., Tuytelaars,

T. (eds.) Computer Vision C ECCV 2014. LNCS, vol. 8694, pp. 446 C461. Springer, Cham (2014). https://doi.org/10.1007/978-3-319-10599-4_29
27. Vaswani, A., et al.: Attention is all you need. In: Advances in Neural Information Processing Systems, vol. 30, pp. 5998–6008 (2017)
28. Dosovitskiy, A., et al.: An image is worth 16x16 words: Transformers for image recognition at scale. arXiv preprint arXiv:2010.11929 (2020)

Topoformer: A Topological Transformer Framework for 3D Shape Classification

Junfeng Long, Yan Guo, Chenwei Sun, Xianggen Liu(✉), and Jiancheng Lv

College of Computer Science, Sichuan University, Chengdu, China
{pepper,guoyan3,sunchenwei}@stu.scu.edu.cn,
{liuxianggen,lvjiancheng}@scu.edu.cn

Abstract. Accurate modeling of CAD models is critical for industrial digitization, yet existing deep learning methods struggle to jointly capture geometric and topological structures across multiple scales. We propose Topoformer, a hierarchical Transformer that captures B-rep topology through multi-scale topological attention while preserving geometric invariance via an unsupervised consistency objective. We also introduce new benchmarks constructed from real-world industrial parts for 3D classification. Experimental results demonstrate that Topoformer outperforms existing methods, achieving 90.17% accuracy on the most challenging benchmark.

Keywords: Transformer · Classification · Boundary representation

1 Introduction

Industrial digitization motivates the development of advanced intelligent interpretation of 3D geometry. Boundary Representation (B-rep) serves as the de facto standard in Computer-Aided Design (CAD), representing mechanical components using parametric surfaces (e.g., NURBS) and explicit topological constraints [1]. While this format ensures high engineering fidelity, its hybrid nature—comprising continuous coordinates and discrete categorical structures—poses significant challenges for deep learning. Specifically, existing neural architectures struggle to process the geometric-topological heterogeneities inherent in B-rep data while maintaining native precision.

Traditional approaches often bypass the complexities of Boundary Representation (B-rep) by discretizing models into point clouds, voxels, or meshes [2–4]. While compatible with standard deep learning frameworks, these approximations introduce significant fidelity gaps—such as quantization errors that obscure fine surface curvatures vital for downstream CAD analysis [5].

To bridge these gaps, recent advancements have shifted toward native B-rep processing via Graph Neural Networks (GNNs), which encode topological elements as graph nodes [7]. Frameworks like UV-Net [8], BRep-BERT [9] and AAGNet [10] demonstrate that preserving connectivity plays an important role in high-quality representation learning. However, current GNN-Transformer

Z. Zhang (Ed.): ISNN 2026, LNCS 16848, pp. 50–61, 2027.
https://doi.org/10.1007/978-981-92-5870-3_5

hybrids [11] often rely on simple layer stacking, which limits their ability to model multi-scale topological dependencies and weakens their invariance to rotation, thereby affecting their robustness and generalization ability.

To address these limitations, we propose Topoformer. The core design objective is to learn global features of 3D models by employing Transformer layers biased by multi-order adjacency matrices. Inspired by multi-resolution analysis [12,13], Topoformer integrates high-order adjacency information into distinct attention channels to jointly capture local geometric patterns and global shape semantics. Our contributions are three-fold:

- **Multi-scale Topological Transformer:** A novel architecture that embeds hierarchical adjacency relationships directly into the Transformer's attention mechanism, enabling the capture of complex 3D surface connectivities.
- **Rotation-Invariant Representation Learning:** An unsupervised regularization strategy that enforces feature consistency across rotated instances, effectively mitigating the model's sensitivity to absolute coordinate values.
- **Industrial B-rep Benchmarks:** Two new large-scale datasets comprising 10,296 realistic manufacturing parts across 12 utility categories, significantly expanding the resources available for B-rep research.

2 Related Work

Early 3D classification primarily relied on polygonal meshes and point clouds. Mesh-based methods, such as MeshCNN [14], preserve combinatorial structures via convolutions on vertices and faces. In contrast, point-based approaches, such as PointNet [2] and PointNet++ [15], aggregate unordered points to capture global and local geometric features. Recent transformer-based methods, such as Point-BERT [16], further improve contextual representation learning. However, point-based methods often sacrifice topological fidelity under aggressive downsampling.

Boundary Representation (B-rep) offers superior geometric and topological detail for CAD models. Recent works like UV-Net [8] and BRepNet [17] utilize face-adjacency graphs and co-edge walks to learn directly from B-rep structures. Nonetheless, these architectures lack intrinsic rotation-equivariant design, limiting model generalization. To address this, we introduce an unsupervised rotational invariance constraint, enhancing robustness compared to prior B-rep-based methods.

Regarding feature propagation, traditional B-rep methods rely on GNNs [18–20], which often suffer from restricted receptive fields and feature dilution [21]. While BRep-BERT [9] integrates GNN embeddings into transformers, our architecture employs a multi-scale graph Transformer. By dynamically integrating adjacency matrices of varying granularities, our design facilitates effective global propagation while preserving fine-grained geometric sensitivity.

3 Method

3.1 Problem Definition

The task of 3D shape classification aims to predict the category of real-world 3D solids. These solids are typically created using CAD software and represented in B-rep format, which consists of surface elements and their topological connections. The surfaces are connected along edges to form a closed volume. Formally, a B-rep file is converted into a graph structure $\mathcal{G}$, where surface features S_f serve as nodes and edge features E_f as edges. Given $\mathcal{G}$, the learning model outputs an integer $O \in \{1, \cdots, K\}$ indicating the shape category, where K is the number of predefined classes.

3.2 Dataset Construction and Processing

To have a comprehensive evaluation of the 3D shape classification models, we build realistic datasets (named Parts-V1 and Parts-V2) by collecting 3D solids from https://www.traceparts.com/. The collected data includes 12 categories of industrial components, such as transmission parts, sealing elements, springs, mold standard parts, bearings, etc. Figure 1 demonstrates that the industrial components exhibit intricate geometric configurations and heterogeneous compositions, posing significant challenges for conventional 3D shape classification paradigms.

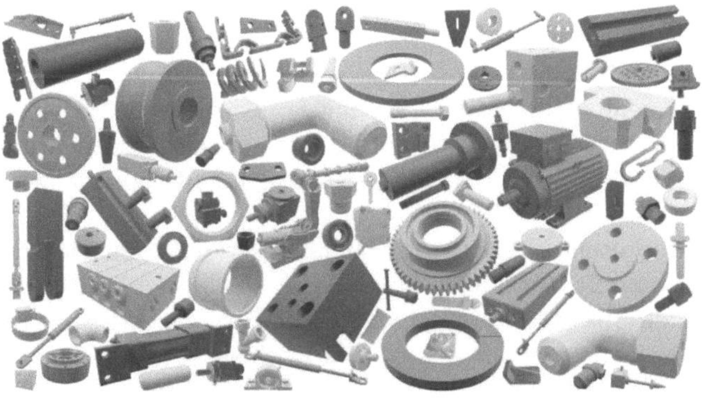

Fig. 1. Overview of the Parts dataset

The B-rep model is represented as a face adjacency graph $\mathcal{G}$ [22] by extracting geometric-topological features from surfaces $S(v, w)$ and edges $C(u)$. We discretize the parametric domains $u \in [u_{\min}, u_{\max}]$ and $(v, w) \in [v_{\min}, v_{\max}] \times [w_{\min}, w_{\max}]$ using uniform sampling. Specifically, each edge is encoded as a feature tensor $E_f \in \mathbb{R}^{6 \times N}$ by concatenating 3D coordinate offsets and unit tangent directions sampled at N parametric points. Each surface is represented as a feature tensor $S_f \in \mathbb{R}^{7 \times M \times N}$, where the first three channels correspond to 3D coordinates, the next three channels correspond to unit surface normals, and the final channel represents a binary visibility mask over an $M \times N$ discretized parametric grid (Fig. 2(a)). For evaluation, we employ two datasets: Parts-V1 (4,696 files, 6 classes) and Parts-V2 (10,296 files, 12 classes) to evaluate performance under different training scales.

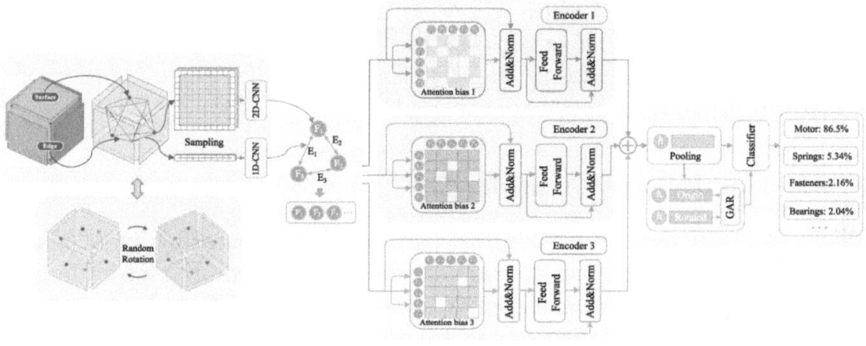

(a) Preprocessing Block (b) Hierarchical GNN (c) Topology-encoded Transformer (d)Postprocessing Block

Fig. 2. Methodology Overview. (a) 3D feature extraction and random rotation augmentation. (b) Neighborhood-based GNN encoding for local topological features. (c) Multi-granularity feature extraction via diverse attention biases. (d) Global feature aggregation and loss computation.

3.3 Hierarchical GNN

To extract initial local features, we design a hierarchical architecture combining CNN and GNN layers. Specifically, the CNN layers first perform initial feature embedding of the raw face-edge characteristics from the topological graph $\mathcal{G}$ (Fig. 2(a)). The 1D-convolution module processes edge features E_f, while the 2D-convolution module handles surface features S_f:

$$h_e = \mathrm{F}_{\mathrm{1D}}(E_f) \ , \tag{1}$$

$$h_s = \mathrm{F}_{\mathrm{2D}}(S_f) \ , \tag{2}$$

where h_e and h_s represent the embedded features for edge e and surface s (hereafter referred to as nodes), respectively. The F_{2D} layer extracts multi-channel surface features $[M, N, 7]$ using a multi-layer CNN with 3×3 kernels to preserve spatial dimensions, with F_{1D} following an analogous design for edges.

Subsequently, as shown in Fig. 2(b), an edge-aware GNN layer is employed to embed edge information from the node's neighborhood into the node features. This ensures that the representations are enriched with local topological context prior to the Transformer layers, facilitating more effective face-to-face attention learning. The update for node s is defined as:

$$\hat{h}_s = \sigma\left((1+\alpha)h_s + \sum_{n\in\mathcal{N}(s)} h_n \odot f(h_{e_{s,n}})\right) , \tag{3}$$

where $\hat{h}_s$ represents the updated feature of nodes, α is a learnable parameter emphasizing the original feature h_s. $\mathcal{N}(s)$ denotes the set of the neighboring nodes and $e_{s,n}$ means the edge between the nodes s and n. The function f is a linear projection that transforms the edge features to align the dimension of the node features h_n, and σ represents the MLP layer.

3.4 Topology-Encoded Transformer

We design multiple parallel transformer encoders, each operating under a unique scope of neighborhood context, to compute attention-based updates for node features. To achieve this, we introduce an attention bias in the Transformer architecture by incorporating the topological distance between node pairs, constrained by a length threshold K. For each Transformer encoder, we have an attention bias defined by the adjacent distance (i.e., the length of the connection path), given by

$$\mathrm{B}_{s,n}^{(k)} = \begin{cases} 0, & \text{if } \mathrm{Dis}(s,n) = k \\ -\infty, & \text{otherwise} \end{cases} \quad k = 1, \cdots, K , \tag{4}$$

where the function $\mathrm{Dis}(s,n)$ measures the shortest path length (in terms of hop count) between nodes s and n in the adjacency graph.

In Topoformer, we leverage the above attention bias to encode the 3D shape topology into the feature extraction of Transformer encoders. We design a multi-scale attention mechanism to model topological structures across different scales, as shown in Fig. 2(c), given by

$$\mathrm{Attention}^{(k)}(\mathbf{Q},\mathbf{K},\mathbf{V},\mathbf{B}^{(k)}) = \mathrm{Softmax}\left(\frac{\mathbf{QK}^T}{\sqrt{d_k}} + \mathbf{B}^{(k)}\right)\mathbf{V} , \tag{5}$$

where the function $\mathrm{Attention}^{(k)}$ performs the topology-encoded self-attention computations in the k-th Transformer encoder. The input features $\mathbf{Q},\mathbf{K},\mathbf{V}$ are the outputs of the hierarchical GNN with individual linear projections. When computing the attention between nodes, the above attention bias is applied. The value of 0 in $\mathrm{B}^{(k)}$ indicates that a node pair is reachable and its attention is retained, whereas a value of $-\infty$ denotes an unreachable node pair and thus masks the corresponding attention. Note that, for a Transformer encoder, multiple self-attention layers are applied, and finally, the updated representation of the node s is obtained, given by

$$h_s^{(k)} = \mathrm{TET}^{(k)}(\hat{h}_s, \mathbf{B}^{(k)}) , \tag{6}$$

where the function $\mathrm{TET}^{(k)}$ stands for the k-th Topology-encoded Transformer encoder.

Based on the updated representations from the topology-encoded Transformer with K layers, we employ a fully connected layer ϕ to fuse them, and use the Max-pooling layer to aggregate the information of all nodes and produce the global topological feature of the B-rep solid $h_{\mathcal{G}}$.

$$h'_s = \phi([h_s^{(1)}, \cdots, h_s^{(K)}]) \ , \tag{7}$$

$$h_{\mathcal{G}} = \text{Max-Pool}\{h'_s \mid s \in \mathcal{G}\} \ . \tag{8}$$

Finally, we leverage an MLP layer to transform the global feature $h_{\mathcal{G}}$ into the predicted probabilities of the predefined categories.

$$p(y|\mathcal{G}) = \text{Softmax}\left(\text{MLP}(h_{\mathcal{G}})\right) \ . \tag{9}$$

The Topoformer is trained using a standard cross-entropy loss over dataset $\mathcal{D}$:

$$\mathcal{L}_{\text{CE}} = - \sum_{(\mathcal{G}, y_c) \in \mathcal{D}} \log p(y_c \mid \mathcal{G}) \ , \tag{10}$$

where $p(y_c \mid \mathcal{G})$ denotes the predicted probability for ground-truth label y_c.

3.5 Geometry-Aware Regularization

To achieve rotation invariance, we implement Geometry-Aware Regularization (GAR) rather than relying on inherent architectural guarantees. By applying stochastic spatial rotations to the absolute coordinates of the original 3D models, GAR explicitly constrains feature consistency through a dedicated regularization loss. This approach enforces the learning of invariant representations, effectively guiding Topoformer to maintain discriminative power across varying spatial orientations.

For each 3D solid represented in B-rep format, we apply random rotations to the geometric features associated with the graph while keeping the underlying topological adjacency unchanged. Specifically, we perform rotations along a fixed axis (e.g., the X-axis) at a set of predefined angles. Concretely, the 3D coordinates and associated geometric attributes of each sampled surface point $S_f[\gamma, \eta]$ and curve point $E_f[\gamma, \tau]$ are transformed by a rotation matrix $\mathrm{R_X}(\theta)$, given by

$$S'_f[\gamma, \eta] = \mathrm{R_X}(\theta) \cdot S_f[\gamma, \eta] \ , \tag{11}$$

$$E'_f[\gamma, \tau] = \mathrm{R_X}(\theta) \cdot E_f[\gamma, \tau] \ , \tag{12}$$

where $\mathrm{R_X}$ represents the randomly chosen X-axis rotation matrix, and θ denotes the random rotation angle. γ is the 3D coordinates and η is the normals of the surface features S_f. Similarly, τ stands for the tangent information of the edge features E_f. The rotation matrix $\mathrm{R_X}$ is

$$\mathrm{R_X}(\theta) = \begin{bmatrix} 1 & 0 & 0 \\ 0 & \cos\theta & -\sin\theta \\ 0 & \sin\theta & \cos\theta \end{bmatrix}, \quad \theta \in \{45°, 90°, \ldots, 315°\}. \tag{13}$$

The newly constructed data $\mathcal{G}' = (S'_f[\gamma, \eta], E'_f[\gamma, \tau])$ is topologically identical to the original graph, differing only by a rotational transformation applied to the feature representations. Consequently, GAR seeks to align the representations of the original and rotated solids. Specifically, we introduce a bounded quadratic cosine similarity loss to encourage this alignment, defined as

$$\mathcal{L}_{\mathrm{GAR}} = \frac{1}{N_{\mathcal{B}}} \sum_{(\mathcal{G},\mathcal{G}')\in\mathcal{B}} \frac{(1 - \mathbf{C}(\mathcal{G},\mathcal{G}'))^2}{(1 - \mathbf{C}(\mathcal{G},\mathcal{G}'))^2 + \beta}, \tag{14}$$

where β is a hyperparameter that modulates the strength of the alignment term. Here, $\mathbf{C}(\mathcal{G},\mathcal{G}')$ denotes the cosine similarity between the two representations, $\mathcal{B}$ represents the training batch, and $N_{\mathcal{B}}$ is the number of paired samples in the batch. The quadratic cosine loss is adopted due to its robustness to outliers, providing a smooth optimization landscape that ensures stable gradient propagation and penalizes low-similarity representations while being more tolerant to highly similar ones.

The overall training objective is defined as a weighted combination of the cross-entropy loss and the GAR loss:

$$\mathcal{L} = \omega_{\mathrm{c}}\mathcal{L}_{\mathrm{CE}} + \omega_{\mathrm{g}}\mathcal{L}_{\mathrm{GAR}}, \tag{15}$$

where ω_{c} and ω_{g} are hyperparameters that balance the contribution of each objective. For simplicity, we set $\omega_{\mathrm{c}} = 1.0$ and $\omega_{\mathrm{g}} = 1.0$ across all experiments, effectively treating the total loss as a direct summation of the two components.

4 Experiments

4.1 Datasets and Experimental Setup

We evaluate Topoformer on three datasets: the SolidLetters [8] (96k alphabet shapes) and our proposed Parts-V1 and Parts-V2 benchmarks, all partitioned using a consistent 8:1:1 split. To assess performance, we compare Topoformer against point-based paradigms (PointNet [2], PointNet++ [15], DGCNN [6]) and CAD-specific frameworks (UV-Net [8], AAGNet [10]). Implementation details include UV sampling at $M = N = 10$, CNN layers $\mathcal{D}_{\mathrm{CNN}} = \{64, 128, 256, 64\}$, and initial GNN weight $\alpha = 0$. Both TET and Transformer encoders utilize 3 layers, with $\mathcal{D}_{\mathrm{Trans}}$ set to 512 for Parts-V1 and 1024 for Parts-V2. We apply a GAR loss weight $\beta = 0.3$ and a final MLP with dimensions $\{64, 128, 256, 64\}$. Training is executed via the Adam optimizer on an NVIDIA A800 GPU, with point-based baselines standardized to 2048 points per shape.

4.2 Classification Task

Table 1 presents the Top-1 and Top-2 classification accuracies of all compared methods. On the SolidLetters dataset, Topoformer consistently achieves the best performance across the evaluated metrics, with a Top-1 accuracy of 97.86%. On the Parts-V1 dataset, some baseline methods are affected by class imbalance and limited training samples. In contrast, Topoformer achieves a Top-1 accuracy of 85.99%. On Parts-V2, although UV-Net and PointNet++ show strong performance, Topoformer surpasses them, achieving a Top-1 accuracy of 90.17%.

Table 1. Classification results on three datasets (Top-1 and Top-2 accuracy).

Dataset	SolidLetters		Parts-V1		Parts-V2	
Model	Top-1	Top-2	Top-1	Top-2	Top-1	Top-2
UV-Net	97.04%	99.48%	76.07%	93.10%	86.70%	**95.93%**
PointNet	97.01%	99.35%	70.46%	93.54%	81.55%	92.80%
PointNet++	97.64%	99.66%	80.14%	94.73%	88.51%	95.47%
DGCNN	94.76%	99.04%	59.53%	76.75%	74.11%	92.18%
AAGNet	97.35%	99.55%	75.86%	92.89%	86.41%	94.94%
Topoformer	**97.86%**	**99.67%**	**85.99%**	**95.04%**	**90.17%**	95.73%

Figure 3 illustrates Topoformer's classification performance. Across SolidLetters, all categories achieve high accuracy. In Parts-V1, Category 2 (springs) shows lower performance due to data scarcity, while in Parts-V2, Category 5 (industrial machinery) remains challenging owing to its structural complexity. Despite these specific cases, Topoformer maintains high classification confidence across most categories.

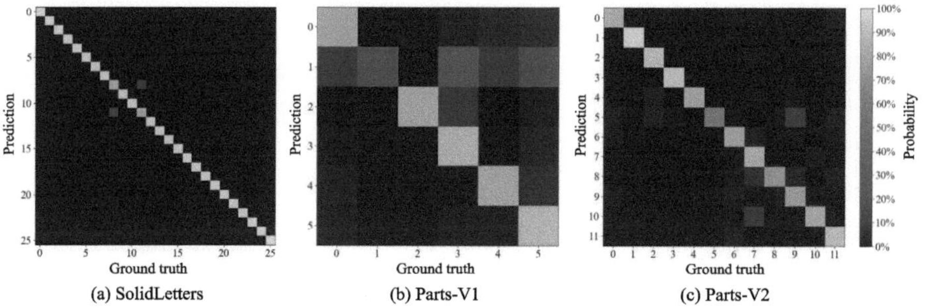

Fig. 3. Confusion matrix of Topoformer on each dataset.

4.3 Ablation Study

We conduct ablation studies on Parts-V1 and Parts-V2 to evaluate the GNN and GAR modules. As shown in Table 2, removing GNN layers decreases accuracy on Parts-V1 (85.99% to 82.97%) and Parts-V2 (90.17% to 89.28%), confirming their role in local representation learning.

To verify the impact of the GAR constraint on feature recognition, we compare Topoformer with and without GAR. Removing GAR significantly reduces performance, particularly on Parts-V1 (77.56%). Replacing GAR with standard MSE loss also yields lower accuracy, demonstrating that GAR enforces stronger geometric consistency than conventional supervision. Furthermore, integrating GAR into UV-Net improves its accuracy from 76.07% to 82.11% on Parts-V1, confirming the module's effectiveness and generalizability.

As shown in Fig. 4, sensitivity analysis shows that a 3-layer TET encoder achieves the best performance. Setting $\beta = 0.3$ yields the highest accuracy, providing stable loss convergence during training.

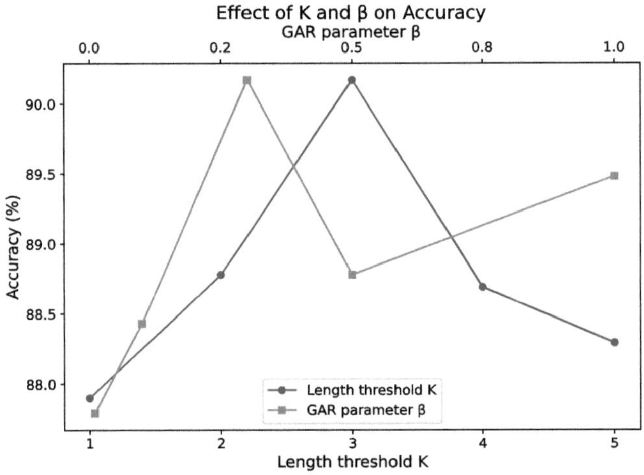

Fig. 4. Effect of the length threshold K and GAR parameter β on classification accuracy on the Parts-V2 dataset.

Table 2. Ablation study on Parts datasets. The best results are highlighted in bold.

Dataset	TF	TF-GAR	TF-GNN	TF+MSE	UV	UV+GAR
Parts-V1	**85.99%**	77.56%	82.97%	85.34%	76.07%	82.11%
Parts-V2	**90.17%**	87.10%	89.28%	89.38%	86.70%	88.98%

4.4 Case Study

To provide a more intuitive qualitative evaluation, we analyze the two categories with the highest misclassification rates in Fig. 3: machine tool accessories and fluid hardware. For each category, we randomly select one typical sample instance for visualization (Fig. 5).

Across all cases, Topoformer consistently assigns the highest posterior probability to the ground-truth class compared with baseline models, demonstrating stronger classification confidence under highly confusing category settings.

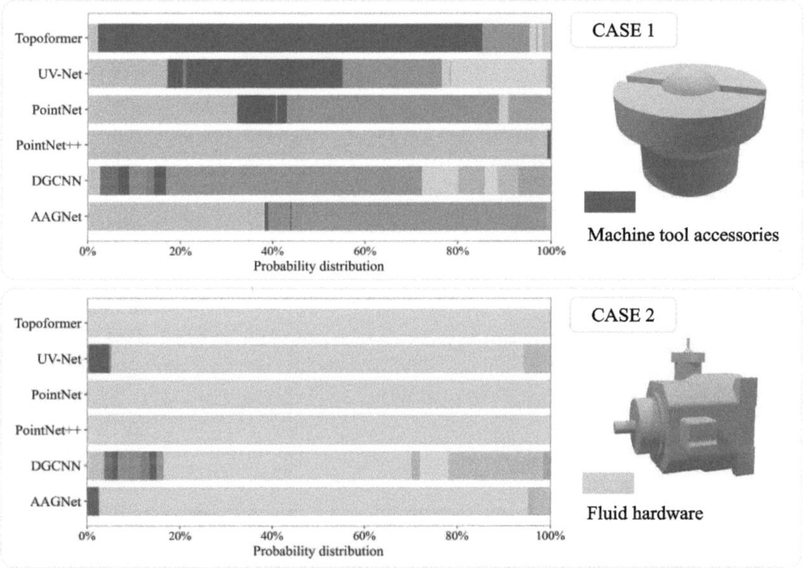

Fig. 5. Predicted class probabilities for Topoformer and baseline models on highly confused categories. The width of each segment is proportional to the predicted probability assigned to each class.

5 Conclusion

In this work, we introduce Topoformer, a model that leverages high-order adjacency matrices to guide the attention mechanism within the Transformer encoder. By propagating surface features across multiple granularities, Topoformer captures both local geometric details and global shape semantics. Additionally, we propose a geometry-aware regularization strategy to enforce rotation-invariant 3D representations. Experimental results demonstrate that integrating Transformer-based global dependency modeling beyond the local receptive fields of traditional GNNs leads to strong performance across standard 3D shape benchmarks.

Acknowledgments. This work was supported by the Science and Technology Major Project of Sichuan Province (2024ZDZX0003, 2025ZDZX0140), and the National Key R&D Program of China (2024YFB3312503). We also acknowledge the support of Sichuan Province Engineering Technology Research Center of Broadband Electronics Intelligent Manufacturing.

References

1. Weiler, K.J.: Topological Structures for Geometric Modeling. Springer, Berlin (1986)
2. Qi, C.R., Su, H., Mo, K., Guibas, L.J.: PointNet: Deep Learning on Point Sets for 3D Classification and Segmentation. In: Proceedings IEEE Conference Computing Visual Pattern Recognition. (CVPR), pp. 652–660 (2017)
3. Wu, Z., et al.: 3D ShapeNets: a Deep Representation for Volumetric Shapes. In: Proceeding IEEE Conference Computing Visual Pattern Recognition. (CVPR), pp. 1912–1920 (2015)
4. Feng, Y., Feng, Y., You, H., Zhao, X., Gao, Y.: MeshNet: mesh Neural Network for 3D Shape Representation. In: Proceedings AAAI Conference Artificial Intelligent (AAAI), vol. 33, pp. 8279–8286 (2019)
5. Yin, G., Xiao, X., Cirak, F.: Topologically Robust CAD Model Generation for Structural Optimisation. Comput. Methods Appl. Mech. Eng. **369**, 113102 (2020)
6. Wang, Y., Sun, Y., Liu, Z., Sarma, S.E., Bronstein, M.M., Solomon, J.M.: Dynamic Graph CNN for Learning on Point Clouds. ACM Trans. Graph. **38**, 1–12 (2019)
7. Ansaldi, S., De Floriani, L., Falcidieno, B.: Geometric Modeling of Solid Objects by Using a Face Adjacency Graph Representation. In: Proceedings SIGGRAPH, pp. 131–139 (1985)
8. Jayaraman, P.K., et al.: UV-Net: learning from Boundary Representations. In: Proceedings IEEE/CVF Conference Computing Visual Pattern Recognition. (CVPR), pp. 11703–11712 (2021)
9. Lou, Y., Li, X., Chen, H., Zhou, X.: BRep-BERT: pre-training Boundary Representation BERT with Sub-graph Node Contrastive Learning. In: Proceedings ACM International Conference Information Knowledge Management (CIKM), pp. 1657–1666 (2023)
10. Wu, H., Lei, R., Peng, Y., Gao, L.: AAGNet: a Graph Neural Network Towards Multi-task Machining Feature Recognition. Robot. Comput.-Integr. Manuf. **86**, 102661 (2024)
11. Zhang, P., et al.: TransGNN: harnessing the Collaborative Power of Transformers and Graph Neural Networks for Recommender Systems. In: Proceedings International ACM SIGIR Conference Research Development Information Retrieval. (SIGIR), pp. 1285–1295 (2024)
12. Wang, H., Li, B., Fan, S., Wu, Y., Liu, X.: TimeSQL: improving Multivariate Time Series Forecasting with Multi-Scale Patching and Smooth Quadratic Loss. Inf. Sci. **671**, 120652 (2024)
13. Song, X., et al.: Evolutionary Multi-Objective Spiking Neural Architecture Search for Image Classification. IEEE Trans. Evol, Comput (2025)
14. Hanocka, R., Hertz, A., Fish, N., Giryes, R., Fleishman, S., Cohen-Or, D.: MeshCNN: a Network with an Edge. ACM Trans. Graph. **38**, 1–12 (2019)

15. Qi, C.R., Yi, L., Su, H., Guibas, L.J.: PointNet++: deep Hierarchical Feature Learning on Point Sets in a Metric Space. In: Advance Neural Information Process Systems (NeurIPS), vol. 30 (2017)
16. Yu, X., Tang, L., Rao, Y., Huang, T., Zhou, J., Lu, J.: Point-BERT: pre-Training 3D Point Cloud Transformers with Masked Point Modeling. In: Proceeding IEEE/CVF Conference Computing Visual Pattern Recognition(CVPR), pp. 19313–19322 (2022)
17. Lambourne, J.G., Willis, K.D.D., Jayaraman, P.K., Sanghi, A., Meltzer, P., Shayani, H.: BRepNet: a Topological Message Passing System for Solid Models. In: Proceedings IEEE/CVF Conference Computing Visual Pattern Recognition (CVPR), pp. 12773–12782 (2021)
18. Kipf, T.N., Welling, M.: Semi-Supervised Classification with Graph Convolutional Networks. In: Proceedings International Conference Learning Representation (ICLR) (2017)
19. Velickovic, P., Cucurull, G., Casanova, A., Romero, A., Lio, P., Bengio, Y.: Graph Attention Networks. In: Proceedings International Conference Learning Representation (ICLR) (2018)
20. Xu, K., Hu, W., Leskovec, J., Jegelka, S.: How Powerful are Graph Neural Networks? In: Proceedings International Conference Learning Representation (ICLR) (2019)
21. Zhou, J., et al.: Graph Neural Networks: a Review of Methods and Applications. AI Open **1**, 57–81 (2020)
22. Kawasaki, T., Jayaraman, P.K., Shida, K., Zheng, J., Maekawa, T.: An Image Processing Approach to Feature-Preserving B-Spline Surface Fairing. Comput.-Aided Des. 99, 1–10 (2018)

Cloud-RAIN: Point Cloud Analysis with Reflectional Invariance

Yiming Cui[1], Lecheng Ruan[2], Hang-Cheng Dong[3], Qiang Li[4], Zhongming Wu[5], Tieyong Zeng[5], and Feng-Lei Fan[5](✉)

[1] Department of Electrical and Computer Engineering, University of Florida, Gainesville, FL, USA
[2] BIGAI, Beijing, China
[3] School of Instrumentation Science and Engineering, Harbin Institute of Technology, Harbin, China
[4] Department of Electrical and Computer Engineering, McMaster University, Hamilton, ON, Canada
[5] Department of Mathematics, The Chinese University of Hong Kong, New Territories, Hong Kong
hitfanfenglei@gmail.com

Abstract. The networks for point cloud tasks are expected to be invariant when the point clouds are affinely transformed such as rotation and reflection. So far, relative to the rotational invariance that has been attracting major research attention in the past years, the reflection invariance is little addressed. Notwithstanding, reflection symmetry can find itself in very common and important scenarios, *e.g.*, static reflection symmetry of structured streets, dynamic reflection symmetry from bidirectional motion of moving objects (such as pedestrians), and left- and right-hand traffic practices in different countries. To the best of our knowledge, unfortunately, no reflection-invariant network has been reported in point cloud analysis till now. To fill this gap, we propose a framework by using quadratic neurons and PCA canonical representation, referred to as Cloud-RAIN, to endow point Cloud models with ReflectionAl INvariance. We prove a theorem to explain why Cloud-RAIN can enjoy reflection symmetry. Furthermore, extensive experiments also corroborate the reflection property of the proposed Cloud-RAIN and show that Cloud-RAIN is superior to data augmentation. Our code is available at https://github.com/YimingCuiCuiCui/Cloud-RAIN.

Keywords: Point Cloud Analysis · Quadratic Networks

1 Introduction

Recently, point clouds have been intensively studied due to their broad applications as a basic representation of 3D models in autonomous driving and robotics. Deep learning techniques are introduced into this field for either direct analysis

Z. Zhang (Ed.): ISNN 2026, LNCS 16848, pp. 62–78, 2027.
https://doi.org/10.1007/978-981-92-5870-3_6

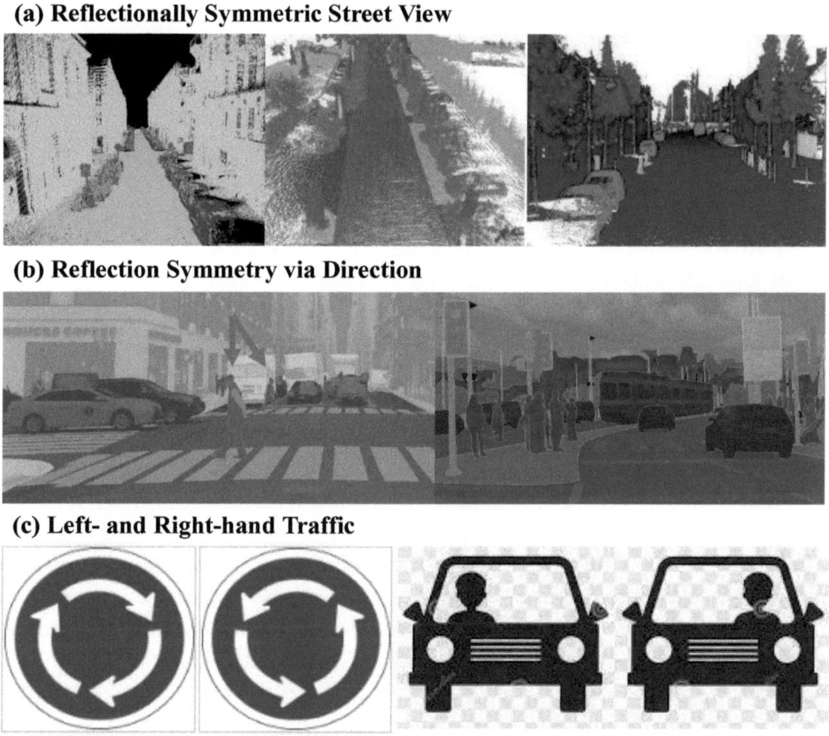

Fig. 1. In self-driving scenarios, front view of cars often exhibits reflection symmetry, which could include (a) static reflection symmetry of structured streets, (b) dynamic reflection symmetry from bidirectional motion of moving objects (such as pedestrians), and (c) left- and right-hand traffic practices in different countries (such as UK and US).

[1–4] or indirect analysis represented by PointNet [5] and its successors [6,7]. As a point cloud is an unordered set, previous direct analysis models focus on the invariance to sample permutation. However, in addition to the invariance to point permutation, the networks are also supposed to infer consistently when the point clouds are affinely transformed such as translation, rotation, and reflection. While the bias caused by translation can be counteracted easily by centralization, past years have witnessed a plethora of works [8–11] proposed to address the invariance to rotation and achieved the encouraging success, *e.g.*, rotation data augmentation, spherical-related convolutions, explicit local rotation-invariant features, and canonical poses.

However, relative to the rotational symmetry, the reflection symmetry is little addressed. Notwithstanding, reflection symmetry can find itself in very common and important scenarios. Often, reflection symmetry appears due to the symmetric city planning. Figure 1(a) shows one such example in autonomous driving, where the driver's front view exhibits reflection symmetry in well-structured traffic scenes; Also, reflection symmetry is obtained by the direction of movement. Figure 1(b) manifests that pedestrians crossing the street to the left and to the right are symmetric in reflection. A devastating car accident may happen if a self-driving car only identifies the pedestrians walking along one direction and fails the opposite; In addition, reflectional symmetry can be given rise by the left- and right-hand traffic practices in different countries, as Fig. 1(c) shows. Furthermore, we argue that reflection symmetry is more common than rotation symmetry in some important scenarios such as driving, where the environment around an autonomous vehicle is unlikely to rotate severely but likely to exhibit reflection symmetry. Therefore, reflectional invariance should be examined dedicatedly in a point cloud model.

However, to the best of our knowledge, no reflection-invariant network has been claimed in point cloud analysis till now. To deal with this issue, here we shift our attention to the quadratic neurons and networks. Recently, motivated by introducing neuronal diversity in deep learning, the so-called quadratic neuron [12] was designed by using a simplified quadratic function to replace the inner product in the conventional neuron. Hereafter, we call a network made of quadratic neurons a quadratic network and a network made of conventional neurons a conventional network. *Why can quadratic networks address reflectional symmetry?* Here, we prove that a quadratic network can approximate a reflectionally symmetric function well compared to conventional networks. The idea is that due to the invariance to the sign flipping of the input, the power term in the quadratic neuron can directly deal with the reflection of the input with respect to axis-aligned planes. Next, we combine quadratic networks with PCA to build a point Cloud method with ReflectionAl INvariance, referred to as Cloud-RAIN. While quadratic neurons deal with the reflectional symmetry due to the sign flipping, the PCA canonical transform derives canonical poses of a point cloud, which can transform the problem of reflection across arbitrary planes into a problem of sign flipping. Thus, Cloud-RAIN can tackle a general arbitrary-plane reflectional invariance problem. Our contributions are threefold:

- We identify reflection invariance as an essential yet overlooked concern in point cloud analysis.
- We prove a theorem showing that a quadratic network is powerful in approximating a reflectionally-invariant function. Then, we propose Cloud-RAIN that combines quadratic neurons and the PCA transform for point cloud applications.
- Empirically, Cloud-RAIN is invariant to reflection and much better than reflection data augmentation.

2 Related Works

Deep Learning-based Point Cloud Analysis. Currently, deep learning methods for point cloud analysis can be divided into two classes: indirect and direct. Indirect methods transform point clouds either to multi-view images or volumetric data followed by a 2D CNN or a 3D CNN, including VoxelNet [1], SPLATNet [13], and PCNN [4]. In contrast, direct methods represented by PointNet [5], PointNet++ [6], RS-CNN [14], PointWeb [15], KPConv [16], PointConv [17], and so on [18–21]. Direct methods are able to capture extrinsic features like 3D corners, edges, and local shape parts. Concurrently, graph-based methods build a graph by considering each point in a point cloud as a vertex and generating directed edges based on the vertices' neighborships. Then, graph networks perform feature learning in spatial or spectral domains. Representative graph models include DGCNN [22], LDGCNN [23], DeepGCNs [24], PointGCN [25], GAC-Net [26], and so on. Recently, Transformers [27] were more and more introduced in the point cloud tasks [28–31]. One common concern regarding deep models is that these models are vulnerable to symmetric transforms when data augmentation is not applied.

Rotation Symmetry. Intensive investigations have been conducted on the rotation-invariant property. These studies are roughly divided into three categories: i) rotation-invariant operations [9,10,32–35], which designs rotation-invariant convolutions or attentions in a network to achieve invariance; ii) explicit local rotation-invariant features [8,11,36–38], which handcrafts rotationally invariant features based on the intrinsic geometric relations of point clouds such as relative distances and angles; and iii) canonical poses [39–41], which transforms point cloud data to their PCA-based canonical poses that preserve shape information to achieve rotational invariance. We think that reflection symmetry is more important than rotation symmetry in some important applications such as driving, where buildings, views, and pedestrians around a normally-moving car are unlikely to encounter severe rotation but very likely to exhibit reflection symmetry. However, unfortunately, endowing a point cloud model with the reflection-invariant property was little studied.

Polynomial Neural Networks. The story of polynomial neurons begins with the Group Method of Data Handling (GMDH [42]), which gradually learns higher-order terms as a feature extractor. Furthermore, the so-called higher-order unit was proposed in [43,44] by adding a nonlinear activation $\sigma(\cdot)$ after GMDH: $y = \sigma(Y(\mathbf{x}_1, \cdots, \mathbf{x}_n))$. To balance the power of high-order units and parametric efficiency, Milenkoiv *et al.* [45] only utilized linear and quadratic terms and proposed to use the annealing technique to find optimal parameters. Recently, higher-order, particularly quadratic units were intensively studied [12,46–49] from the perspective of neuronal diversity. The neurons in [47,49–51] are a special case of [52]. [47] excludes the power term, and [50] only has the power term, which renders an incomplete quadratic representation. This work intends to make the first attempt to introduce quadratic neurons into point cloud analysis to realize reflection invariance.

3 Power of Quadratic Networks in Reflectional Invariance

Here, we illustrate why quadratic neurons are suitable for point cloud tasks requiring reflection invariance. We provide a theorem that a quadratic network is powerful in approximating a symmetric function, *e.g.*, using much fewer parameters than a conventional one. This suggests that for tasks requiring a model to be reflectionally symmetric, quadratic networks are favored over conventional ones, which makes the employment of quadratic networks in reflectional invariance well justified. The core idea of this theorem is that the power term in a quadratic neuron are straightforwardly reflection-invariant, while it is costly for the conventional to learn power terms.

Quadratic Neurons. Given an input vector $\mathbf{f} = (f_1, f_2, \cdots, f_n)$, a conventional neuron computes an inner product plus a bias b followed by a nonlinear activation σ:

$$\sigma(\mathbf{f}^\top \mathbf{w} + b), \tag{1}$$

while a quadratic neuron computes two inner products and one power term and then feeds them into a nonlinear activation function. Mathematically, its inner-working is

$$\sigma\big((\mathbf{f}^\top \mathbf{w}_1 + b_1)(\mathbf{f}^\top \mathbf{w}_2 + b_2) + (\mathbf{f} \circ \mathbf{f})^\top \mathbf{w}_3 + b_3\big), \tag{2}$$

where $\circ$ is a Hadamard product, $\mathbf{w}_1$, $\mathbf{w}_2$, and $\mathbf{w}_3$ are parameters. Compared to the designs of quadratic neurons in [46,53] with a complexity $\mathcal{O}(n^2)$, Eq. 2 is much more compact with the complexity of $\mathcal{O}(3n)$, scaling linearly with the number of features n.

Definition 1 (reflectionally invariant function). *We call a function* $h(\mathbf{x}) = h(\mathbf{x}_1, \mathbf{x}_2, \cdots, \mathbf{x}_d) : (\mathbb{R}^N)^d \to \mathbb{R}$ *reflectionally invariant, if for any* $\pi \in \{-1, 1\}^d$, $h(\mathbf{x})$ *satisfies*

$$h(\mathbf{x}_1, \mathbf{x}_2, \cdots, \mathbf{x}_d) = h(\pi_1\mathbf{x}_1, \pi_2\mathbf{x}_2, \cdots, \pi_d\mathbf{x}_d), \tag{3}$$

where N is the number of samples, d is the number of features, $\mathbf{x}_i \in \mathbb{R}^N$, and $\prod_{i=1}^d \pi_i = -1$.

When $\prod_{i=1}^d \pi_i = -1$, $\pi_1\mathbf{x}_1, \pi_2\mathbf{x}_2, \cdots, \pi_d\mathbf{x}_d$ is a reflection of $\mathbf{x}_1, \mathbf{x}_2, \cdots, \mathbf{x}_d$. When $\prod_{i=1}^d \pi_i = 1$, $\pi_1\mathbf{x}_1, \pi_2\mathbf{x}_2, \cdots, \pi_d\mathbf{x}_d$ is a rotation of $\mathbf{x}_1, \mathbf{x}_2, \cdots, \mathbf{x}_d$. It is also important for a model to maintain rotation invariance over samples. However, previous literature has studied rotation invariance. Therefore, rotation invariance is not the focus of our draft.

Proposition 1 (Proposition 2 in [54]). *Given $M > 0$ and $\epsilon \in (0,1)$, there is a ReLU network with two inputs that implements a function $\widetilde{\times} : \mathbb{R}^2 \to \mathbb{R}$ so that a) for any inputs x, y, if $|x| \le M$ and $|y| \le M$, then $|\widetilde{\times}(x,y) - xy| \le \epsilon$; b) if $x = 0$ or $y = 0$, then $\widetilde{\times}(x,y) = 0$; c) the number of computation units in this ReLU network is $O(\ln(1/\epsilon))$.*

Proposition 2 (Lemma 1 in [55]). *There exists a one-hidden-layer quadratic ReLU network that can implement a mapping $\tilde{\times} : \mathbb{R}^2 \to \mathbb{R}$, satisfying that i) $\tilde{\times}(x,y) = xy$; ii) $\tilde{\times}(x,y)$ has 2 quadratic neurons and accordingly 4 parameters.*

Theorem 1. *Let* $f : \left(\mathbb{R}^N\right)^d \to \mathbb{R}$ *be a continuously differentiable and totally symmetric function, where* Ω *is a compact subset of* $\mathbb{R}^d$*. Let* $\delta > 0$*, then there exists* $\tilde{\mathbf{F}} : \left(\mathbb{R}^N\right)^d \to \mathbb{R}$*, such that for any* X*, we have*

$$\sup_X \left| f(X) - \tilde{\mathbf{F}}(X) \right| \leq \delta, \tag{4}$$

where $\tilde{\mathbf{F}}(X) = \sum_{l=1}^{L} c_l \left(\tilde{\times}_{i=1}^{N} \tilde{\times}_{\alpha=1}^{d} X_{i,\alpha}^{\phi_{i,\alpha}(l)} \right)$*, and* L *is a function of* δ*,* c_l *is a coefficient for different* l*, and* $\phi_{i,\alpha}(l)$ *is the power function dependent on* i *and* α*. To represent* $\tilde{\mathbf{F}}$ *in the sense of the max norm, a conventional network needs* $NdLO(\ln(1/\delta))$ *parameters, while a quadratic network only needs* $4NdL$ *parameters.*

The Sketch of Proof: We first show a polynomial approximation scheme to the symmetric function f, which is $F(\mathbf{x})$. Then, we construct the quadratic and the conventional networks to represent $F(\mathbf{x})$, respectively, with a comparison of the approximation error and the number of parameters.

Proof. **Step 1.** We first consider the case of $N = 1$. According to the Stone-Weierstrass theorem, a polynomial can approximate any continuous function defined on a closed interval. Similarly, based on the fundamental theorem of symmetric polynomials [56], f can be naturally approximated by a polynomial of elementary symmetric polynomials made of power terms. Mathematically, for any symmetric polynomial P, we have $P(X) = Q\left(e_1(X), \ldots, e_d(X)\right)$, where Q is some polynomial, and e_k is as the following:

$$e_k(X) = \sum_{1 \leq j_1 < j_2 < \cdots < j_k \leq d} x_{j_1}^2 x_{j_2}^2 \cdots x_{j_k}^2. \tag{5}$$

To extend the result to the case of $N > 1$, the monomial takes the form

$$F(X) = \prod_{i=1}^{d} \prod_{\alpha=1}^{N} X_{i,\alpha}^{\phi_{i,\alpha}}, \tag{6}$$

where $\phi_{i,\alpha}$ is an even integer, which ensures that this monomial is a reflectionally invariant function. $\phi_{i,\alpha}$ is determined by the target function and the index α, i. The approximation using a symmetric polynomial can be a linear combination of L symmetrized monomials.

$$\mathbf{F}(X) = \sum_{l=1}^{L} c_l F_i^{(l)}(X) = \sum_{l=1}^{L} c_l \left(\prod_{i=1}^{d} \prod_{\alpha=1}^{N} X_{i,\alpha}^{\phi_{i,\alpha}(l)} \right), \tag{7}$$

where L is a function of δ, such that

$$\sup_X |f(X) - \mathbf{F}(X)| \leq \delta/2. \tag{8}$$

Step 2. Now, let us use a conventional and a quadratic neural network to approximate $\mathbf{F}(X)$.

Approximation by a Conventional Network: We can approximate this product by a conventional neural network with the help of Proposition 1. Let $\widetilde{\times}$ be the approximate multiplication. Applying $\widetilde{\times}$ in a compositional manner,

$$\sup_X \left| \widetilde{\times}_{i=1}^{N} \widetilde{\times}_{\alpha=1}^{d} X_{i,\alpha}^{\phi_{i,\alpha}} - \prod_{i=1}^{N} \prod_{\phi=1}^{d} X_{i,\alpha}^{\phi_{i,\alpha}} \right| \leq \delta/2. \tag{9}$$

Let $\tilde{\mathbf{F}} = \sum_{l=1}^{L} c_l \left(\widetilde{\times}_{i=1}^{N} \widetilde{\times}_{\alpha=1}^{d} X_{i,\alpha}^{\phi_{i,\alpha}(l)} \right)$, it can also approximate $f(X)$ based on the triangle inequality:

$$\sup_X \left| f(X) - \tilde{\mathbf{F}}(X) \right| \leq \delta. \tag{10}$$

The number of parameters used to construct $\tilde{\mathbf{F}}$ is $NdLO(\ln(1/\delta))$.

Approximation by a Quadratic Network: A quadratic neuron can precisely approximate a product function without incorporating extra error like a conventional network. To express $\tilde{\mathbf{F}} = \sum_{l=1}^{L} \widetilde{\times}_{i=1}^{N} \widetilde{\times}_{\alpha=1}^{d} X_{i,\alpha}^{\phi_{i,\alpha}(l)}$, the number of parameters used in a quadratic network is $4NdL$.

Remark 1. The above constructive theorem informs us that a conventional network is hard to learn reflectionally invariant functions, *i.e.*, much more parameters, but it is much easier for quadratic networks because quadratic neurons can easily represent a power term and a multiplication operation. Our main theorem heavily relies on [57]. But our highlight is to permute the dimension, while their result permutes samples. Moreover, we intend to use quadratic neurons to reimplement this scheme to show their superiority in representing reflectional symmetry. One may ask if there are other kinds of constructs. Actually, one can also partition the domain into a lattice with a small grid size to approximate $f(X)$ instead of using the linear combination. But still, the quadratic network will show the advantages because the power term in the quadratic neuron is essentially suitable for reflectional invariance.

4 Cloud-RAIN

Inspired by the above encouraging theoretical result, here we propose a point Cloud method which enjoys the ReflectionAl INvariance (Cloud-RAIN) and whose spotlight is the incorporation of quadratic neurons. In addition, we also equip Cloud-RAIN with the PCA canonical transform, because the quadratic neuron alone can only tackle reflection across axis-aligned planes. When combining the PCA canonical transform, quadratic neurons can deal with reflection with respect to arbitrary planes.

4.1 Quadratic Neurons

Given a set of points $\mathbf{X} \in \mathbb{R}^{N \times C}$, where N represents the number of points, and F is the number of features. Usually, $C = 3 + c$ denotes (x, y, z) position in the geometric space, and the feature number (*e.g.*, $c = 3$ for points only with color; $c = 6$ for those with color and normals). Existing point cloud analysis methods usually focus on how to engineer $\mathbf{w}$ or $\mathbf{f}$ to fulfill the needs and characters of the problems. For example, RS-CNN [14] restricts $\mathbf{w}$ to learn the relations between p_i and p_j in geometric space and extends a regular rigid CNN to the irregular configuration. DGCNN [22] proposes the EdgeConv that operates on the edges between the central point and its neighbors and treats $\mathbf{f}$ as the edge features. Point Transformer [28] uses the vector self-attention operator to aggregate local features and the subtraction relation to generate $\mathbf{w}$. Despite these arts, they all use conventional neurons which integrate $\mathbf{w}$ and $\mathbf{f}$ in a linear fashion, as Eq. (1) shows.

Here, the proposed Cloud-RAIN replaces the simple linear relation between $\mathbf{f}$ and $\mathbf{w}$ with a quadratic interaction by quadratic neurons. Theorem 1 shows that quadratic neurons are powerful in learning a reflectionally invariant function.

4.2 PCA Canonical Transform

Note $(\mathbf{x}_1, \mathbf{x}_2, \cdots, \mathbf{x}_D) \rightarrow (\pi_1\mathbf{x}_1, \pi_2\mathbf{x}_2, \cdots, \pi_D\mathbf{x}_D)$ is a special case of reflection symmetry, where the planes of symmetry align with the axes. Here, we introduce PCA canonical transform to empower the quadratic aggregation to deal with reflection across an arbitrary plane. We first show how the PCA transform is performed on the point cloud so as to obtain its canonical representation. Without loss of generality, let the point cloud be $\mathbf{X} \in \mathbb{R}^{N \times 3}$, $\mathbf{X}_i \in \mathbb{R}^{3 \times 1}$ be the i-th point of $\mathbf{X}$, and $\overline{\mathbf{X}} = (\sum_{i=1}^{N} \mathbf{X}_i)/N \in \mathbb{R}^{3 \times 1}$ be the geometric center of $\mathbf{X}$, we perform the PCA decomposition for $\mathbf{X}$ as follows:

$$\frac{\sum_{i=1}^{N} \left(\mathbf{X}_i - \overline{\mathbf{X}}\right)\left(\mathbf{X}_i - \overline{\mathbf{X}}\right)^T}{N} = \mathbf{V}\mathbf{\Lambda}\mathbf{V}^\top, \tag{11}$$

where $\mathbf{V} \in \mathbb{R}^{3 \times 3}$ is the matrix composed of eigenvectors $(\mathbf{v}_1, \mathbf{v}_2, \mathbf{v}_3)$ and $\mathbf{\Lambda} = \text{diag}(\lambda_1, \lambda_2, \lambda_3)$ are the corresponding eigenvalues. By multiplying the point cloud with the eigenvector matrix, we obtain the canonical representation of the point cloud $\mathbf{X}$ as $\mathbf{X}_{\text{can}} = \mathbf{X}\mathbf{V}$.

The reflection-invariant property of $\mathbf{X}_{\text{can}}$ is illustrated in the following: Applying a random reflection matrix $\mathbf{F} \in \mathbb{R}^{3 \times 3}$ on the point cloud $\mathbf{X}$ to get a reflected point cloud $\mathbf{X}\mathbf{F}^\top$ and performing PCA decomposition for $\mathbf{F}\mathbf{X}$ as shown in Eq. (11), we have the following:

$$\begin{aligned}
&\frac{\sum \left(\mathbf{F}\mathbf{X}_i - \mathbf{F}\overline{\mathbf{X}}\right)\left(\mathbf{F}\mathbf{X}_i - \mathbf{F}\overline{\mathbf{X}}\right)^\top}{N} \\
=&\mathbf{F}\left(\frac{\sum \left(\mathbf{X}_i - \overline{\mathbf{X}}\right)\left(\mathbf{X}_i - \overline{\mathbf{X}}\right)^\top}{N}\right)\mathbf{F}^\top \\
=&(\mathbf{F}\mathbf{V})\mathbf{\Lambda}(\mathbf{F}\mathbf{V})^\top,
\end{aligned} \tag{12}$$

where $\mathbf{FV}$ is the eigenvector matrix of the reflected point cloud $\mathbf{XF}^\top$. Therefore, the canonical representation $\mathbf{XF}^\top$ is $\left(\mathbf{XF}^\top\right)_{\text{can}} = \mathbf{XF}^T \cdot \mathbf{FV} = \mathbf{XV}$. Thus, the effect of the reflection is counteracted in $\mathbf{X}_{\text{can}}$.

Table 1. Experiments on S3DIS and ScanNet.

Dataset	S3DIS		ScanNet	
Methods	mAcc	mIOU	mAcc	mIOU
PointNet [5]	53.1	45.0	44.3	41.9
PointNet [5]+Ours	56.5	48.1	45.6	43.7
PointNet++ [6]	62.0	53.9	53.4	54.4
PointNet++ [6]+Ours	64.4	56.0	57.3	57.6
DGCNN [22]	57.0	49.3	48.0	46.7
DGCNN [22]+Ours	61.9	50.6	51.0	49.8
LDGCNN [23]	59.1	49.9	50.6	49.9
LDGCNN [23]+Ours	61.4	50.6	51.6	51.8
GSNet [58]	52.7	44.5	43.7	45.2
GSNet [58]+Ours	54.5	46.2	44.9	46.4
ShellNet [59]	58.3	49.5	49.7	49.0
ShellNet [59]+Ours	61.2	50.4	50.8	51.7
PointMLP [20]	80.3	70.1	68.9	69.3
PointMLP [20] + Ours	81.5	70.9	70.1	70.5
PointMixer [60]	80.5	70.7	69.7	70.6
PointMixer [60] + Ours	81.7	71.8	70.9	71.8

Although employing the canonical representation can realize the reflection invariance with respect to arbitrary planes, it brings up the issue of sign ambiguity. Given an eigenvector $\mathbf{v}$, both choosing $+\mathbf{v}$ and $-\mathbf{v}$ fulfills the rule of PCA transform. What's undesirable is we don't know if it is $+\mathbf{v}$ or $-\mathbf{v}$ being selected. Because $\mathbf{V} = [\mathbf{v}_1, \mathbf{v}_2, \mathbf{v}_3]$, there are eight possible combinations ($[\pm\mathbf{v}_1, \pm\mathbf{v}_2, \pm\mathbf{v}_3]$) that cause sign ambiguity in the canonical representation of $\mathbf{X}_{\text{can}}$. This is the intrinsic problem of using the canonical representation. However, quadratic aggregation can overcome the sign ambiguities due to its power term based on our earlier analyses.

Remark 2. (Theoretical guarantee). The canonical representation alone can deal with reflection across the arbitrary plane but suffers sign ambiguity, while the quadratic aggregation alone is invariant to the sign flipping but only works for reflection across axis-aligned planes. It is the combination between the canonical representation and quadratic aggregation that realizes the invariance to reflection across arbitrary planes. The advantage of such a combination is that it has a theoretical guarantee to be invariant to reflection, which will be confirmed by our experiments later.

5 Experiments

To validate the reflection invariance of the proposed Cloud-RAIN, we conduct extensive experiments on point cloud semantic segmentation with multiple recent networks as a backbone. Other supplementary results such as model complexity analysis are put into the supplementary materials.

Table 2. Performance drops on different reflection operations.

Model	x axis		y axis		z axis		xyz axis	
	ΔmAcc	ΔmIOU	ΔmAcc	ΔmIOU	ΔmAcc	ΔmIOU	ΔmAcc	ΔmIOU
PointNet [5]	↓ 3.5%	↓ 4.1%	↓ 2.6%	↓ 3.1%	↓ 45.0%	↓ 43.8%	↓ 45.8%	↓ 44.2%
PointNet [5] + Ours	↓ 0%	↓ 0%	↓ 0%	↓ 0%	↓ 0%	↓ 0%	↓ 0%	↓ 0%
DGCNN [22]	↓ 3.8%	↓ 4.4%	↓ 3.5%	↓ 3.6%	↓ 44.3%	↓ 49.2%	↓ 45.3%	↓ 49.3%
DGCNN [22] + Ours	↓ 0%	↓ 0%	↓ 0%	↓ 0%	↓ 0%	↓ 0%	↓ 0%	↓ 0%
PointMLP [20]	↓ 4.3%	↓ 5.1%	↓ 3.9%	↓ 4.1%	↓ 64.7%	↓ 68.9%	↓ 66.1%	↓ 69.2%
PointMLP [20] + Ours	↓ 0%	↓ 0%	↓ 0%	↓ 0%	↓ 0%	↓ 0%	↓ 0%	↓ 0%
PointMixer [60]	↓ 4.9%	↓ 5.7%	↓ 4.4%	↓ 4.9%	↓ 65.1%	↓ 69.4%	↓ 66.5%	↓ 69.9%
PointMixer [60] + Ours	↓ 0%	↓ 0%	↓ 0%	↓ 0%	↓ 0%	↓ 0%	↓ 0%	↓ 0%

5.1 Experimental Setups

We mainly use S3DIS [61] and ScanNet [62], the mainstream point cloud semantic segmentation benchmarks, to evaluate our proposed methods. We use the widely-adopted mean accuracy (mAcc) and IOU (mIOU) as evaluation metrics. For S3DIS, following the protocol of previous work [5,6,22,59,63,64], we train for 100 epochs with 4 T V100 GPUs. The batch size is set to 32. Following the standard practice, the raw input points are firstly grid-sampled to generate 4,096 points. Unless otherwise specified, we use scale and jitter as data augmentation. For ScanNet, we train for 100 epochs with weight decay and batch size set to 0.1 and 32, respectively. The number of input points is set to be 8,192 by sampling. Except for random jitter, the data augmentation of the ScanNet is the same as that of the S3DIS. We re-implement the original models for a fair comparison and only replace the linear aggregation with the quadratic one without changing other elements of the model.

5.2 Segmentation Results

We report the basic segmentation results in Table 1. All compared models are either classic models or recently published in flagship venues. The spotlight is that our proposed Cloud-RAIN can universally boost the existing models' performance by a reasonably large margin. In detail, integrated with our Cloud-RAIN, the current models' mAcc and mIOU scores are improved by at least 1.8% and 0.7% on the S3DIS benchmark, respectively. For a simple model like PointNet, our proposed Cloud-RAIN can boost the performance on mAcc and mIOU by 3.4% and 3.1%. On the ScanNet benchmark, the improvement over ScanNet is not on par with that on the S3DIS dataset. However, our proposed methods can still escalate the mAcc and mIOU scores by at least 1.0% relative to the existing methods. We think that if normal features over the ScanNet benchmark are considered, the improvement could be more significant. Although reflection symmetry is the primary goal of this manuscript, we would like to note to readers that the Cloud-RAIN is an effective drop-in replacement to escalate the performance of the original models.

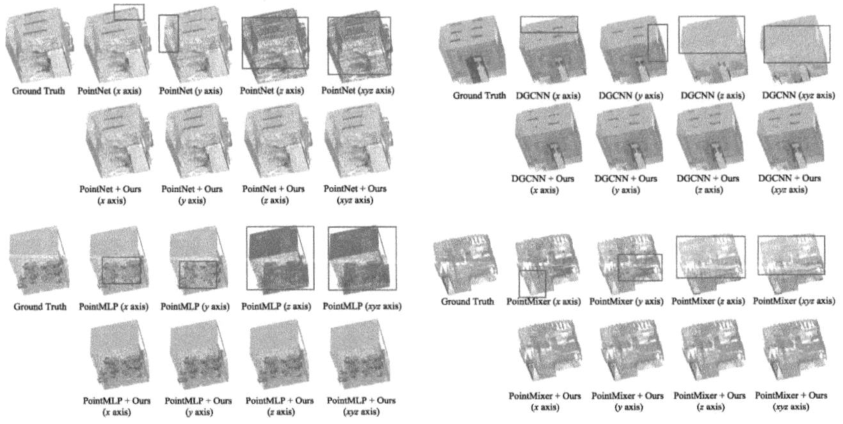

Fig. 2. Comparison of the segmentation results with different reflections.

5.3 Reflection Invariance Analysis

Now, we provide detailed experiments to analyze the reflection-invariance properties of our proposed Cloud-RAIN. In the first part, we concentrate on reflections across axis-aligned planes that are very common in realistic scenarios such as symmetric street views. No PCA transform needs to be used because quadratic aggregation alone can deal with such reflection operations. In the second part, reflection across arbitrary planes is tested.

Reflection Invariance (Axis-Aligned Planes). To analyze the effect of data reflection, we select PointNet, DGCNN, PointMLP, and PointMixer as examples to conduct experiments on the S3DIS benchmark. In the experiment, we flip the point clouds along the x, y, z, and xyz axis, respectively, and test the segmentation performance of the well-trained models with the reflected inputs, accordingly. During the training process, reflection augmentation is not applied for the original models and those integrated with our proposed methods. The results are listed in Table 2.

As seen in Table 2, our proposed Cloud-RAIN can keep the performance for the reflected inputs thanks to the effect of power terms. This is mathematically guaranteed because power terms return the same value for the flipped inputs. In contrast, the existing methods cannot withstand the input reflection, evidenced by the dramatic performance drop. Especially when the reflection is applied to the z axis, the performance drop is devastating. This might be because the dynamic range of positions along the z axis is far more extensive than that along the x and y axes. Thus, the destruction of flipping the z axis to the model is severer.

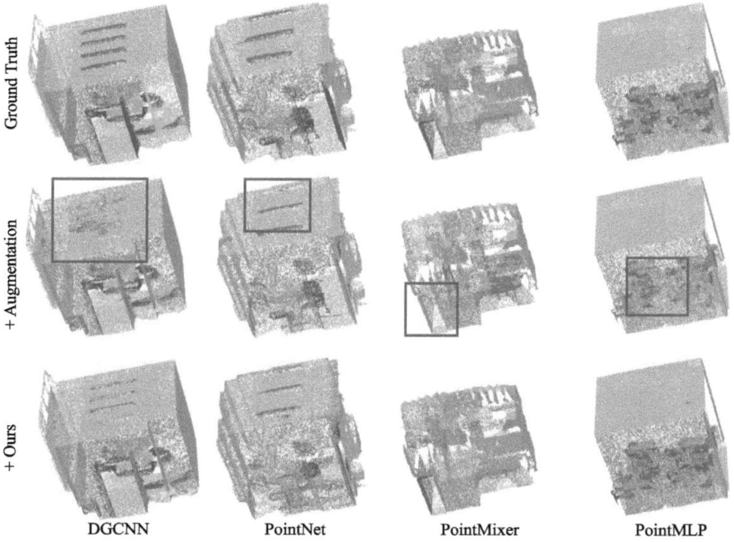

Fig. 3. Comparison of the segmentation results from different models with reflection across randomly-selected planes.

Figure 2 offers semantic segmentation results in different reflection settings. For better visualization, we view the semantic segmentation results from the same angle regardless of their reflection. An interesting finding from Fig. 2 is that when the point clouds are reflected along the z axis, the original model is handicapped in distinguishing the categories of ceiling and floors. Many points with the ceiling category are misclassified into the floor class, indicating that the model just learns a reflectionally asymmetric mapping.

Table 3. Performance of PointNet and DGCNN drops dramatically for reflection across arbitrary planes.

Model	ΔmAcc	Δ mIOU
PointNet	↓ 46.3%(±0.52%)	↓ 42.7(±0.32%)
PointNet + Aug	↓ 2.6%(±0.46%)	↓ 2.4%(±0.28%)
PointNet + PCA	↓ 1.7%(±0.38%)	↓ 1.9%(±0.32%)
PointNet + Ours	↓ 0%	↓ 0%
DGCNN	↓ 45.9%(±0.22%)	↓ 40.4%(±0.18%)
DGCNN + Aug	↓ 8.0%(±0.28%)	↓ 8.5%(±0.32%)
DGCNN + PCA	↓ 6.8%(±0.22%)	↓ 7.7%(±0.26%)
DGCNN + Ours	↓ 0%	↓ 0%
PointMLP	↓ 66.9%(±0.22%)	↓ 69.8%(±0.16%)
PointMLP + Aug	↓ 10.2%(±0.32%)	↓ 11.3%(±0.38%)
PointMLP + PCA	↓ 7.4%(±0.20%)	↓ 8.6%(±0.22%)
PointMLP + Ours	↓ 0%	↓ 0%
PointMixer	↓ 67.1%(±0.38%)	↓ 70.0%(±0.42%)
PointMixer + Aug	↓ 11.0%(±0.30%)	↓ 11.9%(±0.34%)
PointMixer + PCA	↓ 8.6%(±0.26%)	↓ 9.4%(±0.28%)
PointMixer + Ours	↓ 0%	↓ 0%

Reflection Invariance (Arbitrary Planes). We conduct experiments with PointNet, DGCNN, PointMLP, and PointMixer on the S3DIS benchmark to study the reflection across arbitrary planes. We randomly generate reflection planes, across which we flip the point clouds. To contrast the effectiveness of the proposed Cloud-RAIN, we apply data augmentation during the training process, where the training data are flipped along randomly generated planes. In the inference stage, we generate a plane randomly and flip the point clouds across the plane. For reliability of results, we run 5 experiments and calculate the mean as well as the standard deviation. We also train the original models aided by PCA transform. All those comparative results are listed in Table 3.

From Table 3, we notice that without reflection data augmentation, all models suffer a severe performance loss for both mAcc and mIOU scores when there is no training augmentation. Even when the models are trained with randomly generated data augmentation, PointNet is still subjected to at least a 2.0% drop on both evaluation metrics, while DGCNN's mAcc and mIOU scores drop at least 8.0%. What is favorable is that the performance of our models does not have any compromise with randomly generated reflection planes, which agrees with our theoretical analyses that Cloud-RAIN has the theoretical guarantee for reflection invariance. Furthermore, aided by PCA transform, the existing models are still subjected to the performance drop, which means these models cannot overcome the sign ambiguity given rise by PCA. This means that quadratic aggregation is necessary for reflectional symmetry over arbitrary planes.

We visualize and compare a representative example from the S3DIS with respect to an arbitrary plane reflection as Fig. 3. We sample 4,096 points from the point clouds of each room and reflect those sampled points across the same randomly generated plane. Then, we merge the segmentation results to be the original room. From Fig. 3, we notice that when the test data are reflected across an arbitrary plane, the models trained without data augmentation cannot even accurately distinguish the categories like walls, ceilings, and so on. When it comes to the model trained with reflection data augmentation, though it is not severely affected by the arbitrary reflection during the inference time, there is still a structural distortion in distinguishing the boundaries compared with our proposed method, as highlighted in the red rectangle.

6 Conclusion

In this manuscript, we have identified the reflection invariance, alongside the rotation invariance, as a practical concern in point cloud applications. Then, we have derived a theorem suggesting that quadratic neurons can learn a reflectionally symmetric function much easier than conventional neurons do. Next, we have introduced quadratic neurons and PCA canonical transform to prototype a reflectionally-invariant model. Lastly, extensive comparison experiments show that embedding quadratic aggregation in the existing models can not only strengthen the reflection invariance as our theoretical analyses illustrate. In the future, more efforts can be invested in exploring polynomial aggregation to unify rotational and reflectional invariance ultimately.

References

1. Zhou, Y., Tuzel, O.: Voxelnet: end-to-end learning for point cloud based 3D object detection. In: CVPR (2018)
2. Xie, J., Dai, G., Zhu, F., Wong, E.K., Fang, Y.: Deepshape: deep-learned shape descriptor for 3D shape retrieval. TPAMI **39**(7), 1335–1345 (2017)
3. Su, H., Maji, S., Kalogerakis, E., Learned-Miller, E.G.: Multi-view convolutional neural networks for 3D shape recognition. In: ICCV (2015)
4. Atzmon, M., Maron, H., Lipman, Y.: Point convolutional neural networks by extension operators. TOG **37**(4), 1–12 (2018)
5. Qi, C.R., Su, H., Mo, K., Guibas, L.J.: Pointnet: deep learning on point sets for 3D classification and segmentation. In: CVPR (2017)
6. Qi, C.R., Yi, L., Su, H., Guibas, L.J.: Pointnet++: deep hierarchical feature learning on point sets in a metric space. In: NIPS (2017)
7. Liu, X., Qi, C.R., Guibas, L.J.: Flownet3D: learning scene flow in 3D point clouds. In: CVPR (2019)

8. Sun, X., Lian, Z., Xiao, J.: Srinet: learning strictly rotation-invariant representations for point cloud classification and segmentation. In: ACM Multimedia (2019)
9. Esteves, C., Allen-Blanchette, C., Makadia, A., Daniilidis, K.: Learning SO(3) equivariant representations with spherical CNNs. In: Ferrari, V., Hebert, M., Sminchisescu, C., Weiss, Y. (eds.) ECCV 2018. LNCS, vol. 11217, pp. 54–70. Springer, Cham (2018). https://doi.org/10.1007/978-3-030-01261-8_4
10. Zhang, Z., Hua, B.-S., Rosen, D.W., Yeung, S.-K.: Rotation invariant convolutions for 3D point clouds deep learning. In: 3DV (2019)
11. Li, X., Li, R., Chen, G., Fu, C.-W., Cohen-Or, D., Heng, P.-A.: A rotation-invariant framework for deep point cloud analysis. TVCG (2021)
12. Fan, F., Cong, W., Wang, G.: A new type of neurons for machine learning. Int. J. Numer. Methods Biomed. Eng. **34**(2), e2920 (2018)
13. Su, H., et al.: SPLATNet: sparse lattice networks for point cloud processing. In: CVPR (2018)
14. Liu, Y., Fan, B., Xiang, S., Pan, C.: Relation-shape convolutional neural network for point cloud analysis. In: CVPR (2019)
15. Zhao, H., Jiang, L., Fu, C.-W., Jia, J.: PointWeb: enhancing local neighborhood features for point cloud processing. In: CVPR (2019)
16. Thomas, H., Qi, C.R., Deschaud, J.-E., Marcotegui, B., Goulette, F., Guibas, L.J.: Kpconv: flexible and deformable convolution for point clouds. In: ICCV (2019)
17. Wu, W., Qi, Z., Fuxin, L.: Pointconv: deep convolutional networks on 3D point clouds. In: CVPR (2019)
18. Hu, Q., et al.: Randla-net: efficient semantic segmentation of large-scale point clouds. In: CVPR (2020)
19. Tang, L., Zhan, Y., Chen, Z., Yu, B., Tao, D.: Contrastive boundary learning for point cloud segmentation. In: CVPR (2022)
20. Ma, X., Qin, C., You, H., Ran, H., Fu, Y.: Rethinking network design and local geometry in point cloud: a simple residual MLP framework. In: ICLR (2022)
21. Nie, D., Lan, R., Wang, L., Ren, X.: Pyramid architecture for multi-scale processing in point cloud segmentation. In: CVPR (2022)
22. Wang, Y., Sun, Y., Liu, Z., Sarma, S.E., Bronstein, M.M., Solomon, J.M.: Dynamic graph CNN for learning on point clouds. TOG **38**(5), 1–12 (2019)
23. Zhang, K., Hao, M., Wang, J., de Silva, C.W., Fu, C.: Linked dynamic graph CNN: learning on point cloud via linking hierarchical features, *[cs]* (2019). arXiv:1904.10014
24. Li, G., Müller, M., Thabet, A., Ghanem, B.: DeepGCNs: can GCNs go as deep as CNNs?. In: ICCV (2019)
25. Zhang, Y., Rabbat, M.: A graph-CNN for 3D point cloud classification. In: ICASSP (2018)
26. Wang, L., Huang, Y., Hou, Y., Zhang, S., Shan, J.: Graph attention convolution for point cloud semantic segmentation. In: CVPR (2019)
27. Vaswani, A., et al.: Attention is all you need. In: NIPS (2017)
28. Zhao, H., Jiang, L., Jia, J., Torr, P.H., Koltun, V.: Point transformer. In: ICCV (2021)
29. Lai, X., et al.: Stratified transformer for 3D point cloud segmentation. In: CVPR (2022)
30. Park, C., Jeong, Y., Cho, M., Park, J.: Fast point transformer. In: CVPR (2022)
31. Zhang, C., Wan, H., Shen, X., Wu, Z.: Patchformer: an efficient point transformer with patch attention. In: CVPR (2022)
32. Poulenard, A., Rakotosaona, M.-J., Ponty, Y., Ovsjanikov, M.: Effective rotation-invariant point CNN with spherical harmonics kernels. In: 3DV (2019)

33. Rao, Y., Lu, J., Zhou, J.: Spherical fractal convolutional neural networks for point cloud recognition. In: CVPR (2019)
34. Fuchs, F., Worrall, D., Fischer, V., Welling, M.: Se (3)-transformers: 3D roto-translation equivariant attention networks. In: NIPS (2020)
35. Deng, C., Litany, O., Duan, Y., Poulenard, A., Tagliasacchi, A., Guibas, L.J.: Vector neurons: a general framework for so (3)-equivariant networks. In: ICCV (2021)
36. Chen, C., Li, G., Xu, R., Chen, T., Wang, M., Lin, L.: Clusternet: deep hierarchical cluster network with rigorously rotation-invariant representation for point cloud analysis. In: CVPR (2019)
37. Xu, J., Tang, X., Zhu, Y., Sun, J., Pu, S.: SGMnet: learning rotation-invariant point cloud representations via sorted gram matrix. In: ICCV (2021)
38. Yu, R., Wei, X., Tombari, F., Sun, J.: Deep positional and relational feature learning for rotation-invariant point cloud analysis. In: Vedaldi, A., Bischof, H., Brox, T., Frahm, J.-M. (eds.) ECCV 2020. LNCS, vol. 12355, pp. 217–233. Springer, Cham (2020). https://doi.org/10.1007/978-3-030-58607-2_13
39. Fujiwara, K., Hashimoto, T.: Neural implicit embedding for point cloud analysis. In: CVPR (2020)
40. Xiao, Z., Lin, H., Li, R., Geng, L., Chao, H., Ding, S.: Endowing deep 3D models with rotation invariance based on principal component analysis. In: ICME (2020)
41. Li, F., Fujiwara, K., Okura, F., Matsushita, Y.: A closer look at rotation-invariant deep point cloud analysis. In: ICCV (2021)
42. Ivakhnenko, A.G.: Polynomial theory of complex systems. TSMC **4**, 364–378 (1971)
43. Poggio, T.: On optimal nonlinear associative recall. Biol. Cybern. **19**(4), 201–209 (1975)
44. Giles, C.L., Maxwell, T.: Learning, invariance, and generalization in high-order neural networks. Appl. Opt. **26**(23), 4972–4978 (1987)
45. Milenkovic, S., Obradovic, Z., Litovski, V.: Annealing based dynamic learning in second-order neural networks. In: ICNN (1996)
46. Zoumpourlis, G., Doumanoglou, A., Vretos, N., Daras, P.: Non-linear convolution filters for CNN-based learning. In: ICCV (2017)
47. Tsapanos, N., Tefas, A., Nikolaidis, N., Pitas, I.: Neurons with paraboloid decision boundaries for improved neural network classification performance. TNNLS **30**(1), 284–294 (2018)
48. Chrysos, G.G., Moschoglou, S., Bouritsas, G., Deng, J., Panagakis, Y., Zafeiriou, S.: Deep polynomial neural networks. TPAMI **44**(8), 4021–4034 (2021)
49. Xu, Z., Yu, F., Xiong, J., Chen, X.: Quadralib: a performant quadratic neural network library for architecture optimization and design exploration. In: MLSys (2022)
50. Goyal, M., Goyal, R., Lall, B.: Improved polynomial neural networks with normalised activations. In: IJCNN (2020)
51. Bu, J., Karpatne, A.: Quadratic residual networks: a new class of neural networks for solving forward and inverse problems in physics involving PDEs. In: SDM (2021)
52. Fan, F., et al.: Quadratic autoencoder (q-ae) for low-dose CT denoising. TMI **39**(6), 2035–2050 (2019)
53. Jiang, Y., Yang, F., Zhu, H., Zhou, D., Zeng, X.: Nonlinear CNN: improving CNNs with quadratic convolutions. Neural Comput. Appl. **32**(12), 8507–8516 (2020)
54. Yarotsky, D.: Error bounds for approximations with deep RELU networks. Neural Netw. **94**, 103–114 (2017)
55. Fan, F.-L., et al.: One neuron saved is one neuron earned: on parametric efficiency of quadratic networks. arXiv preprint arXiv:2303.06316 (2023)

56. Macdonald, I.G.: Symmetric Functions and Hall Polynomials. Oxford University Press (1998)
57. Han, J., Li, Y., Lin, L., Lu, J., Zhang, J., Zhang, L.: Universal approximation of symmetric and anti-symmetric functions. Commun. Math. Sci. **20**(5), 1397–1408 (2022)
58. Xu, M., Zhou, Z., Qiao, Y.: Geometry sharing network for 3D point cloud classification and segmentation. In: AAAI (2020)
59. Zhang, Z., Hua, B.-S., Yeung, S.-K.: Shellnet: efficient point cloud convolutional neural networks using concentric shells statistics. In: ICCV (2019)
60. Choe, J., Park, C., Rameau, F., Park, J., Kweon, I.S.: Pointmixer: MLP-mixer for point cloud understanding. In: Avidan, S., Brostow, G., Cissé, M., Farinella, G.M., Hassner, T. (eds.) ECCV 2022. LNCS, vol. 13687, pp. 620–640. Springer, Cham (2022). https://doi.org/10.1007/978-3-031-19812-0_36
61. Armeni, I., et al.: 3D semantic parsing of large-scale indoor spaces. In: CVPR (2016)
62. Dai, A., Chang, A.X., Savva, M., Halber, M., Funkhouser, T., Nießner, M.: Scannet: richly-annotated 3D reconstructions of indoor scenes. In: CVPR (2017)
63. Cui, Y., Liu, X., Liu, H., Zhang, J., Zare, A., Fan, B.: Geometric attentional dynamic graph convolutional neural networks for point cloud analysis. Neurocomputing **432**, 300–310 (2021)
64. Wang, Y., Derr, T.: Tree decomposed graph neural network. In: CIKM (2021)

U-IMANet: U-Shape Interactive Multimodal Attention Network for Audio-Visual Scene Classification

Tianwei Liang and Ming Yin(✉)

School of Electronic Science and Engineering, South China Normal University, Foshan, China
m.yin@scnu.edu.cn

Abstract. Scene classification, as one of the hallmark tasks of computer vision, has been widely used in intelligent environment monitoring, including commercial area, industrial area and residential area. Due to the distribution difference of modalities being quite large, the direct concatenation in the feature-level fusion may not result in promising performance. To this end, in this paper, a novel U-shape Interactive Multimodal Attention Network, namely U-IMANet, is proposed for audio-visual scene classification. In particular, the features of the audio and video data are first extracted using a pre-trained model. Frame-Level Cross Attention (FCA) is then developed to preserve the appearance and structure of the objects and background as well as identities, by carrying over from the first frame to subsequent frames. Finally, a four-layer U-shape Multimodal Interactive Module (UMIM) is designed to mine the interaction between multi-modalities. Experimental results on two benchmark datasets demonstrate that the proposed U-IMANet outperforms state-of-the-art models, with competitive performance.

Keywords: Multimodal learning · Modality interactive · Scene classification · Audio-visual fusion · Attention mechanism

1 Introduction

Due to the superiority of attention mechanism in many fields, multimodal fusion based on attention mechanism for AVSC has become a mainstream scheme. Hou *et al.* proposed a scene classification system based on a cross-attention mechanism to effectively integrate audio and visual features [1]. Considering the fine-grained relationship between audio and visual modal within the intra-class scene, an AVSC model equipped with contrastive event-object alignment and semantic-based fusion is proposed in a follow-up study [2]. Lei *et al.* proposed a multimodal graph fusion model by constructing the sample-to-sample cross similarity graphs and then introducing the graph convolution network (GCN) to fuse the graphs from different modalities [3]. However, the distribution of features extracted from each modality is different to some extent. In aforementioned

Z. Zhang (Ed.): ISNN 2026, LNCS 16848, pp. 79–88, 2027.
https://doi.org/10.1007/978-981-92-5870-3_7

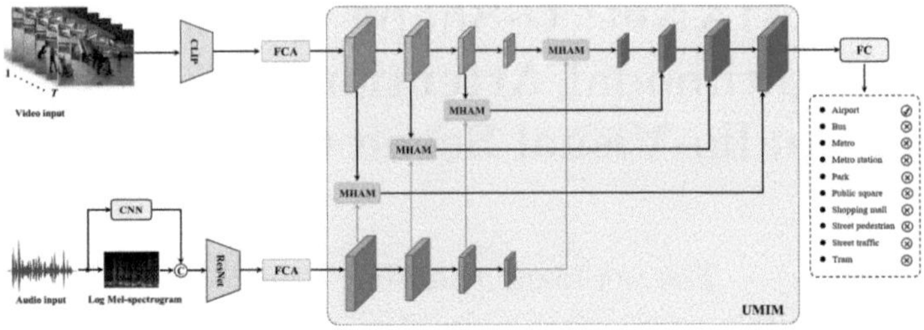

Fig. 1. An illustration of the proposed network U-IMANet. Frame-Level Cross Attention (FCA) aims to enhance the consistency of feature sequences. U-shape Multimodal Interactive Module (UMIM) is applied to fuse the audio features and visual features. ⓒ stands for the concatenation.

methods, neither concatenation nor fusion based on attention mechanism can fully exploit the consistency and complementarity of multimodal data, while the GCN-based approach requires additional specific graph structure in advance.

In addition, beyond the fusion, the temporal correlation among different modalities is worth paying attention to the AVSC task. In particular, the existing methods fail to take full advantage of the temporal correlation information for audio and visual modalities. To this end, this work proposes a novel U-shape Interactive Multimodal Attention Network (U-IMANet) for AVSC. From Fig. 1, the modal features are firstly extracted individually through a pre-trained model and fed into FCA blocks. In FCA blocks, the features of the first frame to subsequent frames are separately interacted through cross-attention, and then fed into four-layer U-shape Multimodal Interactive Module (UMIM) where the audio and visual features are fused by sharing attention weight to achieve the final scene classification result.

To recap, the main contributions of our work are on fourfold:

- Frame-level cross-attention (FCA) is designed to increase the time consistency of the sequences by utilizing the normal attention mechanism. FCA is to further improve the expressiveness of features on frame level.
- A U-shape Multimodal Interactive Module, termed by UMIM, is designed to capture complementary relationship between audio and visual modalities.
- Finally a U-shape Interactive Multimodal Attention Network, namely U-IMANet, equipped with FCA and UMIM is proposed for audio-visual scene classification.
- Experimental results demonstrate that our approach outperforms state-of-the-arts on the TAU and AVE datasets.

2 Related Work

2.1 Multimodal Fusion

How to integrate multiple modalities effectively has been a non-trivial issue in Artificial Intelligence (AI) field, with the continuously increasing of data. Multimodal fusion aims to integrate the information from different single modality to perform the down-stream task. In general, multimodal fusion is usually divided into early fusion and late fusion [4,5].

2.2 Scene Classification

Recently, Multimodal scene classification (MSC) is of increasing interest to researchers in recent years [6]. MSC can exploit complementary sources of different modalities simultaneously to further improve the performance of scene classification. In order to integrate information from multiple modalities, the efficient fusion method plays a key role in MSC. Additionally, external sound knowledge and uncertainty estimation network are introduced by the works [7] [8] to enhance the effectiveness of acoustic modality and the reliability of prediction results, respectively. Saurabh *et al.* proposed a method to integrate local context information into the self-attention model to improve scene classification, using the local and global time relationship between video frames to obtain more comprehensive context representation information of a single frame [9]. The Attention Cluster network [10] incorporates an attention mechanism that operates autonomously within each modality, subsequently integrating the vectors of each modality through concatenation. Both DMRN [11] and MAFnet [12] engage in a dynamic interaction between audio and visual modalities. Specifically, DMRN highlights image pixels based on sound information, while MAFnet focuses audio features maps based on visual information. Most multimodal fusion strategies cannot effectively capture the complementary relationship between modalities, and the obtained fused feature representations have some redundancy.

3 U-Shape Interactive Multimodal Attention Network

In order to address the aforementioned issues, in this paper, we present a novel U-shape Interactive Multimodal Attention Network, namely U-IMANet, is proposed for audio-visual scene classification, shown in Fig. 1.

3.1 Feature Extractor

Unlike the existing studies, we apply the CLIP model (pre-trained on ImageNet) without fine-tuning on the target dataset to extract the visual features $\mathcal{X}_\alpha \in \mathbb{R}^{B\times T\times d}$. The B represents the batch size, the terms T and d denote the time dimension and feature dimension, respectively. As for Audio modality, instead of learning raw audio waves directly, we first used the Librosa Library to attain log-mel spectrograms. To extract the audio feature $\mathcal{X}_\beta \in \mathbb{R}^{B\times T\times d}$, we concatenate the Log-Mel spectrogram and the Wavegram and then input in ResNet (pre-trained on ImageNet).

3.2 Frame-Level Cross Attention (FCA)

How to exploit the sequence correlation for AVSC is non-trivial. Thus, in Fig. 2, we develop a Frame-Level Cross Attention block that is inspired by self-attention mechanism to explore the temporal relationship within each modality. This module aims to preserve the correlation in sequence that consists of several frames, especially related to the basis information. To be concrete, the input tokens $\mathbf{X} = [x^1, ..., x^T] \in \mathbb{R}^{T\times d}$ are linearly projected to quires $Q \in \mathbb{R}^{T\times d}$, keys $K \in \mathbb{R}^{T\times d}$, and values $V \in \mathbb{R}^{T\times d}$. The self-attention is calculated by the scaled dot product, which can be formulated as follows:

$$SA(Q, K, V) = Softmax(\frac{QK^T}{\sqrt{d}})V, \tag{1}$$

where T and d denote the sequence length and dimension, respectively.

Distinct from SA, however, FCA is used to handle the relationship between each frame and the first frame. It is formulated by,

$$FCA(Q^m, K^1, V^1) = Softmax(\frac{Q^m(K^1)^T}{\sqrt{d}})V^1, \; m = 1, ..., T, \tag{2}$$

where Q^m is defined by the m-th frame x^m, meanwhile K^1 and V^1 are generated by the first frame x^1.

3.3 U-Shape Multimodal Interactive Module (UMIM)

Furthermore, in order to exploit the fine-grained information of audio events and visual objects, a UMIM with multiple layers is proposed in this work. The detailed network structure is presented in Fig. 3.

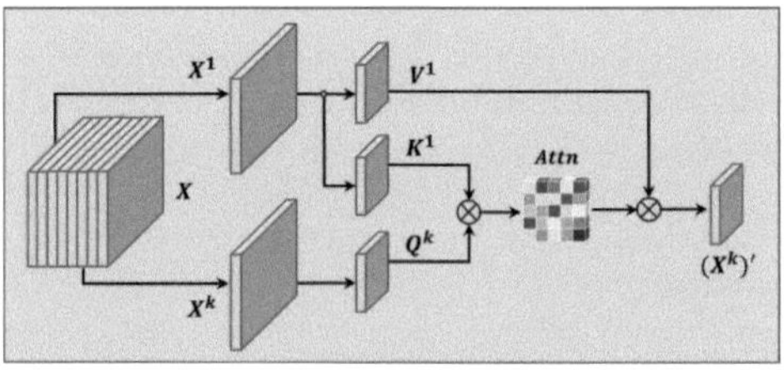

Fig. 2. Frame-Level Cross Attention. ⊗ stands for the matrix multiplications.

Given two input streams of one layer, $\mathbf{X}_\alpha \in \mathbb{R}^{B\times T\times d}$ and $\mathbf{X}_\beta \in \mathbb{R}^{B\times T\times d}$, which are linearly projected to queries (Q_α and Q_β), keys (K_α and K_β), and values (V_α and V_β). The output of Self-Attention branch $\mathbf{Z}_\alpha$ and $\mathbf{Z}_\beta$ can be formulated as follow:

$$\begin{aligned}\hat{\mathbf{Z}}_\alpha &= LN(\mathbf{X}_\alpha + SA(Q_\alpha, K_\alpha, V_\alpha)), \; \hat{\mathbf{Z}}_\beta = LN(\mathbf{X}_\beta + SA(Q_\beta, K_\beta, V_\beta)), \\ \mathbf{Z}_\alpha &= LN(\hat{\mathbf{Z}}_\alpha + FFN(\hat{\mathbf{Z}}_\alpha)), \; \mathbf{Z}_\beta = LN(\hat{\mathbf{Z}}_\beta + FFN(\hat{\mathbf{Z}}_\beta)),\end{aligned} \tag{3}$$

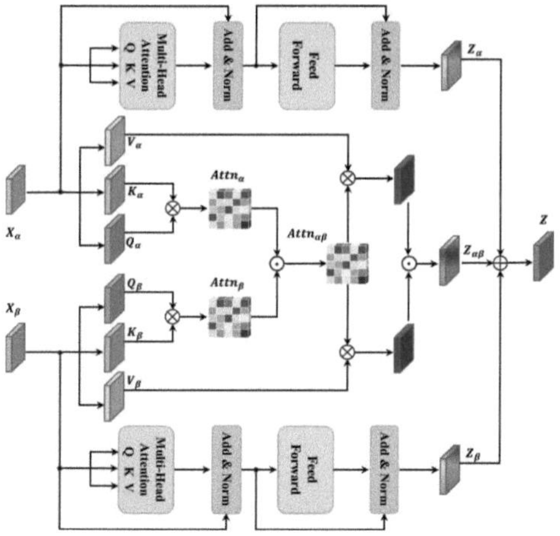

Fig. 3. Multimodal-Hybrid Attention Module. ⊙, ⊗, and ⊕ stand for the element-wise multiplications, matrix multiplications, and element-wise addition, respectively.

where *LN* and *FFN* is the layer normalization and feedforward network, respectively.

In the WFCA branch, Q and the transposes of K are multiplied, and activated using the softmax function to obtain the attention map of each modality. The information contained in audio and visual modality is different and mutually independent and complementary. The joint attention map $Attn_{\alpha\beta}$ is expressed as follows:

$$Attn_{\alpha} = Softmax(\frac{Q_{\alpha}(K_{\alpha})^T}{\sqrt{d}}),\ Attn_{\beta} = Softmax(\frac{Q_{\beta}(K_{\beta})^T}{\sqrt{d}}),\ Attn_{\alpha\beta} = Attn_{\alpha} \odot Attn_{\beta}, \tag{4}$$

where $\odot$ denotes element-wise multiplications. Then, the joint attention map $Attn_{\alpha\beta}$ is multiplied by V_{α} and V_{β} respectively. Similarly, using Hadamard product to calculate the joint feature map $\mathbf{Z}_{\alpha\beta}$. Next, the output of this layer is expressed as follows:

$$\mathbf{Z}_{\alpha\beta} = Attn_{\alpha\beta} V_{\alpha} \odot Attn_{\alpha\beta} V_{\beta},\ \mathbf{Z} = \mathbf{Z}_{\alpha} + \mathbf{Z}_{\beta} + \mathbf{Z}_{\alpha\beta}. \tag{5}$$

Finally, we concatenate the output of each layer to form the feature $\mathbf{Z}$ with the most relevant information, which can be classified through the last Fully Connected (FC) layer with the softmax activation function. The output probability vector $\boldsymbol{p} = [p_1, p_2, ..., p_c]$ is a c-dimensional vector, corresponding to the number of classes. The model is trained by minimizing the cross-entropy loss which is defined by, $\mathcal{L} = -\sum_i^c y_i \log(p_i)$, where y_i denotes the ground truth label and p_i denotes the predicted probability.

4 Experiments and Results

4.1 Dataset and Metric

We evaluate our model on two public available datasets, such as Audio-Visual Urban Scenes 2021 (TAU) dataset [13] and Audio-Visual Event (AVE) dataset [11]. According to previous works, log multi-class cross-entropy loss (Log-loss) and classification accuracy are used to evaluate the performance of the proposed model.

Table 1. The performance comparisons on the TAU dataset in terms of Log-loss and Accuracy (%).

#	Method	Audio Backbone	Visual Backbone	Aug. Method	Log-loss	Accuracy (%)
1	Official baseline [13]	OpenL3	OpenL3	None	0.658	77.00
2	SAGCN [14]	VGGish	ResNet50	None	-	93.10
3	AVSM [15]	VGGish	ResNet50	8 types	0.187	93.50
4	CNN_Transformer [16]	VGGish	Transformer	4 types	0.223	93.90
5	2-stage fine-tuning [17]	ResNet	EfficientNet	5 types	0.159	94.10
6	CEOA-SF [2]	Transformer	ConvNeXt	None	-	94.10
7	Joint mixup [18]	VGGish	DenseNet-161	15 types	0.192	94.20
8	Joint training [19]	ResNet	EffNetV2-S	4 types	0.152	94.59
9	U-IMANet (Ours)	ResNet	CLIP/ResNet50	None	**0.151**	**95.33**

4.2 Implementation Details

For each 10-second audio file, STFT is applied to extract Log-Mel spectrogram on the raw signal every 171 ms over 512 ms windows with 512 Mel filter banks, which generates a feature tensor shape of $59 \times 2 \times 512$. Additionally, 1D-CNN is adopted for Wavegram extraction, which also generates a feature tensor shape of $59 \times 2 \times 512$. The input of the visual modality is an image sequence consisting of 20 images. Pre-trained models ResNet and CLIP/ResNet50 are directly applied to extract audio and visual feature without any data augmentation tricks.

4.3 Results

Table 1 demonstrates the performance of U-IMANet and the state-of-the-art models for AVSC on TAU dataset. Notably, the model in #1 is the baseline used in DCASE 2021 Task 1B. Similar to all other compared methods, the proposed network also uses the weights from pre-trained models. Specifically, the weights are learned by ResNet and CLIP image encoder to extract audio and visual feature, respectively. In this work, the proposed model does not use any ensemble methods. Thus, to be fair, the comparison is conducted to the best result achieved by single model of other methods. Except for Baseline, the methods of #2, #6, and the proposed model, others use diverse visual or audio data augmentation tricks. Additionally, the model of #5 uses additional large scene dataset ImageNet and Places365 for training.

From the Table 1, all methods utilizing the pre-trained model as their encoder demonstrated exceptional performance, leading to accuracy rate of 93% and a Log-loss value below 0.25. In terms of the accuracy metric, our approach achieves the best score of 95.33%, with a gain of 18.33% against the official baseline and 0.74% compared to the second best-performing system. Furthermore, our approach achieves the lowest Log-loss with the value of 0.151.

Table 2. The performance comparisons on the AVE dataset in terms of Accuracy (%).

#	Method	Audio Backbone	Visual Backbone	Accuracy
1	Attention Cluster [10]	VGGish	DenseNet	80.72
2	DMRN [11]	VGGish	DenseNet	85.02
3	MESL [20]	CNN	ResNet	87.80
4	HAD [21]	VGGish	ResNet	88.90
5	MAFnet [12]	VGGish	DenseNet	90.86
6	U-IMANet (Ours)	ResNet	CLIP/ResNet50	93.75

Table 2 shows the event recognition results on the AVE dataset. As illustrated, U-IMANet demonstrates a notable superiority over other methods on the AVE dataset. In particular, our method achieves an accuracy rate of 93. 75%, exceeding the MAFnet methods by a margin of 2.89%.

4.4 Ablation Study

For simplicity, FCA block, UMIM and their combinations are evaluated for the impact on the classification performance. The results are given in Table 3. As can be seen, both FCA and UMIM have a positive influence for the final prediction. The feature is effectively enhanced by FCA block via attention mechanism, meanwhile high level representations are further enhanced via UMIM. Noted that the best performance in the single part comparison is achieved by UMIM with 94.77% on accuracy metric, improved by 1.66% compared to the basic backbone.

Since the number of UMIM layers is the main hyperparameter affecting the performance of the model, experiments are also carried out. As shown in Table 4, by varying the number of UMIM layers from 1 to 5, it is observed that the proposed model attains its optimal performance when equipped with 4 UMIM layers. Furthermore, due to Joint Training is the previous state-of-the-art method, in Table 5 we give a comprehensive comparison using various combinations of modality. It is worth noting that UMIM is inherently designed for multimodal feature interaction, such that it cannot be used for only one modality.

Table 3. The ablation study of different modules on the TAU dataset.

Audio Backbone	Visual Backbone	FCA	UMIM	Accuracy (%)
ResNet	CLIP/ResNet50	×	×	93.11
ResNet	CLIP/ResNet50	✓	×	93.99
ResNet	CLIP/ResNet50	×	✓	94.77
ResNet	CLIP/ResNet50	✓	✓	**95.33**

Table 4. Results of our model with different number of UMIM layers on the TAU dataset.

Audio Backbone	Visual Backbone	FCA	UMIM Layer	Accuracy (%)
ResNet	CLIP/ResNet50	✓	1	95.06
ResNet	CLIP/ResNet50	✓	2	95.02
ResNet	CLIP/ResNet50	✓	3	95.15
ResNet	CLIP/ResNet50	✓	4	**95.33**
ResNet	CLIP/ResNet50	✓	5	95.08

In summary, our model achieves the best scores except for the audio modality. This may lie in the assumption that the visual modality can provide more information than audio one. In addition, it has been verified that the CLIP image encoder, as the visual extractor, exhibits remarkable performance.

Table 5. The performance comparison between our method and Joint Training on different modality combinations on the TAU dataset.

#	Audio	Visual	Accuracy(%)	
			Ours	Joint Training
1	✓	×	73.26	**77.98**
2	×	✓	**92.28**	90.91
3	✓	✓	**95.33**	94.59

4.5 Visualizations

To gain a deeper insight into the effectiveness of our proposed approach, we used t-SNE to visualize the low-dimensional embeddings of the test samples on the TAU dataset. As shown in Fig. 4, the visualization of the embeddings showcases the impact of FCA blocks and UMIM. That is, the visualization corresponds to the results of the experiments 1 and 4 in Table 3. In Fig. 4(a), it can be seen that samples from tram class are easily confused with those from Bus and

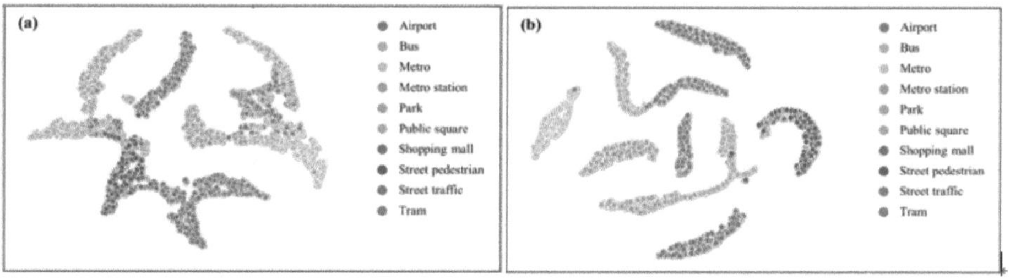

Fig. 4. The t-SNE visualization of embedding distributions on the TAU dataset before (a) and after (b) the usage of FCA and UMIM.

Metro classes. Samples from Public square are mixed with samples from Street pedestrian, while those from Shopping mall are mixed with samples from Airport. Through Fig. 4(b), we can observe clear boundaries on the clusters of 10 scene categories within the proposed modules. The intra-class distances are reduced while the inter-class distances are enlarged at the same time. In essence, our approach leverages the power of FCA and UMIM to sculpt a discriminative feature space that is optimized for effective and efficient scene classification.

5 Conclusion

In this work, we proposed a novel multimodal feature fusion network for AVSC, which utilizes the consistency and complementarity information within modalities through an attention weight sharing strategy, so as to enhance the representation of fusion feature. Experimental results showcase the superior performance of our model over the state-of-the-art methods on the TAU dataset, and competitive accomplishment on the AVE dataset.

Acknowledgements. This work is supported by the National Natural Science Foundation of China (No. 62376101), in part by Guangdong Basic and Applied Basic Research Foundation (No.2024A1515240036).

References

1. Hou, Y., et al.: CNN-based dual-stream network for audio-visual scene classification. Technical report, DCASE2021 Challenge (2021)
2. Hou, Y., Kang, B., Botteldooren, D.: Audio-visual scene classification via contrastive event-object alignment and semantic-based fusion. In: IEEE 24th International Workshop on Multimedia Signal Processing (MMSP), pp. 1–6. IEEE (2022)
3. Lei, H., Chen, N.: Audio-visual scene classification based on multi-modal graph fusion. In: INTERSPEECH 2022, pp. 4157–4161 (2022)
4. Snoek, C.G., Worring, M., Smeulders, A.W.: Early versus late fusion in semantic video analysis. In: Proceedings of the 13th Annual ACM International Conference on Multimedia, pp. 399–402 (2005)

5. Gunes, H., Piccardi, M.: Affect recognition from face and body: early fusion vs. late fusion. In: IEEE International Conference on Systems, Man and Cybernetics, vol. 4, pp. 3437–3443. IEEE (2005)
6. Zeng, D., et al.: Deep learning for scene classification: a survey. arXiv preprint arXiv:2101.10531 (2021)
7. Nie, L., et al.: Enhancing micro-video understanding by harnessing external sounds. In: Proceedings of the 25th ACM International Conference on Multimedia, pp. 1192–1200 (2017)
8. Wang, B., Huang, X., Cao, G., Yang, L., Wei, X., Tao, Z.: Attention-enhanced and trusted multimodal learning for micro-video venue recognition. Comput. Electr. Eng. **102**, 108127 (2022)
9. Sahu, S., Goyal, P.: Leveraging local temporal information for multimodal scene classification. In: ICASSP 2022-2022 IEEE International Conference on Acoustics, Speech and Signal Processing (ICASSP), pp. 1830–1834. IEEE (2022)
10. Long, X., Gan, C., De Melo, G., Wu, J., Liu, X., Wen, S.: Attention clusters: purely attention based local feature integration for video classification. In: Proceedings of the IEEE Conference on Computer Vision and Pattern Recognition, pp. 7834–7843 (2018)
11. Tian, Y., Shi, J., Li, B., Duan, Z., Xu, C.: Audio-visual event localization in unconstrained videos. In: Ferrari, V., Hebert, M., Sminchisescu, C., Weiss, Y. (eds.) ECCV 2018. LNCS, vol. 11206, pp. 252–268. Springer, Cham (2018). https://doi.org/10.1007/978-3-030-01216-8_16
12. Brousmiche, M., Rouat, J., Dupont, S.: Multimodal attentive fusion network for audio-visual event recognition. Inf. Fusion **85**, 52–59 (2022)
13. Wang, S., Mesaros, A., Heittola, T., Virtanen, T.: A curated dataset of urban scenes for audio-visual scene analysis. In: ICASSP 2021-2021 IEEE International Conference on Acoustics, Speech and Signal Processing (ICASSP), pp. 626–630. IEEE (2021)
14. Wang, Z., Wu, Y., Wang, Y., Gong, W., Gonzàlez, J.: Audio-visual scene recognition using attention-based graph convolutional model. Multimedia Tools Appl., 1–25 (2024)
15. Wang, Q., et al.: A model ensemble approach for audio-visual scene classification. In: Proceedings of the 6th Workshop on Detection and Classification of Acoustic Scenes and Events (2021)
16. Yang, Y., Luo, Y.: Scene classification using acoustic and visual feature. Technical report, DCASE2021 Challenge (2021)
17. Wang, M., Chen, C., Xie, Y., Chen, H., Liu, Y., Zhang, P.: Audio-visual scene classification using transfer learning and hybrid fusion strategy. In: Proceedings of the Detection and Classification of Acoustic Scenes and Events, DCASE (2021)
18. Wang, Q., et al.: A study on joint modeling and data augmentation of multi-modalities for audio-visual scene classification. In: 13th International Symposium on Chinese Spoken Language Processing (ISCSLP), pp. 453–457. IEEE (2022)
19. Chen, C., Wang, M., Zhang, P.: Audio-visual scene classification using a transfer learning based joint optimization strategy. arXiv preprint arXiv:2204.11420 (2022)
20. Senocak, A., Kim, J., Oh, T.-H., Ryu, H., Li, D., Kweon, I.S.: Audio-visual fusion layers for event type aware video recognition. arXiv preprint arXiv:2202.05961 (2022)
21. Zuo, Y., Yao, H., Zhuang, L., Xu, C.: Hierarchical augmentation and distillation for class incremental audio-visual video recognition. IEEE Trans. Pattern Anal. Mach. Intell. **46**(11), 7348–7362 (2024)

Memristive Circuit Design of Nonlinear Enhancement for Audio-Visual Modalities

Moshu Xing, Xiaoping Wang(✉), and Ruoxin Wu

School of Artificial Intelligence and Automation, Huazhong University of Science and Technology, Wuhan 430074, China
{moshuxing,wangxiaoping,ruoxinwu}@hust.edu.cn

Abstract. This paper proposes a novel memristive audio-visual fusion circuit inspired by the nonlinear enhancement mechanism observed in biological multisensory integration. The core innovation lies in the synergistic integration of volatile and nonvolatile memristors combined with millisecond-level response speed, enabling efficient single-modal signal processing, semantic consistency detection, and intensity-dependent nonlinear enhancement within a single circuit. Crucially, the circuit incorporates external safety/danger correction signals to dynamically modulate risk assessment. Simulation results demonstrate that the system not only emulates the integration characteristics of the brain under varying stimulus intensities but also exhibits response timing consistent with biological behavior. This work provides a reference for constructing low-power neuromorphic systems with brain-inspired perception and decision-making capabilities.

Keywords: modal fusion · nonlinear enhancement · memristor · bionic circuit

1 Introduction

Multisensory integration is a core mechanism through which the biological nervous system achieves efficient environmental perception and rapid decision-making. Extensive research in neuroscience has demonstrated that when stimuli from different modalities, such as vision and audition, are spatiotemporally congruent, multimodal neurons in the cerebral cortex exhibit a nonlinear enhancement of their response. Under conditions of weak signals, this manifests as superadditivity, where the neural response to multimodal stimuli exceeds the sum of the individual single-modal responses. Conversely, under strong signal conditions, the integration tends toward subadditivity, with the multimodal response becoming less than the sum of its unimodal components [1–3]. This dynamic and adaptive integration strategy significantly reduces the latency from sensory coding to motor command generation, thereby enhancing the efficiency of decisions critical for survival.

Memristors, as a new type of electronic device characterized by resistance state memory, have garnered widespread attention due to their advantages such

Z. Zhang (Ed.): ISNN 2026, LNCS 16848, pp. 89–103, 2027.
https://doi.org/10.1007/978-981-92-5870-3_8

as nonlinearity, low power consumption, and synapse-like plasticity [4–6]. Volatile memristors rapidly revert to their initial resistance state after power-off, making them suitable for simulating the dynamic response of neural circuits. In contrast, non-volatile memristors can stably maintain their resistance state without an extra bias, which is often utilized for long-term information storage and synaptic weight regulation. These unique properties establish memristors as a core component in brain-inspired bionic circuit design [7,8]. In recent years, progress has been made in research on multisensory fusion based on memristors. Jiang et al. proposed a bio-inspired memristor-based neuromorphic circuit, which bridges multiple unisensory channels with a multisensory mutual associative memory (MMAM) unit [9]. Similarly, Zhang et al. proposed a multisensory mutual associative memory networks framework and memristive circuit to mimic the ability of the biological brain to make associations of information received simultaneously, applied to cross-modal association of image and audio recognition results [10]. Tao et al. proposed a memristor-based neural network circuit of emotional grading in intelligence decision system [11]. Furthermore, existing studies have predominantly focused on cross - modal associations, rarely addressing modal fusion and the dynamic regulation of fusion effects under combinations of different signal strengths. Additionally, the response time characteristics of the system have not been thoroughly analyzed [12,13]. These research gaps constrain the further application of memristor-based multisensory fusion systems in complex intelligent scenarios. To address the aforementioned issues, this paper designs an audio-visual modal fusion and decision-making circuit based on memristors, with the main contributions outlined as follows:

1) A memristor-based nonlinear enhancement circuit exhibiting superadditivity and subadditivity is proposed, which can flexibly respond to auditory and visual signals of varying intensities: strong, moderate, and weak.
2) An external safety or danger correction signal is introduced to simulate the brain's decision-making correction process, thereby improving the rationality of risk assessment.
3) The response time characteristics of decision execution are analyzed, verifying the consistency between the designed circuit and biological behavioral patterns.

2 Memristor Model

Based on their ability to retain resistance states after power-off, memristors can be categorized into two types: volatile and non-volatile. The former rapidly relaxes to its initial resistance value once the excitation is removed, while the latter can maintain its resistance state stably for extended periods without an extra bias.

The volatile memristor model employed in this paper is adopted from [14]. The parameters of the volatile memristor model employed in this study are listed in Table 1. α is the barrier height in the Schottky barrier region, β is the depletion

Table 1. Parameters of volatile memristors VM1-VM5

Parameters	α	β	γ	δ	η_1	η_2	λ	σ	θ
Settings	1e−4	0.2	1e−3	1	4	2	0.05	1e−4	0.01

Table 2. Parameters of non-volatile memristors M1–M6

Parameters	R_{on} (kΩ)	R_{off} (kΩ)	V_{on} (V)	V_{off} (V)	k_{on}	k_{off}	α	p
M1, M4	5	100	0.6	−0.2	1e4	1e3	0.28	1.58
M2, M5	5	100	0.4	−0.2	2e4	1e3	0.28	1.58
M3, M6	5	100	0.2	−0.2	5e4	1e3	0.28	1.58

width of the Schottky barrier region, γ is the tunneling barrier height, and δ is the effective tunneling distance in the conductive region. η_1 and η_2 represent the interface effects under positive and negative voltages, respectively. λ is a positive constant that controls the rate of change of x, σ and θ are positive constants determined by the material.

The non-volatile memristor model used in this paper is adopted from [15]. The parameters for the non-volatile memristors used in this study are listed in Table 2. V_{on} and V_{off} denote the positive and negative threshold voltages, respectively. The parameters k_{on}, k_{off}, α, and p are adjustable fitting parameters used to characterize different memristor devices. R_{on} and R_{off} correspond to the minimum and maximum memristance values. A change in memristance occurs when the voltage across the memristor exceeds its threshold.

3 Experimental Design for Audio Visual Modality Fusion Based on Memristor

As shown in Fig. 1a, a woman and a cat perceive the approach of a dog through visual and auditory cues. When the dog is close, representing a dangerous state, an escape response is required. Conversely, no escape is needed when the dog is farther away. When the individual sensory signals are weak (indicating the dog is distant), the neural computation underlying their integration exhibits superadditivity, exceeding not only the strongest unimodal response but also their arithmetic sum. As the dog moves closer, the cues become more effective, the unimodal responses grow stronger, and the integration strength gradually diminishes. The computation progressively shifts to an additive (middle) and then a subadditive (bottom) regime [1]. Multisensory integration can shorten the interval between sensory coding and the formation of motor commands, thereby accelerating the decision-making process.

Three experiments were designed to simulate the above process, as illustrated in Fig. 1b. Experiment 1 tested the system's response when a unimodal signal weakened from strong to weak, simulating the brain's unimodal processing. Experiment 2 simulated the fusion of visual and auditory signals under strong, moderate, and weak states to achieve nonlinearity. Experiment 3 introduced an external supplementary signal to correct hazard judgment and decision-making responses. The meanings of the various input signals in the experiments are as follows: The visual input signal V_{in1} represents the size of the target dog, while the auditory input signal V_{in2} corresponds to the loudness of the dog's bark. External safety information V_{e1} indicates that the dog is leashed, muzzled,

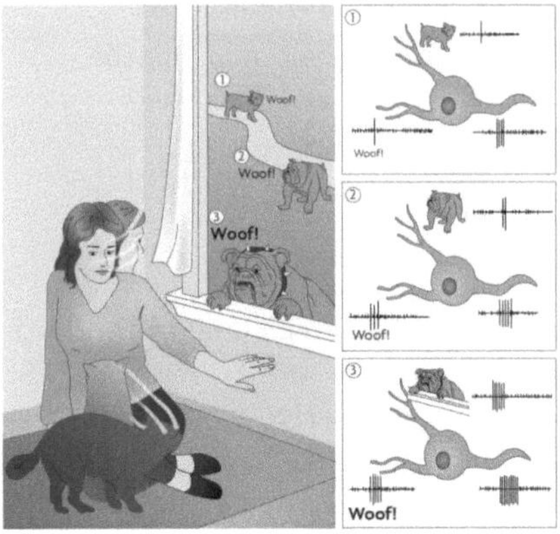

(a)

Experiment 1: Test Experiment

$\begin{cases} \text{Visual Signal } V_{in1} : \text{Strong} - \text{Moderate} - \text{Weak} - \text{None} - \text{None} - \text{None} \\ \text{Auditory Signal } V_{in2} : \text{None} - \text{None} - \text{None} - \text{Strong} - \text{Moderate} - \text{Weak} \end{cases}$

Experiment 2: Multimodal Fusion

$\begin{cases} \text{Visual Signal } V_{in1} : \text{Strong} \\ \text{Auditory Signal } V_{in2} : \text{Strong} \end{cases}$

$\begin{cases} \text{Visual Signal } V_{in1} : \text{Moderate} \\ \text{Auditory Signal } V_{in2} : \text{Moderate} \end{cases}$

$\begin{cases} \text{Visual Signal } V_{in1} : \text{Weak} \\ \text{Auditory Signal } V_{in2} : \text{Weak} \end{cases}$

Experiment 3: Fusion Calibration Experiment

$\begin{cases} \text{Visual Signal } V_{in1} : \text{Strong} \\ \text{Auditory Signal } V_{in2} : \text{Strong} \\ \text{External Correction Signal } V_{e1} : \text{Negative Signal} \end{cases}$

$\begin{cases} \text{Visual Signal } V_{in1} : \text{Weak} \\ \text{Auditory Signal } V_{in2} : \text{Weak} \\ \text{External Correction Signal } V_{e2} : \text{Positive Signal} \end{cases}$

(b)

Fig. 1. Experimental design for audio visual modality fusion based on memristor: (a) nonlinear enhancement of neurons in the superior colliculus, (b) experimental design scheme of this paper

or has a gentle temperament, posing no danger and thus requiring no escape. External danger information V_{e2} signifies that dogs are fighting, have a history of aggression, or carry diseases, representing a high level of danger that necessitates leaving the area promptly. The following variables are observed in the experiment: the danger signal v_f, which quantifies the level of threat, with higher values indicating greater danger. When this danger exceeds a certain threshold, both the woman and the cat must flee the room. The escape signal V_{out} indicates whether the woman has initiated movement to escape.

4 Circuit Design

4.1 Designed Circuit of Each Module

The brain contains a large number of multimodal neurons capable of responding simultaneously to both visual and auditory stimuli [16]. The multisensory neuron circuit adopted in this paper is shown in Fig. 2a [17]. The input voltage V_{in} represents the intensity of a visual stimulus (seeing a dog) or an auditory stimulus (hearing a dog bark). V_{in} is applied to the positive terminal of the memristor M, while a bias of $-1.5\,\text{V}$ is used for the purpose of neuron circuit recovery. When $V_{in} > V_{on}$, the resistance of M drops sharply, causing the voltage at point B to decrease. Once this voltage falls below the turn-on voltage of the PMOS transistor, the MOS transistor conducts, and the circuit outputs 1 V, simulating neuronal firing. When $0.1\,\text{V} < V_{in} < V_{on}$, the resistive state of the memristor remains unchanged, and the circuit maintains its current state. When $V_{in} < 0.1\,\text{V}$, the voltage at point A becomes $-1.5\,\text{V}$, which is lower than the negative threshold voltage of M. This causes the resistance of M to increase rapidly, raising the voltage at point B. When this voltage exceeds the turn-on voltage of the PMOS transistor, the MOS transistor turns off, the output returns to zero, and the circuit resets to its resting state, simulating the neuronal recovery process.

This circuit fully simulates the entire process of neuronal firing and resting. By adjusting the threshold voltage of the memristor M, the firing threshold of the neuron can be flexibly modulated, enabling it to respond to multimodal input signals of varying intensities. As shown in Fig. 2b, the audiovisual fusion module takes the output of the preceding multisensory neuronal circuit as its input, which corresponds to the intensity levels of the audiovisual stimuli. Here, v_{11}/v_{21}, v_{12}/v_{22}, and v_{13}/v_{23} represent strong, medium, and weak levels of audiovisual signals, respectively. A gating circuit is employed to determine whether the input signals belong to the same intensity level: high-level outputs from v_{p1}, v_{p2}, and v_{p3} indicate matching results for strong, medium, and weak signals, respectively.

$$v_{p1} = v_{11} \wedge v_{12} \wedge v_{13} \wedge v_{21} \wedge v_{22} \wedge v_{23} \tag{1}$$

$$v_{p2} = \overline{v_{11}} \wedge v_{12} \wedge v_{13} \wedge \overline{v_{21}} \wedge v_{22} \wedge v_{23} \tag{2}$$

$$v_{p3} = \overline{v_{11}} \wedge \overline{v_{12}} \wedge v_{13} \wedge \overline{v_{21}} \wedge \overline{v_{22}} \wedge v_{23} \tag{3}$$

After amplification, the high-level signal is applied across the memristor, causing its voltage to exceed the threshold. This results in a sharp drop in resistance, which subsequently leads to a decrease in v_{m1} until the comparator switches states and v_{c1} outputs a high-level signal. This module incorporates three parallel fusion pathways, corresponding to strong, medium, and weak signals respectively. The outputs v_{c1}, v_{c2}, and v_{c3} indicate the occurrence of fusion and its intensity. Based on the nonlinear characteristics of signal enhancement, a stronger input yields lower gain, while a weaker input produces more significant gain. The output voltages for the three fusion pathways are set to 2.2 V, 2.7 V, and 4.9 V respectively. When only a single visual or auditory signal is present, the unimodal cognitive pathway becomes activated. As is shown in Fig. 2c, upon detecting a unimodal input, the XOR gate outputs a high-level signal. This signal is then applied to the negative terminal of the memristor via an inverting amplifier, creating a voltage difference across the memristor that exceeds its threshold. Consequently, the memristance decreases, leading to a reduction in the voltage at the right terminal of the memristor until the comparator switches states, outputting a 2 V level and thereby closing the analog switch. In reference to neuroscience research, the visual pathway conducts signals more slowly than the auditory pathway, resulting in a correspondingly lower perceived level of danger. Therefore, in this work, the gain for the visual signal is set lower than that for the auditory signal [18]. After being weighted and inverted, the two signals are output through the analog switch as the unimodal danger signal v_u. When the input consists of bimodal signals or no signal, the XOR gate outputs a low level, the memristor retains its original resistance, P4 remains off, the analog switch stays open, and no output is produced at v_u. The dynamic resistance of memristor VM simulates the transmission and processing of unimodal signals in the brain, with its rate of change corresponding to the speed of information processing.

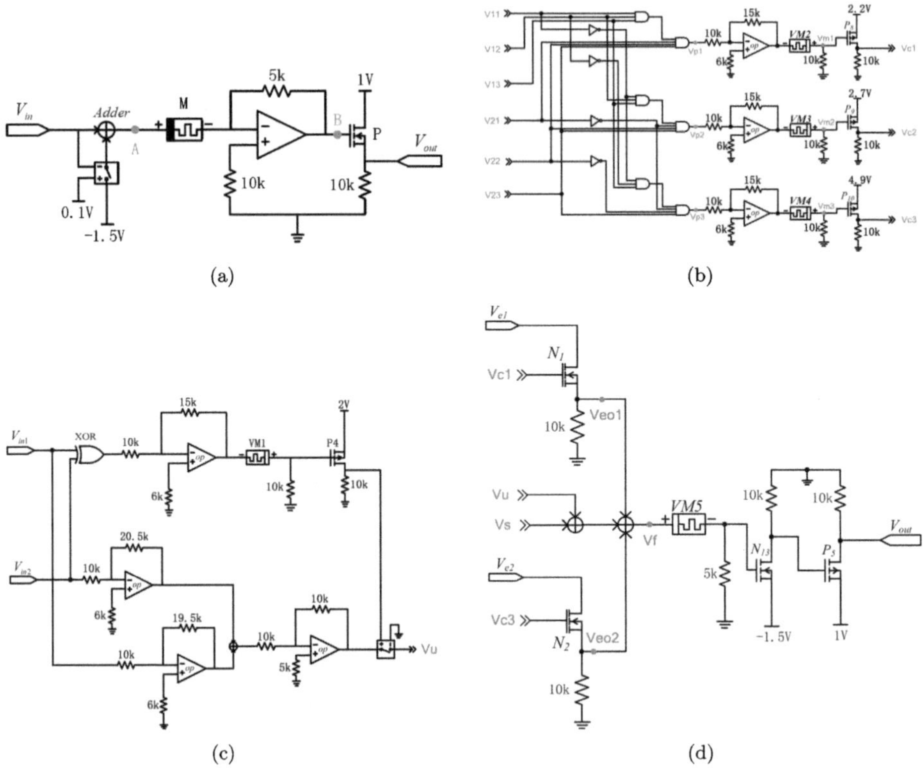

Fig. 2. The designed circuit of each module: (a) multisensory neuronal circuit in [17], (b) memristive circuit design of audiovisual fusion, (c) the designed circuit of unimodal cognitive pathway, (d) the designed circuit of decision module.

As shown in Fig. 2d, the decision module receives six input signals: an external safety signal V_{e1}, an external hazard signal V_{e2}, outputs from the audiovisual integration module (v_{c1} or v_{c3}), output from the unimodal cognitive pathway (v_u), and the audiovisual integration Cderived hazard level (v_s). Its final output, v_f, represents the brain s assessment of the overall hazard level of the scene. In the absence of external supplementary signals (i.e., $V_{e1} = V_{e2} = 0$), v_f is determined by the summation of v_u and v_s, thereby integrating both unimodal and multimodal information. When external signals are present, two distinct scenarios arise: 1) if both incoming signals are strong, the safety signal V_{e1} is added to v_f, resulting in a marked reduction in perceived hazard level; 2) if both incoming signals are weak, the hazard signal V_{e2} is added to v_f, leading to a pronounced increase in perceived hazard level.

A higher v_f indicates a greater environmental hazard level and triggers more urgent escape behavior. The MOS transistor N13, together with the MOS transistor P5 and a resistor, forms a comparator circuit. When v_f exceeds the threshold voltage of the memristor VM5, the resistance of VM5 decreases rapidly, causing the voltage on the right side to rise above the turn-on voltage of N13. As a result, both MOS transistors conduct, driving the circuit output to a high logic level of 1 V and thereby initiating the escape response. The resistance variation of the memristor VM5 reflects the brain s assessment of hazard severity and its subsequent decision-making process.

4.2 Complete Circuit Design

As illustrated in Fig. 3, the memristor-based nonlinear enhancement circuit of audiovisual modalities comprises four functional modules: an audiovisual signal perception module, an audiovisual modality fusion module, a unimodal cognitive pathway, and a decision-making module. The perception module consists of six multimodal neurons, three of which respond specifically to visual stimuli and the other three to auditory stimuli. To accommodate the three signal intensity levels employed in this experiment—strong, moderate, and weak—the threshold voltage(von) of the memristors M1/M4, M2/M5, and M3/M6 are set to 0.6 V, 0.4 V, and 0.2 V, respectively. Consequently, the input signal strength is categorized as follows: $V_{in} < 0.2\,\mathrm{V}$(no signal), $0.2V < V_{in} < 0.4\,\mathrm{V}$(weak), $0.4V < V_{in} < 0.6\,\mathrm{V}$(moderate), and $V_{in} > 0.6\,\mathrm{V}$(strong).

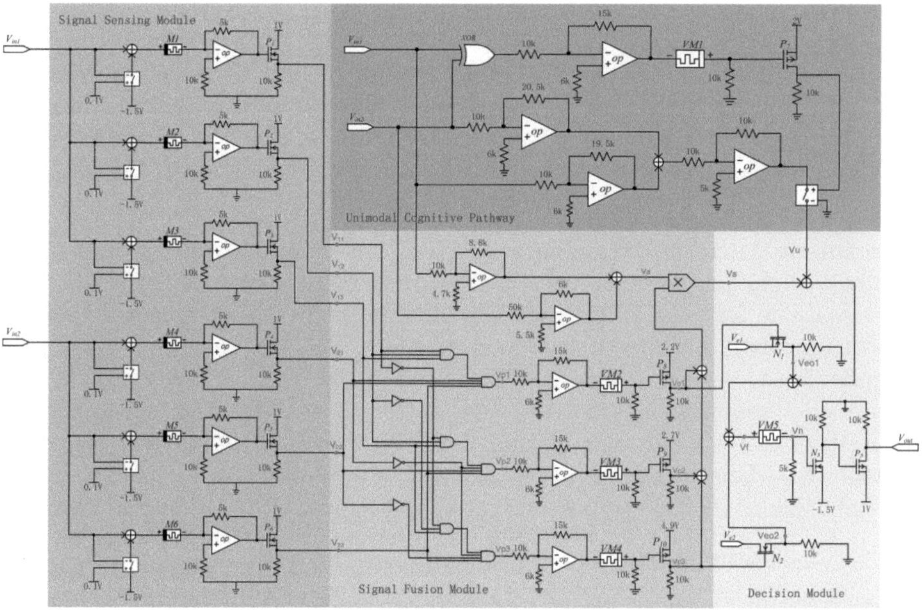

Fig. 3. Memristor-based nonlinear enhancement circuit of audiovisual Modalities

Upon perceiving audiovisual stimuli of varying intensities, multimodal neurons generate six distinct output signals v_{11} through v_{23}, which are subsequently fed into a fusion module to integrate signals across multiple intensity levels. Research has demonstrated that visual stimuli are more effective than auditory stimuli alone; therefore, the amplification factor for visual signals is set higher than that for auditory signals [19]. After summation, the combined signal is routed to a multiplier, where it is multiplied by the fused voltages v_{c1}, v_{c2}, and v_{c3}, yielding the multimodal hazard signal v_s.

The unimodal cognitive pathway operates upon input of either a single visual or auditory signal and generates a unimodal hazard signal, denoted as v_u. The fusion module activates only when bimodal inputs are present and produces a fused hazard signal, denoted as v_s. External information signals V_{e1} and V_{e2} govern the assessment of hazard severity through a correction mechanism. The final hazard signal v_f is generated by superimposing v_u, v_s, V_{e1} and V_{e2} and is applied to the memristor VM5. The execution stage comprises a voltage-divider circuit formed by VM5 and a resistor, together with a MOS transistor comparator circuit. This stage implements the escape response. When v_f exceeds the threshold voltage of VM5, the resistance of VM5 decreases and the node voltage v_n increases. If v_n exceeds the turn-on voltage of transistor N3, the circuit outputs a high-level signal $V_{out} = 1\,\text{V}$, triggering the escape behavior; otherwise, N3 remains off and the circuit outputs $V_{out} = 0$, maintaining quiescence.

5 Experimental Simulation Result and Analysis

The experiments in this paper were conducted in LTspice. This paper designed three experiments to achieve audiovisual modality fusion, with the following signal settings: strong signal at 0.75 V, moderate signal at 0.5 V, and weak signal at 0.25 V; external safety signal $V_{e1} = -0.5\,\text{V}$ and danger signal $V_{e2} = 1\,\text{V}$.

Experiment 1: Test Experiment. When the woman perceives only a dog visually without hearing any vocalization, the apparent size of the dog serves as a visual cue for proximity: a larger perceived area indicates closer distance and consequently a higher risk of being bitten. This study defines three risk categories based on the visually estimated size of the dog: low risk when the dog appears small requiring no evasive action; moderate risk when the dog appears of intermediate size warranting cautious observation without immediate response; and high risk when the dog appears large necessitating immediate escape. Under this framework, auditory input is set to zero, while visual input is modeled as strong, moderate, or weak signals, respectively. The corresponding system responses are illustrated in Fig. 4.

As shown in Fig. 4a, from 0 to 2 s, only a strong visual stimulus is applied; thus, the unimodal cognitive pathway operates exclusively. The XOR gate outputs a high logic level, which is inverted and amplified before being applied to the negative terminal of memristor VM1. As the voltage across VM1 exceeds its threshold, its resistance decreases. This reduction causes the voltage on the right-hand side to drop, thereby turning on transistor P7, which in turn activates the

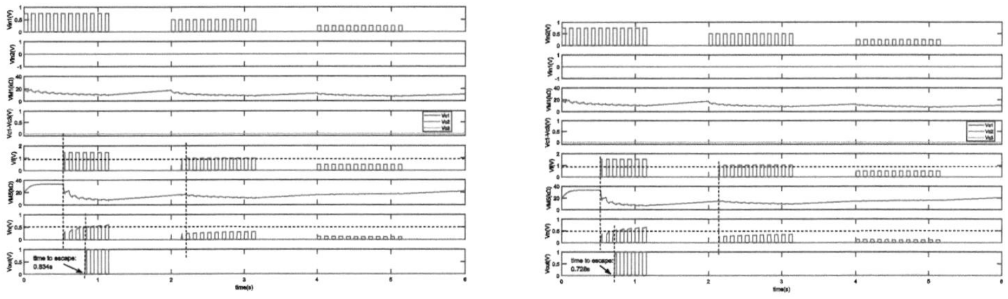

(a) (b)

Fig. 4. Simulation results of experiment 1: (a) visual unimodal input, (b) auditory unimodal input

analog switch. Consequently, the visual signal is amplified and routed to the decision module, generating a unimodal hazard signal v_u. Since only a strong visual stimulus is present, only three visual neurons fire, yielding $v_{11} = v_{12} = v_{13} = 1\,\text{V}$. All three AND gates therefore output low logic levels: $v_{c1} = v_{c2} = v_{c3} = 0$. As a result, the fusion module remains inactive, and the fused output voltage v_f equals the unimodal output voltage v_u. At this point, $v_f = 1.46\,\text{V}$, exceeding the threshold voltage of VM5 (0.9 V), causing VM5 s resistance to decrease and the node voltage v_n to rise. At $t = 0.834s$, v_n surpasses the comparator s reference voltage of 0.5 V, prompting the circuit to output a high logic level ($V_{out} = 1\,\text{V}$) and thereby triggering the escape response.

During the 24 s interval when only the medium-intensity visual signal is input, the system continues to process the signal via the unimodal pathway. At this time, $v_f = v_u = 0.975\,\text{V}$, which is slightly higher than the threshold voltage of VM5(0.9 V). This causes the resistance of VM5 to decrease and the voltage at v_n to rise, though it remains below the turn-on voltage of N3(0.5 V). Consequently, the circuit output stays at a low level, with $V_{out} = 0$, indicating that the system detects a certain hazard but has not reached the threshold required to trigger an escape behavior.

During the 4 to 6-s interval, only a weak visual signal was input, simulating the switch being turned on. Due to the input of only a weak visual signal, only the third neuron fired, resulting in $v_{13} = 1\,\text{V}$. At this time, v_f was 0.488 V, which was below the threshold voltage of VM5(0.9 V). Consequently, the resistance of VM5 remained unchanged, and N3 was turned off. V_{out} remained at 0, indicating a very low level of danger, and the system maintained its stationary position. No escape behavior was triggered, and the static state was preserved.

When only auditory input (dog barking) is present without visual input, sound intensity directly reflects distance and risk level: the louder the sound, the higher the risk, and the more urgent the need to escape; a moderate sound prompts temporary observation; a faint sound leads to remaining still. Based on this principle, the visual input was set to 0 in the experiment, and strong, moderate, and weak levels of auditory signals were sequentially input. The system response is shown in Fig. 4b.

Similar to the scenario with only visual input, during the 0–2 s interval, only a strong auditory signal is input. The unimodal cognitive pathway operates, and the amplified auditory signal is transmitted through the conducting analog switch to the decision module. At this time,$v_f = v_u = 1.538$ V, which exceeds the threshold of VM5, causing its resistance to decrease and vn to rise. vn surpasses the comparator threshold at 0.728 s, resulting in a circuit output $V_{out} = 1$ V. This is judged as a high risk and triggers an escape behavior. During the 2–4 s interval, only a moderate auditory signal is input. Here, $v_f = v_u = 1.025$ V, which is higher than the threshold voltage of VM5, leading to a reduction in the resistance of VM5 and an increase in the v_n voltage. However, v_n remains below the turn-on voltage of 0.5 V for the MOS transistor N3, so N3 does not conduct, and V_{out} stays at 0. This indicates a certain level of hazard, but no escape action is executed. From 4 to 6 s, only a weak visual signal is input. v_f is 0.513 V, which is below the threshold voltage of VM5. Consequently, the resistance of VM5 remains unchanged, and N3 does not turn on. V_{out} stays at 0 throughout this period, indicating a very low hazard level, and the system maintains its stationary position.

Experiment 2: Multimodal Fusion Experiment. As shown in Fig. 5a, When both audio and visual modalities receive strong input signals ($V_{in1} = V_{in2} = 0.75$ V) and no external correction is applied ($V_{e1} = V_{e2} = 0$), the voltages across memristors M1 through M6 all exceed their thresholds, causing their resistances to drop sharply. Consequently, all neurons fire and output 1 V. The logic gate determines this as a strong audio-visual signal, resulting in $v_{p1} = 1$ V and $v_{p2} = v_{p3} = 0$, which triggers the fusion enhancement mechanism. The resistance of VM2 continues to decrease, and once the voltage at its right-hand side falls below the threshold of the subsequent comparator, the output v_{c1} becomes 2.2 V. This reflects the brain s process of integrating congruent signals. After the input signals are weighted, amplified, and multiplied by the fusion signal, a fused danger signal $v_s = 1.65$ V is obtained. At this point, the unimodal pathways are shut down, giving $v_u = 0$, and with no external correction applied, the final danger signal v_f equals v_s, i.e., 1.65 V. v_f exceeds the threshold voltage of VM5, which is 0.9 V, causing its resistance to decrease. Consequently, the node voltage v_n rises and reaches the turn-on voltage of N3, 0.5 V, at 0.648 s. This triggers the comparator to switch, resulting in a system output V_{out} of 1 V and the execution of an escape behavior.

As shown in Fig. 5b, when both the audio and visual signals are at medium intensity ($V_{in1} = V_{in2} = 0.5$ V) and no external correction is applied ($V_{e1} = V_{e2} = 0$), the four corresponding neurons fire, resulting in $v_{12} = v_{13} = v_{22} = v_{23} = 1$ V. The gate circuit identifies this as a medium-intensity audio-visual signal ($v_{p2} = 1$ V), which triggers the fusion pathway. The resistance of VM3 decreases, causing the voltage at its right terminal to fall below the threshold of the subsequent comparator. Consequently, the comparator outputs $v_{c2} = 2.7$ V.

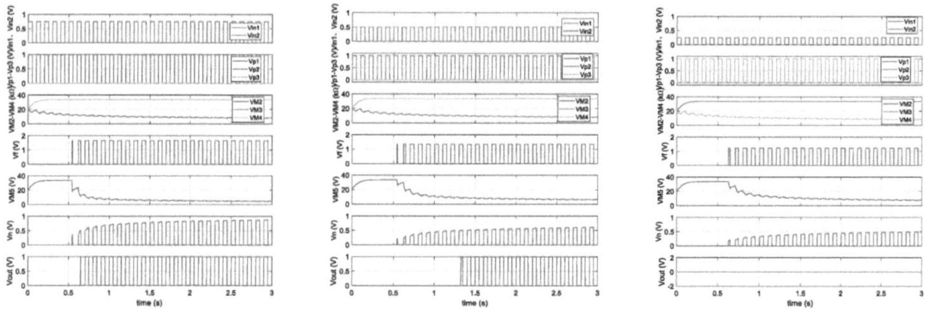

(a) (b) (c)

Fig. 5. Simulation results of experiment 2: (a) fusion of strong visual and auditory signals, (b) fusion of moderate visual and auditory signals, (c) fusion of weak visual and auditory signals

After weighted fusion, the signal voltage becomes $v_s = 1.35\,\text{V}$. At this point, the unimodal pathways are shut off, and the final hazard signal is $v_f = v_s = 1.35\,\text{V}$. This voltage exceeds the threshold of VM5, reducing its resistance. As a result, the node voltage v_n rises and, at 1.33 s, turns on the subsequent MOS transistor. The system then produces an output of $V_{out} = 1\,\text{V}$, which corresponds to the implementation of an escape behavior.

As shown in Fig. 5c, when both the auditory and visual signals are of weak intensity and no external calibration is applied, with inputs $V_{in1} = V_{in2} = 0.25\,\text{V}$ and $V_{e1} = V_{e2} = 0$, the corresponding two neurons fire, resulting in $v_{13} = v_{23} = 1\,\text{V}$. The fusion module identifies these as weak audiovisual signals and, due to their semantic consistency, triggers fusion, producing $v_{p3} = 1\,\text{V}$. The resistance of VM4 decreases, leading to an output $v_{c3} = 4.9\,\text{V}$. After weighted fusion, v_s is calculated as $1.225\,\text{V}$, so $v_f = v_s = 1.225\,\text{V}$. Although this voltage exceeds the threshold of VM5, causing its resistance to drop and v_n to rise, v_n remains consistently below the turn-on voltage of the subsequent NMOS transistor N3. Consequently, the system output remains at $V_{out} = 0$, and no escape behavior is triggered.

Experiment 3: Fusion Calibration Experiment. When the woman sees a large dog and hears loud barking, her initial instinct would be to flee immediately. However, if she simultaneously receives external safety cues, such as the dog being on a leash, wearing a muzzle, or the owner informing her that the dog has a gentle temperament, her judgment is corrected. Consequently, the perceived level of danger drops sharply, leading to the assessment that the current situation is relatively safe and does not require escape.

As illustrated in Fig. 6a, with V_{in1} and V_{in2} both set to $0.75\,\text{V}$, V_{e1} at $-0.5\,\text{V}$, and V_{e2} at 0, all six neurons fire simultaneously. The gate circuit identifies this as a dual strong-consistency signal, resulting in $v_{p1} = 1\,\text{V}$. This triggers the

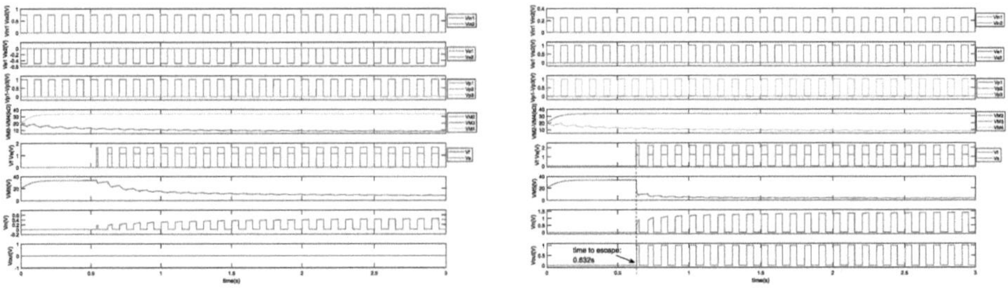

(a) (b)

Fig. 6. Simulation results of experiment 3: (a) strong visual signal combined with strong auditory signal and external safety information, (b) weak visual signal combined with weak auditory signal and hazardous external information

fusion pathway, producing an output $v_{c1} = 2.2\,\text{V}$, which subsequently generates a fused danger signal $v_s = 1.65\,\text{V}$. Without correction, this voltage would drive the system to output an escape command. However, the external safety signal V_{e1} is superimposed onto vs by turning on the MOS transistor N1, causing the final danger signal v_f to drop to 1.15 V. Although this voltage still exceeds the threshold of VM5 and induces a change in its resistance, it is insufficient for the node voltage v_n to surpass the threshold of the subsequent comparator. Consequently, the system output remains at $V_{out} = 0$, indicating that the current scenario is relatively safe and no escape behavior is triggered. This process demonstrates the critical role of external information in calibrating multimodal hazard assessment.

Similarly, as shown in Fig. 6b, when observing a distant small dog and hearing faint barking, but receiving external hazard information (such as the dog being engaged in a fight, having a history of aggression, or carrying a disease), the hazard judgment is primarily dominated by external information. Here, $V_{in1} = V_{in2} = 0.25\,\text{V}$, $V_{e1} = 0$, and $V_{e2} = 1\,\text{V}$. Only two discharges occur at this moment. The gate circuit, after evaluation, identifies two weak signals, resulting in $v_{p3} = 1\,\text{V}$. The resistance of VM4 decreases rapidly, reducing the voltage at the right terminal and turning P10 on. This yields a fused output signal $v_{c3} = 4.9\,\text{V}$. The weighted multiplication of the audio-visual signals produces $v_s = 1.225\,\text{V}$. As known from Experiment 2, this value is insufficient to trigger a system response. However, N2 is now conducting, causing v_f to increase significantly to 2.25 V. Consequently, v_n rises rapidly, flipping the comparator. The system outputs $V_{out} = 1\,\text{V}$, indicating that an escape action is implemented to quickly leave the scene.

Table 3 compares the relationship between the hazard level v_f and the escape response V_{out} across three experiments. A condition where $v_f \neq 0$ indicates the presence of a hazard, while the triggering of the escape action depends on the change in the resistance of the memristor VM5, specifically whether the voltage at its right terminal rises to the turn-on threshold of the MOS transistor. Table 3 reveals several key patterns verified by the experiments:

Table 3. Input signals and corresponding actuation times

input signals	$v_f(V)$	$v_{out}(V)$	Actuation time (s)
Strong Visual Signal	1.46	1	0.834
Moderate Visual Signal	0.98	0	
Weak Visual Signal	0.49	0	
Strong Auditory Signal	1.54	1	0.728
Moderate Auditory Signal	1.03	0	
Weak Auditory Signal	0.51	0	
Strong Audio-visual Signal	1.65	1	0.648
Moderate Audio-visual Signal	1.35	1	1.330
Weak Audio-visual Signal	1.23	0	
Strong Audio-visual Signal+External Safety Signal	1.15	0	
Weak Audio-visual Signal+External Danger Signal	2.25	1	0.632

1) When a single modality provides a moderate signal input, the system chooses to remain vigilant without initiating escape. In contrast, when two moderate signals from different modalities are simultaneously input, the system executes an escape action. This indicates that multimodal information, when semantically consistent, produces a significant perceptual enhancement effect, thereby improving the recognition of and response intensity to potential hazards [1].
2) The system exhibited subadditive integration under strong signal inputs: the fused $v_f(1.65\,\text{V})$ was significantly smaller than the sum of the unimodal responses $(1.46\,\text{V}+1.54\,\text{V}=3\,\text{V})$. Conversely, under weak signal inputs, it demonstrated superadditive integration: the fused $v_f(1.23\,\text{V})$ exceeded the sum of the unimodal responses $(0.49\,\text{V}+0.51\,\text{V}=1\,\text{V})$. This pattern reflects the principle of inverse effectiveness in multisensory integration.
3) The system exhibited significant differences in escape response times under various conditions. The slowest response occurred with moderate-intensity bimodal input (1.33 s), which aligns with the behavioral characteristic of cautious decision-making by humans in ambiguous situations. In contrast, the fastest response was observed in the presence of a clear external danger signal (0.632 s). The response time following the fusion of strong audiovisual signals (0.648 s) was shorter than that for any single-modality input (visual: 0.834 s, auditory: 0.728 s), demonstrating a fusion enhancement effect. Additionally, a single strong auditory input elicited a faster response than a single strong visual input.

6 Conclusion

This paper presents a memristor-based circuit for audiovisual modality fusion and decision-making. Its correctness is validated through three experiments, and the experimental results are systematically analyzed. The circuit comprises four functional modules. The multimodal perception module determines the relative intensities of audio and visual stimuli. The signal fusion module performs nonlinear enhancement of audiovisual signals with matched intensity levels. The unimodal cognitive pathway emulates the brain s processing of individual sensory inputs. The decision-making module evaluates the degree of danger by integrating baseline information with corrective feedback and subsequently triggers an escape response. The circuit achieves semantic-level fusion and amplification of congruent audiovisual information, thereby providing a viable solution for hazard-related decision-making. It holds significant potential for applications in environmental perception and safety-critical decision-making for intelligent robots, multimodal hazard warning systems in autonomous driving, and immersive interaction with integrated safety mechanisms in augmented reality and virtual reality environments.

References

1. Stein, B.E., Stanford, T.R.: Multisensory integration: current issues from the perspective of the single neuron. Nat. Rev. Neurosci. **9**(4), 255–266 (2008)
2. Meredith, M.A., Nemitz, J.W., Stein, B.E.: Determinants of multisensory integration in superior colliculus neurons. I. Temporal factors. J. Neurosci. **7**(10), 3215–3229 (1983)
3. Meredith, M.A., Stein, B.E.: Interactions among converging sensory inputs in the superior colliculus. Science **221**(4608), 389–391 (1983)
4. Chua, L.O.: Memristor-The missing circuit element. IEEE Trans. Circ. Theory **18**(5), 507–519 (1971)
5. Strukov, D.B., Snider, G.S., Stewart, D.R., et al.: The missing memristor found. Nature **453**(7191), 80–83 (2008)
6. Yang, J., Cho, H., Ryu, H., et al.: Tunable synaptic characteristics of a Ti/TiO2/Si memory device for reservoir computing. ACS Appl. Mater. Interfaces. **13**(28), 33244–33252 (2021)
7. Zhang, J., Xu, K., Lu, C., et al.: One-step annealing-configured Hf0.2Zr0.8O2 memristive CAntiferroelectric devices for bioinspired CSNN neuromorphic computing. Nano Lett. **25**(50), 17266–17274(2025)
8. Indiveri, G., Linares-Barranco, et al.: Memristive devices for computing. Nat. Nanotechnol. **8**(2), 87–96 (2013)
9. Jiang, M., Zhang, Y., Jiang, N., et al.: Bio-inspired neuromorphic circuit design of nonassociative learning for multisensory enhancement and depression. IEEE Trans. Circ. Syst. I Regular Papers **72**(10), 5517–5530 (2025)
10. Zhang, Y., Lv, J., Zeng, Z.: The framework and memristive circuit design for multisensory mutual associative memory networks. IEEE Trans. Cybernet. **53**(12), 7844–7857 (2023)

11. Tao, K., Zhou, H., Wang, Y., et al.: Memristor-based neural network circuit of emotional grading in intelligence decision system. In: International Conference on Neuromorphic Computing (ICNC), Chongqing, China, pp. 1–4 (2024)
12. Liu, J., Xiong, F., Zhou, Y., et al.: Bioinspired memristive neural network circuit design of cross-modal associative memory. IEEE Trans. Cogn. Dev. Syst. **16**(2), 794–808 (2024)
13. Ji, X., Dong, Z., Zhou, G., et al.: MLG-NCS: multimodal local CGlobal neuromorphic computing system for affective video content analysis. IEEE Trans. Syst. Man Cybernet. Syst. **54**(8), 5137–5149 (2024)
14. Chen, L., Li, C.D., Huang, T.W., et al.: The bipolar and unipolar reversible behavior on the forgetting memristor model. Neurocomputing **171**, 1637–1643 (2016)
15. Yan, R.N., Hong, Q.H., Wang, C.H., et al.: Multilayer memristive neural network circuit based on online learning for license plate detection. IEEE Trans. Comput.-Aided Des. Integr. Circ. Syst. **41**, 3000–3011 (2022)
16. Quiroga, R.Q., Kraskov, A., Koch, C., et al.: Explicit encoding of multimodal percepts by single neurons in the human brain. Curr. Biol. **19**(15), 1308–1313 (2009)
17. Ding, M.H., Wang, X.P., Zeng, Z.G.: Memristive circuit design and implementation of olfactory perception and cross-modal association (in Chinese). Sci. Sin. Inform. **53**, 2561–2576 (2023)
18. Woelfle, R., Grahn, J.A.: Auditory and visual interhemispheric communication in musicians and non-musicians. PLoS ONE **8**(12), e84446 (2013)
19. Yasin, B., Mustafa, F., Permatasari, R.: How much videos win over audios in listening instruction for EFL learners. Turk. Online J. Educ. Technol. **17**(1), 92–100 (2018)

Deep Learning

ClusterSuit: Accelerating In-Run Data Shapley with Cluster-Smoothed Layer Selection

Mengkai Han(✉)

Department of Data Science, City University of Hong Kong, 83 Tat Chee Avenue, Kowloon, Hong Kong SAR
mengkahan3-c@my.cityu.edu.hk

Abstract. Data valuation is essential for trustworthy AI, but Shapley-based methods remain computationally prohibitive across training. We propose ClusterSuit, combining profiling-guided hotspot layer selection with cluster smoothing post-processing. On ResNet-18, top-3 Layer4 layers account for 80% of layer time; hotspot selection improves throughput by 55% versus depth-based selection with comparable Shapley quality. Cluster smoothing boosts mislabeled-data AUROC from 0.5371 to 0.7355 (+37% relative, +8.5% absolute over In-Run Data Shapley). Experiments on CIFAR-10 and GPT-2 show improved accuracy and reduced computational overhead.

Keywords: data valuation · Shapley value · mislabeled data detection · neural network interpretability · cluster smoothing

1 Introduction

Data valuation supports data cleaning [1], active learning [2], and detecting mislabeled or outlier samples [3]. Its goal is to quantify each training point's contribution. The Shapley value [4] is a principled choice for this task, offering fairness, efficiency, and additivity [5].

Classical Data Shapley requires retraining on many subsets and is thus prohibitive for large models [1,3]. In-Run Data Shapley [6] avoids retraining via Taylor approximation in a single run, but remains costly for deep networks: even with "ghost dot-product" and "ghost Hessian-gradient product" tricks, second-order variants can take roughly 2× training time. We identify two key limitations:

Computational Inefficiency: Existing methods treat all layers uniformly, computing gradients over all parameters despite uneven impact on valuation. In deep nets, certain "hotspot layers" dominate forward/backward cost and carry more attribution signal; uniform computation wastes effort on low-impact layers.

Limited Discriminative Power: Raw gradient dot products weakly separate mislabeled data. In-Run Data Shapley reaches AUROC = 0.678 on CIFAR-10,

Z. Zhang (Ed.): ISNN 2026, LNCS 16848, pp. 107–118, 2027.
https://doi.org/10.1007/978-981-92-5870-3_9

partly because similar gradient signatures can correspond to different label quality—motivating post-processing that exploits gradient-space similarity.

We propose **ClusterSuit**, accelerating In-Run Data Shapley via profiling-guided layer selection and cluster-smoothing post-processing. Our contributions are: (1) **Hotspot Layer Identification**—lightweight profiling pinpoints expensive, informative layers for targeted optimization; (2) **Selective Gradient Computation**—valuation gradients focus on hotspot layers to cut cost while preserving key signal; and (3) **Cluster Smoothing Post-Processing**—clustering by gradient signatures and propagating quality scores within clusters, raising mislabeled-data AUROC from 0.5371 to 0.7355.

Experiments on noisy-label CIFAR-10 and language modeling (ResNet-18, GPT-2) show improved valuation accuracy and large speedups over full-layer computation.

2 Background and Related Work

We summarize data valuation, In-Run Data Shapley, efficient gradient computation, and layer-wise gradient heterogeneity to motivate ClusterSuit.

2.1 Data Valuation and Shapley Value

The Problem of Data Valuation. Data valuation quantifies how each training sample affects model performance—critical as models scale on heterogeneous data [1]. It enables data cleaning [3], mislabeled-sample detection [7], and fair compensation for contributors [8].

Shapley Value for Data Attribution. The Shapley value [4] provides a principled attribution framework. For $\mathcal{D} = \{z_i\}_{i=1}^N$ and utility $\mathcal{U}(S)$ on subset S, the value of z_i is

$$\phi_{z_i}(\mathcal{U}) = \sum_{S \subseteq \mathcal{D} \setminus \{z_i\}} \frac{|S|!(N-|S|-1)!}{N!} [\mathcal{U}(S \cup \{z_i\}) - \mathcal{U}(S)]. \tag{1}$$

It uniquely satisfies Efficiency, Symmetry, Dummy, and Linearity [5].

Traditional Data Shapley and Its Limitations. Data Shapley [1] sets $\mathcal{U}(S) = \mathrm{Perf}(\mathcal{A}(S))$ for algorithm $\mathcal{A}$ and validation performance Perf. Exact Shapley requires 2^N subset evaluations; Monte Carlo approximations [1,3] still demand many retrainings—prohibitive at foundation-model scale [6].

2.2 In-Run Data Shapley

Concept and Formulation. In-Run Data Shapley [6] computes values in a single training run via Taylor approximation of small per-step loss changes. At iteration t, the local utility for using subset S of the batch on validation point $z^{(\text{val})}$ is

$$\mathcal{U}_t^{(t)}(S : w_t) \triangleq \ell(w_{t+1}(S), z^{(\text{val})}) - \ell(w_t, z^{(\text{val})}), \tag{2}$$

where $w_{t+1}(S) \triangleq w_t - \eta_t \sum_{z \in S} \nabla \ell(w_t, z)$ and η_t is the learning rate.

Taylor Approximation and Closed-Form Computation. First-order expansion yields

$$\mathcal{U}_t^{(t)}(S : w_t) \approx -\eta_t \sum_{z \in S} \nabla \ell(w_t, z^{(\text{val})}) \cdot \nabla \ell(w_t, z), \tag{3}$$

and the first-order In-Run Data Shapley

$$\phi_{z_i}^{(1)} \approx -\frac{1}{T} \sum_{t=0}^{T-1} \eta_t \nabla \ell(w_t, z_i) \cdot \nabla \ell(w_t, z^{(\text{val})}). \tag{4}$$

The dot product measures alignment between training and validation gradients at w_t.

Second-Order Extension. Second-order In-Run Data Shapley adds gradient-Hessian-gradient terms for batch interactions:

$$\phi_{z_i}^{(2)} \approx \phi_{z_i}^{(1)} + \frac{1}{T} \sum_{t=0}^{T-1} \frac{\eta_t^2}{2|B_t|} \sum_{z_j \in B_t, z_j \neq z_i} g_{z_j}^\top H_{z^{(\text{val})}} g_{z_i}, \tag{5}$$

with $g_z = \nabla \ell(w_t, z)$ and $H_{\text{val}} = \nabla^2 \ell(w_t, z^{(\text{val})})$.

2.3 Efficient Gradient Computation

The Per-Sample Gradient Problem. In-Run Shapley needs training-validation gradient dot products over all parameters. Naive per-sample backprop ($|B_t|$ passes) is $|B_t|$-times slower than standard training (e.g., 32× for batch size 32).

Ghost Clipping from Differential Privacy. Lee & Kifer [9] "ghost clipping" computes per-sample gradient norms in one backward pass from activations and output gradients, without forming full per-sample vectors.

Ghost Dot-Product Technique. Wang et al. [6] extend this to dot products. For linear layer $W \in \mathbb{R}^{d_{\text{in}} \times d_{\text{out}}}$, activations a, and output gradients $\frac{\partial \ell}{\partial s}$,

$$\frac{\partial \ell}{\partial W}[i] = \frac{\partial \ell}{\partial s}[i]^\top \cdot a[i], \tag{6}$$

and

$$\frac{\partial \ell}{\partial W}[i] \cdot \frac{\partial \ell}{\partial W}[j] = \left(\frac{\partial \ell}{\partial s}[i] \cdot \frac{\partial \ell}{\partial s}[j] \right) \cdot (a[i] \cdot a[j]), \tag{7}$$

computable within a single backward pass. Associativity further avoids materializing $O(B \cdot d_{\text{in}} \cdot d_{\text{out}})$ gradients, reducing memory to $O((d_{\text{in}} + d_{\text{out}}) \cdot B)$.

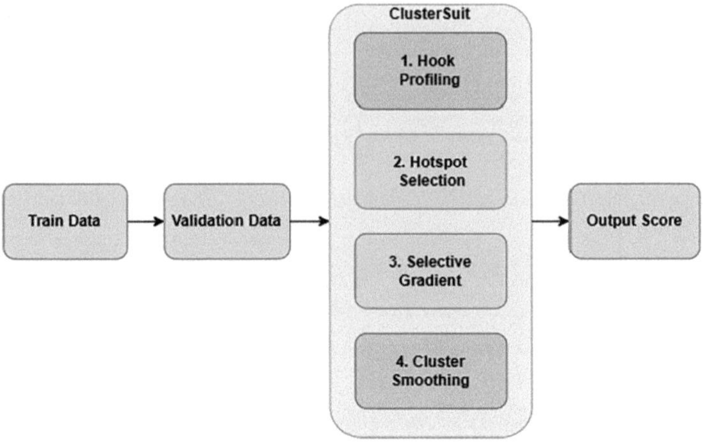

Fig. 1. ClusterSuit Framework Architecture. Four phases: profiling, selective hotspot computation, In-Run Shapley, and cluster smoothing. Hotspots use full ghost dot-product; non-hotspots are skipped.

2.4 Layer-Wise Gradient Analysis

Heterogeneity Across Network Layers. Layers differ in gradient magnitude, sensitivity, and task-relevant information [10,11]; some "critical" layers dominate linear-probe and ablation evidence.

Gradient Magnitude and Computational Cost. Gradient cost scales with layer size (e.g., $d_{\text{in}} \times d_{\text{out}}$ for linear layers). In transformers, attention and FFN layers dominate cost [12]; embeddings behave differently despite fewer parameters.

Selective Computation Opportunities. Layers contribute unevenly to valuation: small layers add little to dot products at high cost, while large layers dominate both time and signal—motivating selective computation on the most impactful layers.

3 Methodology

ClusterSuit comprises four components: (1) profiling-based hotspot layer identification; (2) selective gradient computation on hotspots; (3) cluster-smoothing post-processing; and (4) theoretical analysis.

3.1 Overview of ClusterSuit Framework

Figure 1 summarizes the four-phase pipeline. ClusterSuit performs efficient gradient-based data valuation during training via three principles: non-invasive auxiliary computation; layer-selective resource allocation; and cluster-aware post-processing for detection accuracy. Operationally, stage one profiles hotspots and computes raw Shapley values on selected layers; stage two applies cluster smoothing using gradient-space similarity.

3.2 Hotspot Layer Identification via Profiling

Design Rationale. Unlike uniform layer treatment in prior In-Run Shapley, we exploit heterogeneous per-layer cost: a few layers dominate forward time.

Profiling Method. PyTorch forward hooks record per-layer time over warmed-up iterations; layers are ranked by average time and top-K are chosen as hotspots. Profiling runs once per architecture and amortizes across runs.

Hotspot Layer Selection

$$\mathcal{H}_K = \text{TopK}(\{\text{Cost}_\ell\}_{\ell \in \mathcal{L}}, K), \tag{8}$$

where Cost_ℓ is layer cost and K is the hotspot count.

Key Insight: On ResNet-18, the top 3 layers (all in Layer4) account for $\sim$80% of layer-wise time.

3.3 Selective Gradient Computation

Layer-Wise Computation Strategy. Gradient dot products run only on hotspots: full ghost dot-product on hotspot linear/conv layers; skip non-hotspot linear/conv; index-based accumulation on hotspot embeddings; full computation on non-hotspot LayerNorm/RMSNorm.

Implementation via Autograd Hooks. Hotspots use Section 2.3.3 ghost dot-products via autograd hooks on activations and gradients. Detached copies avoid interfering with training gradients.

Gradient Validation Caching. With cache size > 0, validation gradients are computed every K steps and reused—helpful on GPT-2 where cost is spread across layers.

3.4 Cluster Smoothing Post-processing

Motivation. Raw first-order dot products weakly detect mislabels (AUROC ≈ 0.54 even with hotspots). Similar gradient signatures often share label quality; mislabels cluster with inconsistent patterns relative to neighbors.

Algorithm. Given raw scores s_i, we (1) build a k-NN graph with cosine weights $w_{ij} = \frac{\nabla\ell(w,z_i)\cdot\nabla\ell(w,z_j)}{\|\nabla\ell(w,z_i)\|\cdot\|\nabla\ell(w,z_j)\|}$, and (2) smooth via label propagation: $s'_i = (1-\alpha)\cdot s_i + \alpha \cdot \frac{\sum_{j\in\mathcal{N}_k(i)} w_{ij}\cdot s_j}{\sum_{j\in\mathcal{N}_k(i)} w_{ij}}$, with $\alpha \in [0,1]$ and $\mathcal{N}_k(i)$ the k-nearest neighbors.

Interpretation. Mislabeled samples get lower smoothed scores from neighbor disagreement; correct labels gain from local consensus.

Parameter Selection. α trades raw values ($\alpha = 0$) vs. cluster consensus ($\alpha = 1$). Section 4.4 shows $\alpha = 0.2$ is optimal (AUROC $= 0.7355$).

3.5 Integration with Training Loop

GhostEngineManager Architecture. GhostEngineManager handles engine init/attach/lifecycle: config-based engine selection, fixed validation batch, hook attachment at optimizer steps, and checkpoint save/load.

Training Loop Integration Pattern. Per iteration: attach batch metadata; compute training gradients and hotspot ghost dot-products; aggregate Shapley values; clear gradients. Hooks avoid changing core loop logic.

Distributed Training Compatibility. Under DDP, dot products are local per worker; aggregation aligns with gradient all-reduce, scaling linearly in GPU count without extra communication.

3.6 Theoretical Analysis

Approximation Bound for Hotspot Selection. Let $G_\ell^{(i)}$ be sample i's gradient at layer ℓ. Full vs. selective dot products:

$$\phi_i^{\text{full}} = \sum_{\ell\in\mathcal{L}} G_\ell^{(\text{val})}\cdot G_\ell^{(i)}, \quad \phi_i^{\text{select}} = \sum_{\ell\in\mathcal{H}_K} G_\ell^{(\text{val})}\cdot G_\ell^{(i)}. \tag{9}$$

Theorem 1 (Hotspot Approximation Bound). For target error $\epsilon > 0$, if

$$\sum_{\ell\notin\mathcal{H}_K} \|G_\ell^{(\text{val})}\|_F \cdot \|G_\ell^{(i)}\|_F \le \epsilon \cdot \sum_{\ell\in\mathcal{L}} \|G_\ell^{(\text{val})}\|_F \cdot \|G_\ell^{(i)}\|_F, \tag{10}$$

then

$$\frac{|\phi_i^{\text{full}} - \phi_i^{\text{select}}|}{|\phi_i^{\text{full}}|} \le \epsilon. \tag{11}$$

Proof Sketch. Cauchy-Schwarz bounds each layer term; summing non-hotspot layers yields the bound.

Cluster Smoothing Convergence. Theorem 2 (Cluster Smoothing Convergence). One-step smoothing from raw vector S converges to fixed point

$$S^* = (1 - \alpha) \cdot (I - \alpha W)^{-1} \cdot S, \tag{12}$$

with normalized similarity matrix W.

Proof Sketch. Cluster smoothing is label propagation on a similarity graph; for connected graphs and $\alpha < 1$, convergence to a unique fixed point is exponential.

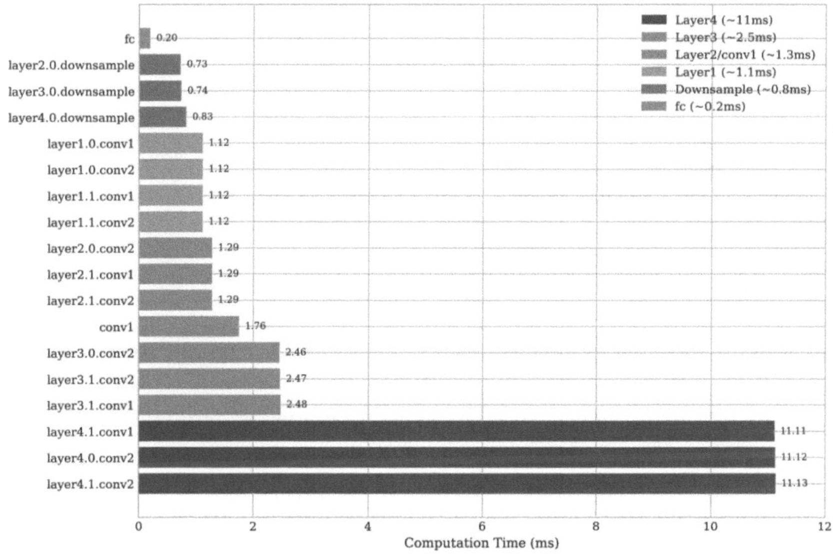

Fig. 2. ResNet-18 layer time distribution. Layer4 conv layers dominate (~11.1 ms), 4.5× Layer3 (~2.5 ms) and 8.6× Layer2 (~1.3 ms). Top-3 layers account for 80% of layer-wise time.

4 Experiments

We evaluate ClusterSuit on CIFAR-10 (noisy labels) and GPT-2 language modeling, measuring speedup for In-Run Shapley and mislabeled-sample detection (AUROC).

4.1 Experimental Setup

Datasets. CIFAR-10 with CIFAR-10N aggregate noise: 50K train/10K test, 4,505 mislabels. GPT-2: local corpus with 2.048M train tokens and 102.4K val/test tokens.

Table 1. ResNet-18 hotspot layers (top 6; full top-18 in appendix)

Rank	Layer Name	Avg Time (ms)	Time Range (ms)
1	layer4.1.conv2	11.126	11.078–11.291
2	layer4.0.conv2	11.123	11.067–11.191
3	layer4.1.conv1	11.107	11.063–11.228
4	layer3.1.conv1	2.479	2.462–5.231
5	layer3.1.conv2	2.472	2.466–2.488
6	layer3.0.conv2	2.457	2.449–2.466

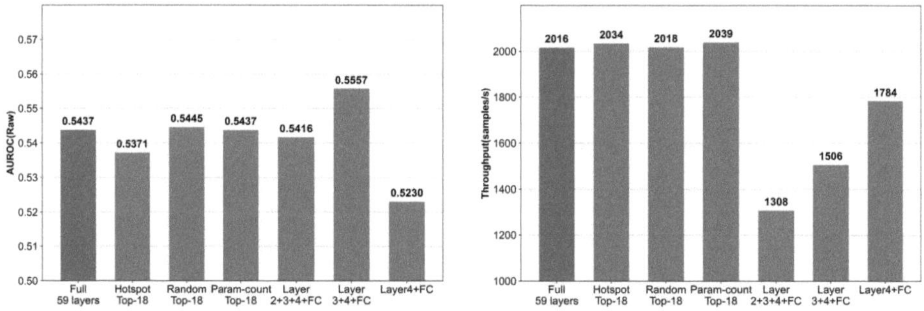

(a) AUROC by layer selection strategy (b) Throughput comparison

Fig. 3. Detection performance: AUROC and throughput by layer selection strategy.

Models. ResNet-18 (59 layers) for CIFAR-10; GPT2-Small (124M params, 12 blocks) for LM.

Hardware. NVIDIA PRO 6000 (CUDA device 1); PyTorch with bfloat16 mixed precision.

4.2 Hotspot Layer Analysis Results

ResNet-18 Layer Computation Time Analysis. We profile per-layer forward time on ResNet-18 (Fig. 2, Table 1).

Layer4 conv layers average ∼11.1 ms—4.5× Layer3 and 8.6× Layer2; top-3 hotspots cover ∼80% of layer-wise time. Downsample layers are comparatively cheap.

GPT-2 Layer Analysis Results. Hotspots are transformer blocks h.[6..11], matching deeper-layer cost concentration in transformers.

4.3 Impact of Layer Selection on Shapley Value Quality

We compare hotspot subsets for mislabel detection (AUROC) under 5K steps, batch 128, LR 0.001, aggregate noise (Fig. 3, Table 2).

Table 2. Detection performance by layer selection

Layer Subset	# Layers	Test Acc	Throughput (samples/s)	AUROC (raw)	AUROC (smoothed)
Full-layer (all 59)	59	61.11%	2016	0.5437	–
Hotspot Top-18 + CS	18	60.84%	2034	0.5371	**0.7355**
Hotspot Top-18 (no smoothing)	18	60.93%	2034	0.5371	–
Random Top-18	18	60.70%	2018	0.5445	–
Param-count Top-18	18	61.30%	2039	0.5437	–
Layer2+3+4+FC	~18	61.30%	1308	0.5416	–
Layer3+4+FC	~12	61.41%	1506	0.5557	–
Layer4+FC	~6	59.70%	1784	0.5230	–

Raw AUROC stays near 0.54 for all selection strategies; cluster smoothing (CS) raises Hotspot Top-18 from 0.5371 to 0.7355 (+37% relative). Hotspot Top-18 attains best smoothed AUROC and highest throughput among 18-layer strategies (2034 samples/s); Layer2+3+4+FC matches raw AUROC but is 36% slower.

4.4 Cluster Smoothing Ablation Study

We vary smoothing strength α (Fig. 4, Table 3).

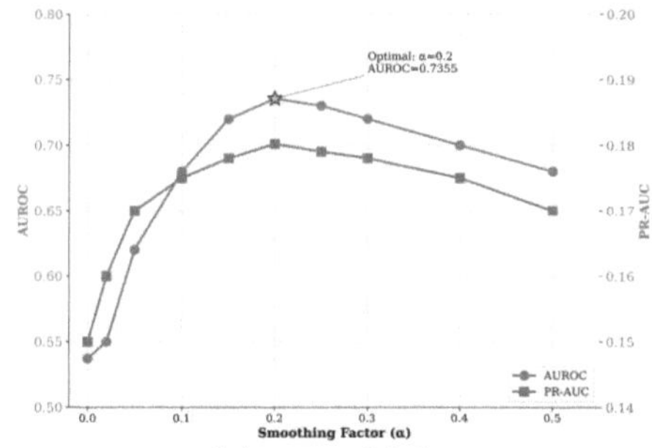

(a) α sensitivity

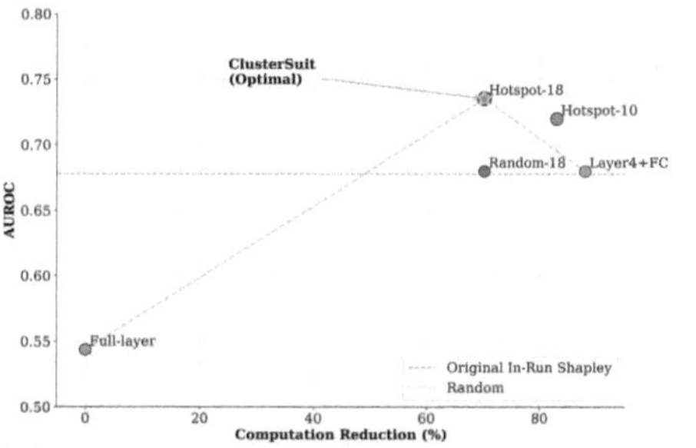

(b) Computation reduction vs. AUROC trade-off

Fig. 4. α sensitivity and computation–AUROC trade-off.

$\alpha = 0.2$ maximizes AUROC (0.7355), beating original In-Run Shapley raw AUROC 0.678 by 8.5% absolute.

4.5 Gradient Cache Optimization

Table 4 reports validation-gradient cache interval K on CIFAR-10.

With CS ($\alpha = 0.2$), all K yield AUROC $\approx$ 0.7355. On ResNet-18, K barely affects throughput (~2030 samples/s); on GPT-2, $K = 2$ gives ~0.6% gain. Smaller val batches raise throughput but lengthen total training.

4.6 Comparison with Baseline Methods

Table 5 summarizes comparisons against prior In-Run Data Shapley and full-layer baselines (layer-strategy details in Table 2).

Note: Original In-Run Shapley from Wang et al. (ICLR 2025), Table 4; our throughput on PRO 6000.

4.7 Summary

Hotspot selection plus CS ($\alpha = 0.2$) reaches AUROC 0.7355 with competitive throughput on ResNet-18; raw dot products alone stay near 0.54 across strategies, so CS is critical. Time-based hotspots beat depth-based picks (e.g., Layer2+3+4+FC: similar raw AUROC, 36% lower throughput). Param-count Top-18 is a close alternative (15/18 overlap with time-based). ResNet-18 hotspots concentrate in Layer4 (top-3 $\approx$ 80% time); GPT-2 is more uniform, yielding only $\sim$0.5% hotspot speedup.

5 Discussion

5.1 Theoretical Implications

Connection to Data Valuation Theory. ClusterSuit connects Shapley-based data valuation [1] to practical training. Shapley values are principled but costly; we show that samples with similar gradient signatures share label-quality patterns, and cluster smoothing exploits this for much better detection. As in network pruning [11], a layer subset suffices for valuation when paired with post-processing.

Relationship with Neural Network Interpretability. Hotspots align with interpretability findings: deeper, higher-cost layers carry more task signal [10]. Their gradients, with cluster smoothing, yield stronger mislabel signals.

5.2 Practical Considerations

ClusterSuit is most beneficial on ResNet-18/CIFAR-10 (55% throughput gain vs. depth-based selection); on GPT-2, gains are modest ($\sim$0.6% with cache). For best detection, use cluster smoothing with $\alpha = 0.2$ (AUROC = 0.7355). We recommend $K = 18$ hotspots on ResNet-18 (59 layers), $\alpha = 0.2$ for smoothing, and cache $K = 2$ or 5 (minimal throughput change on CIFAR-10; slight GPT-2 benefit).

Limitations. Profiling adds one-time cost per architecture (amortized across runs). Hotspots are architecture- and task-specific. Our profiling targets gradient-based valuation; extension to influence functions and TRAK [13,14] is open.

Table 3. Cluster smoothing ablation (Hotspot Top-18)

Alpha	AUROC	PR-AUC
0.00	0.5371	0.15
0.10	0.68	0.175
0.20	**0.7355**	**0.1802**
0.30	0.72	0.178
0.50	0.68	0.17

Table 4. Gradient cache K on CIFAR-10

Cache K	Test Acc	AUROC (raw)	Throughput	Std Dev
0	60.93%	0.5371	2033.68	150.47
2	61.64%	0.5379	2028.07	143.24
5	61.76%	0.5429	2026.48	144.75

Table 5. Baseline comparison

Method	# Layers	AUROC (raw)	AUROC (smoothed)	Throughput
Original In-Run Data Shapley	59	**0.678**	N/A	N/A (A100)
Full-layer (all 59)	59	0.5437	–	2016
Hotspot Top-18 + CS (Ours)	18	0.5371	**0.7355**	2034 (+55% vs. L2+3+4+FC)

5.3 Comparison with Related Approaches

Comparison with Original In-Run Data Shapley. We compare ClusterSuit to Wang et al. [6] (Table 6):

With $\alpha = 0.2$, we reach AUROC 0.7355 (+8.5% absolute over 0.678) while computing on ~30% of layers via hotspot selection.

vs. Full-Layer In-Run Shapley. Comparable throughput with far better smoothed AUROC.

vs. TracIn and Influence Functions. TracIn [13] and influence functions [14] score perturbation-based influence; we accelerate In-Run Shapley. Hotspot selection may complement those methods.

5.4 Broader Impact and Applications

Efficient valuation supports mislabel cleaning, active learning, and fair data compensation [1]. Shapley scores also aid OOD detection, concept-drift monitoring, and fairness auditing. In federated settings, they can guide client scheduling and distributed quality control.

Table 6. Comparison with Original In-Run Data Shapley

Aspect	In-Run Data Shapley (Original)	ClusterSuit (Ours)
Hardware	A100 GPU	NVIDIA PRO 6000 GPU
Layer Selection	Full-layer	Hotspot top-18
Post-processing	None	Cluster smoothing ($\alpha = 0.2$)
CIFAR-10 AUROC (raw)	**0.678**	0.5371
CIFAR-10 AUROC (smoothed)	N/A	**0.7355**
Computation	100% layers	~30% (18/59)

5.5 Conclusion and Future Work

We present ClusterSuit: profiling-guided hotspot selection plus cluster smoothing for faster In-Run Shapley and stronger mislabel detection ($\alpha = 0.2$, AUROC = 0.7355 vs. 0.678 raw baseline). Top-18 hotspots raise throughput from 1308 to 2034 samples/s (+55%) using ~30% of ResNet-18 layers. Combining efficiency (hotspots) and accuracy (smoothing) moves data valuation toward practical pipelines for quality assessment and debugging.

Future work: theory for cluster smoothing; adaptive/multi-scale smoothing; scaling to larger LMs, ViTs, and real-world data.

References

1. Ghorbani, A., Zou, J.: Data shapley: equitable valuation of data for machine learning. In: ICML, pp. 2242–2251 (2019)
2. Wei, K., Iyer, R., Bilmes, J.: Submodularity in data subset selection and active learning. In: ICML, pp. 1954–1963 (2015)
3. Jia, R., Dao, D., Wang, B., et al.: Towards efficient data valuation based on the shapley value. In: ICML, pp. 1167–1176 (2019)
4. Shapley, L.S.: A value for n-person games. In: Contributions to the Theory of Games, vol. 2, no. 28, pp. 307–317 (1953)
5. Young, H.P.: Monotonic solutions of cooperative games. IJGT **14**(2), 65–72 (1985)
6. Wang, J.T., Mittal, P., Song, D., Jia, R.: Data shapley in one training run. In: ICLR (2025)
7. Ghorbani, A., Kim, M., Zou, J.: A distributional framework for data valuation. In: ICML, pp. 3535–3544 (2020)
8. Henderson, P., et al.: Foundation models and fair use. arXiv:2303.15715 (2023)
9. Lee, J., Kifer, D.: Concentrated differentially private gradient descent with adaptive per-iteration privacy budget. In: KDD, pp. 1656–1665 (2021)
10. Alain, G., Bengio, Y.: Understanding intermediate layers using linear classifier probes. In: ICLR Workshop (2016)
11. Morcos, A.S., et al.: On the importance of single directions for generalization. In: ICLR (2018)
12. Vaswani, A., et al.: Attention is all you need. In: NeurIPS, pp. 5998–6008 (2017)
13. Park, S., et al.: TRAK: attributing model behavior at scale. In: ICML, pp. 17064–17083 (2023)
14. Koh, P.W., Liang, P.: Understanding black-box predictions via influence functions. In: ICML, pp. 1885–1894 (2017)

DFNet: Depthwise-Frequency Network for Multivariate Time Series Forecasting

Guang-fei Yang[1], Zhen-dong Liu[2(✉)], Li-li Geng[3], and Er-biao Yuan[1]

[1] School of Economics and Management, Dalian University of Technology, Dalian 116024, China

[2] School of Computer Science and Technology, Dalian University of Technology, Dalian 116024, China
lzd@mail.dlut.edu.cn

[3] Department of Infectious Diseases, Central Hospital of Dalian University of Technology, Dalian 116000, China

Abstract. Multivariate time series forecasting remains a fundamental challenge due to the complex interplay between local temporal dependencies and global periodic patterns. Existing methods often focus exclusively on either the time domain or the frequency domain, limiting their ability to capture heterogeneous dynamics in a unified manner. To address this, we propose DFNet, a novel end-to-end framework that jointly models time and frequency representations through two synergistic modules: (1) Time-domain Learning Structure (TLS) based on depthwise separable convolutions for efficient cross-channel local dependency modeling, and (2) Frequency-domain Learning Structure (FLS) equipped with learnable spectral filters for adaptive emphasis of critical frequency components. DFNet fuses these representations early and trains them jointly, enabling dynamic complementarity between local and global patterns. Extensive experiments on real-world datasets show that DFNet outperforms the current state-of-the-art methods across multiple benchmarks.

Keywords: Multivariate Time Series Forecasting · Spectral-temporal Fusion · Frequency-domain Analysis

1 Introduction

Multivariate time series forecasting underpins critical applications ranging from energy load prediction [17] and climate analysis [13] to financial risk assessment [1] and traffic management [7]. Unlike univariate forecasting, multivariate approaches must model complex interdependencies between multiple correlated variables, presenting significant challenges in capturing both local aperiodic patterns and global periodic structures that coexist in real-world time series.

As illustrated in Fig. 1, real-world time series exhibit heterogeneous periodic characteristics. Some sequences demonstrate strong periodic patterns while others display predominantly aperiodic behavior. This heterogeneity necessitates a

Z. Zhang (Ed.): ISNN 2026, LNCS 16848, pp. 119–132, 2027.
https://doi.org/10.1007/978-981-92-5870-3_10

comprehensive modeling approach that can simultaneously capture both local dependencies where values depend primarily on recent observations, and global dependencies where values are influenced by observations at regular intervals across the entire sequence.

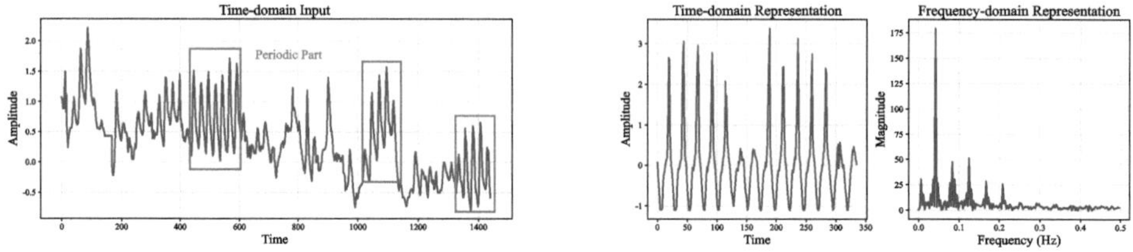

(a) (b)

Fig. 1. Real-world time series with distinct periodic patterns. (a) Aperiodic time series. The values show only short-term dependencies without a consistent long-term pattern. (b) Strongly periodic time series. The data exhibits a clear and predictable repeating pattern over time.

Time-domain and frequency-domain representations offer complementary perspectives for time series analysis. Time-domain methods excel at modeling local temporal dynamics and transient behaviors through direct observation of amplitude variations over time. In contrast, frequency-domain methods decompose signals into constituent frequencies, revealing underlying periodic structures and enabling energy concentration in dominant spectral components. This spectral concentration provides noise reduction capabilities and facilitates modeling of long-range dependencies that are challenging to capture in the time domain.

Despite their complementary strengths, effectively integrating time-domain and frequency-domain representations remains challenging for three fundamental reasons: (Q1) How to design an end-to-end framework that synergistically combines time-domain and frequency-domain learning without redundant computation? (Q2) How to efficiently model cross-channel dependencies in the time domain while maintaining computational efficiency? (Q3) How to identify and leverage the most informative frequency components while filtering out noise and irrelevant spectral information?

To address these challenges, we propose DFNet (Depthwise-Frequency Network), a novel architecture with the following contributions:

1. We propose an end-to-end collaborative framework for multivariate time series forecasting that unifies time-domain and frequency-domain representations through a feature fusion mechanism, establishing a new paradigm for capturing heterogeneous temporal patterns.
2. We design an efficient Time-domain Learning Structure (TLS) based on depthwise separable convolutions. It effectively models local-cross-channel dependencies (LCCD) with minimal parameter overhead, enabling efficient

modeling of cross-channel dependencies in the time domain while maintaining low computational complexity.

3. We develop a Frequency-domain Learning Structure (FLS) with adaptive spectral filtering. It automatically identifies and emphasizes informative frequency components through learnable filters, thereby enhancing model robustness to noise and non-stationarity.

Comprehensive experiments on multiple benchmarks demonstrate that our approach achieves state-of-the-art performance, validating the effectiveness of the proposed time-frequency collaborative paradigm.

2 Related Work

Deep Learning for Multivariate Time Series Forecasting. Multivariate time series forecasting has evolved from classical statistical models to deep neural architectures capable of capturing nonlinear inter-variable dynamics. Early recurrent approaches (LSTMs [5], GRUs [3]) and dilated convolutions (TCNs [2]) often struggle with long-range dependencies and scalability.

Recent advances favor simpler yet effective designs. MLP-based models like DLinear [18] demonstrate that a hierarchical feed-forward network (FFN) can achieve competitive performance by explicitly separating trend and residual components or leveraging direct input-output mappings. SOFTS [4] proposes a centralized module to aggregate global information before redistributing it across channels, improving efficiency over pairwise attention. While efficient, these approaches largely operate in the time domain and lack mechanisms to explicitly exploit global periodic structures, which is a key limitation when dealing with heterogeneous temporal patterns.

Transformer-based models [12,19] introduce attention mechanisms for long-range interactions. PatchTST [11] treats time series patches as tokens and enforces channel independence to preserve per-variable dynamics. iTransformer [10] inverts the attention axis to model cross-channel correlations directly. Nevertheless, most remain confined to time domain or use attention as a black-box correlation learner, lacking explicit separation of periodic and aperiodic components, failing to fully resolve the heterogeneity challenge highlighted in our work.

Frequency-Domain Modeling in Time Series Forecasting. Recognizing the complementary nature of spectral representations, a growing line of research integrates frequency-domain techniques into deep forecasting pipelines. The core insight is that real-world signals often exhibit energy concentration in a few dominant frequencies, enabling noise suppression and efficient long-range modeling.

Autoformer [12] replaces self-attention with an autocorrelation mechanism implemented via Fast Fourier Transform (FFT), effectively identifying lagged periodicities. FEDformer [20] decomposes queries and keys into low-rank frequency subspaces to compute attention in the spectral domain. More recent works push this direction further: FreTS [16] designs frequency-domain MLPs to model both temporal and channel dependencies; FITS [14] applies low-pass filtering followed by complex-valued linear mapping. Most notably, FilterNet [15] introduces learnable frequency filters that adaptively pass or attenuate spectral components, aligning closely with our goal of adaptive spectral selection. However, existing frequency-based models typically operate independently from time-domain dynamics or employ fixed filtering strategies. They lack unified end-to-end frameworks that jointly optimize local temporal interactions and global spectral structures, while cross-channel modeling remains either neglected or computationally prohibitive.

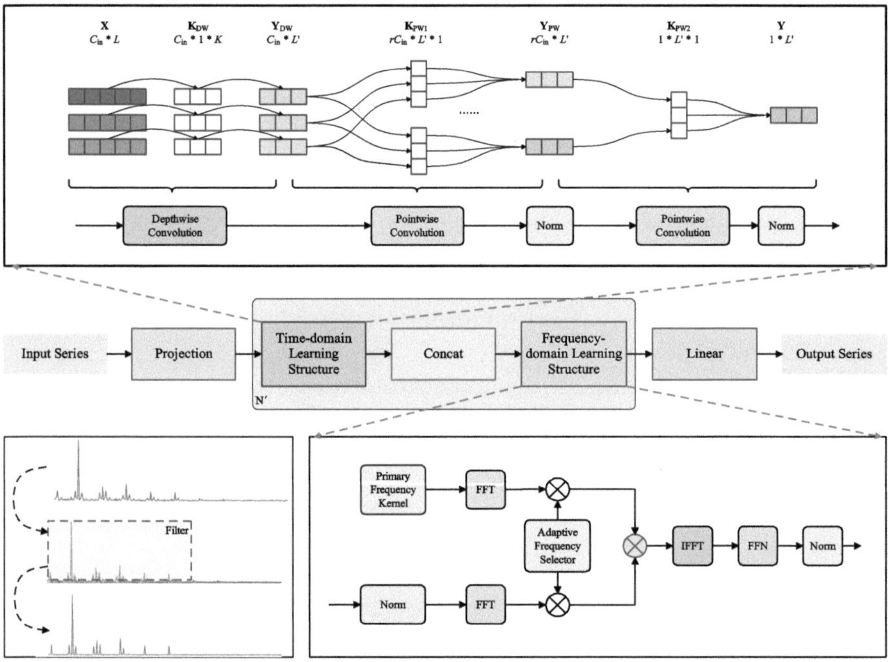

Fig. 2. Overall architecture of DFNet. (i) A depthwise separable convolution-based temporal learning module to learn temporal features and capture cross-channel relationships. (ii) Time-domain feature injection through concatenation operation. (iii) Frequency-domain learning module based on Fourier transform, which is used to learn global features and capture key frequencies.

3 Methodology

3.1 Overview

Figure 2 illustrates the overall architecture of DFNet. The core philosophy is to establish an end-to-end collaborative framework between the time and frequency domains, enabling simultaneous capture of LCCD and global periodic dependencies. The pipeline consists of three key stages: time-domain feature extraction, time-domain feature injection and frequency-domain learning & prediction.

3.2 LCCD Extraction with TLS

To address challenge Q2, we design the TLS, a lightweight module that explicitly captures LCCD with minimal parameters. Given input $\mathbf{X} \in \mathbb{R}^{B \times T \times C}$, where T is the sequence length and C is the number of variates, TLS operates in two stages to jointly model intra-channel temporal dynamics and inter-channel correlations.

Channel-Wise Depthwise Convolution for Local Dynamics. We first apply a depthwise convolution with kernel size K, stride s, and padding p, which processes each channel independently using channel-specific convolution kernels:

$$\mathbf{Z}_{b,t,c} = \sum_{\tau=0}^{K-1} \mathbf{X}_{b,t+\tau,c} \cdot \mathbf{W}^{dw}_{c,\tau}, \quad \forall b \in \{1, \ldots, B\}, c \in \{1, \ldots, C\} \tag{1}$$

where b, t, c denote batch, time step and channel index respectively. $\mathbf{W}^{dw} \in \mathbb{R}^{C \times K}$ represents the depthwise convolution kernels. The output is $\mathbf{Z} \in \mathbb{R}^{B \times T' \times C}$, where the sequence length $T' = \lfloor \frac{T+2p-K}{s} \rfloor + 1$. This isolates local temporal patterns in each univariate series like short-term trends, spikes, or irregular fluctuations, avoiding premature mixing of heterogeneous channel behaviors. This step preserves the distinct local dynamics across variables, which is essential for modeling real-world time series with mixed periodic and aperiodic components.

Pointwise Expand-and-Compress FFN for Cross-Channel Fusion. To model cross-channel dependencies efficiently, we employ a pointwise FFN with an expand-and-compress bottleneck architecture. Given feature $\mathbf{Z} \in \mathbb{R}^{B \times T' \times C}$ from the previous depthwise convolution stage, we first expand it to a higher-dimensional representation $\mathbf{H} \in \mathbb{R}^{B \times T' \times rC}$ for rich cross-channel interactions, then compress it back to obtain the LCCD representation $\mathbf{V}_{\text{LCCD}} \in \mathbb{R}^{B \times T' \times C}$:

$$\mathbf{H} = \text{ReLU}(\text{BN}(\mathbf{Z} * \mathbf{W}^{(1)})), \quad \mathbf{V}_{\text{LCCD}} = \text{BN}(\mathbf{H} * \mathbf{W}^{(2)}) \tag{2}$$

where $\mathbf{W}^{(1)} \in \mathbb{R}^{C \times rC}$ and $\mathbf{W}^{(2)} \in \mathbb{R}^{rC \times C}$ are learnable weights. The expansion factor r is a hyperparameter. BN is batch normalization [6] and is applied to stabilize training and mitigate distribution shifts across sequences. The expansion stage enables nonlinear feature mixing across channels, while the compression stage distills these interactions into a compact LCCD representation that captures both local dynamics and cross-channel correlations. The TLS module requires only $O(C \cdot K + rC^2)$ parameters and provides fine-grained, time-localized context complementary to FLS that spectral methods may overlook.

3.3 Trend Forecasting with Frequency-Domain Learning Structure

To address challenge Q3, we design the FLS, a lightweight module that automatically identifies and emphasizes informative frequency components, enhancing model robustness to noise and non-stationarity.

LCCD Injection. The LCCD vector $\mathbf{V}_{\text{LCCD}} \in \mathbb{R}^{B \times T' \times C}$ encodes the local and cross-channel information of all channels. We inject it into the local sequence representation $\mathbf{S} \in \mathbb{R}^{B \times C \times T}$ via concatenation. This enables indirect interaction between channels. This centralized approach reduces complexity while leveraging comprehensive LCCD information to enhance local channel representation:

$$\mathbf{S}_{\text{fuse}} = \text{Concat}(\mathbf{S}, \mathbf{V}_{\text{LCCD}}^{\top}) \in \mathbb{R}^{B \times C \times (T+T')} \tag{3}$$

Normalization and FFT Transformation. Given the fused representation $\mathbf{S}_{\text{fuse}} \in \mathbb{R}^{B \times C \times (T+T')}$, we apply Reversible Instance Normalization (RevIN) [8] to address distribution shift, which is a common challenge in time series forecasting. Unlike batch normalization that computes statistics across the batch, RevIN normalizes each instance independently along the temporal dimension:

$$\mathbf{Y} = \text{RevIN}(\mathbf{S}_{\text{fuse}}) = \frac{\mathbf{S}_{\text{fuse}} - \boldsymbol{\mu}}{\boldsymbol{\sigma}} \tag{4}$$

where $\boldsymbol{\mu}, \boldsymbol{\sigma} \in \mathbb{R}^{B \times C \times 1}$ are the per-instance mean and standard deviation. The key advantage of RevIN is its reversibility: the original scale can be precisely recovered via $\text{RevIN}^{-1}(\cdot)$ using the stored statistics, which is essential for accurate forecasting.

Then, the normalized sequence $\mathbf{Y} \in \mathbb{R}^{B \times C \times (T+T')}$ and a learnable primary frequency kernel $\mathbf{w} \in \mathbb{R}^{1 \times (T+T')}$ which is initialized from a standard normal distribution, are transformed into the frequency domain using the Real Fast Fourier Transform (rFFT):

$$\begin{aligned} \hat{\mathbf{Y}} &= \text{rFFT}(\mathbf{Y}) \in \mathbb{C}^{B \times C \times (\lfloor (T+T')/2 \rfloor + 1)} \\ \hat{\mathbf{w}} &= \text{rFFT}(\mathbf{w}) \in \mathbb{C}^{1 \times (\lfloor (T+T')/2 \rfloor + 1)} \end{aligned} \tag{5}$$

Adaptive Frequency Selector. To perform adaptive frequency filtering, a compact set of learnable parameters $\boldsymbol{\alpha} \in \mathbb{R}^{1 \times r_\alpha}$ is introduced, where $r_\alpha \ll T+T'$. These parameters are passed through a sigmoid function to generate frequency-specific attention coefficients:

$$\boldsymbol{\beta} = \sigma(\boldsymbol{\alpha}) \in \mathbb{R}^{1 \times r_\alpha} \tag{6}$$

where $\sigma(\cdot)$ denotes the sigmoid function. The coefficients $\boldsymbol{\beta}$ are then zero-padded to match the frequency domain dimensionality and applied multiplicatively to both the input spectrum $\hat{\mathbf{Y}}$ and the kernel spectrum $\hat{\mathbf{w}}$:

$$\tilde{\mathbf{Y}} = \hat{\mathbf{Y}} \odot \text{pad}(\boldsymbol{\beta}), \quad \tilde{\mathbf{w}} = \hat{\mathbf{w}} \odot \text{pad}(\boldsymbol{\beta}) \tag{7}$$

Here $\odot$ represents element-wise multiplication, and pad($\cdot$) zero-pads the vector to dimension $\lfloor (T+T')/2 \rfloor+1$ by appending zeros, enabling selective amplification or suppression of specific frequency components.

Spectral Convolution and Reconstruction. The filtered spectra $\tilde{\mathbf{Y}}$ and $\tilde{\mathbf{w}}$ are combined element-wise in the frequency domain:

$$\hat{\mathbf{Z}}_{\text{freq}} = \tilde{\mathbf{Y}} \odot \tilde{\mathbf{w}} \in \mathbb{C}^{B\times C\times(\lfloor (T+T')/2 \rfloor+1)} \tag{8}$$

Then the result is transformed back to the time domain via inverse rFFT:

$$\mathbf{Z}_{\text{freq}} = \text{rFFT}^{-1}(\hat{\mathbf{Z}}_{\text{freq}}) \in \mathbb{R}^{B\times C\times(T+T')} \tag{9}$$

This process selectively emphasizes informative frequency components while suppressing noise and irrelevant patterns. As illustrated in Fig. 3, frequency filtering effectively removes high-frequency noise while preserving dominant periodic components, demonstrating the value of spectral-domain analysis for enhancing signal quality and capturing meaningful temporal patterns.

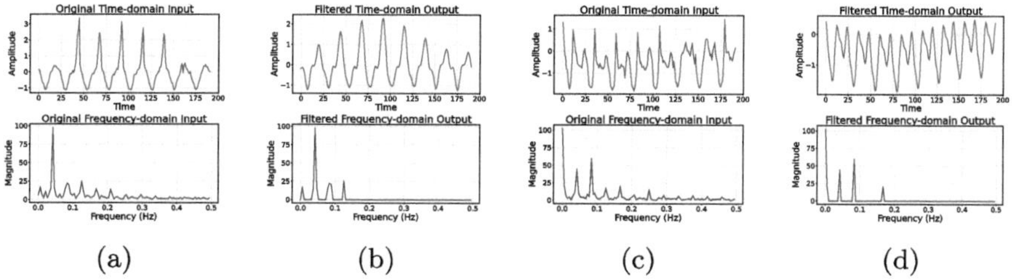

Fig. 3. Demonstration of filtering effect of key frequency features. This filtering process effectively removes noise while enhancing the model's ability to capture global temporal patterns.

Feature Refinement and Denormalization. The reconstructed time-domain signal $\mathbf{Z}_{\text{freq}}$ is further refined through a lightweight FFN comprising two linear projection layers with LeakyReLU activation:

$$\mathbf{F} = \text{FFN}(\mathbf{Z}_{\text{freq}}) \in \mathbb{R}^{B\times C\times H} \tag{10}$$

where H is the forecasting horizon. Finally, the refined feature $\mathbf{F}$ undergoes inverse RevIN transformation to recover the original data scale and distribution:

$$\mathbf{O} = \text{RevIN}^{-1}(\mathbf{F}) \in \mathbb{R}^{B\times C\times H} \tag{11}$$

where $\mathbf{O}$ is the final output of the FLS module, preserving both the temporal structure and channel-wise characteristics for subsequent forecasting tasks.

4 Experiments

4.1 Experimental Setup

Datasets. To thoroughly evaluate the performance of our proposed DFNet, we conduct extensive experiments on 7 widely used, real-world datasets: ETT [19] comprises two hourly-level datasets (ETTh) and two 15-min-level datasets (ETTm), each containing seven oil temperature and load features; Electricity (ECL) collects the hourly electricity consumption of 321 clients. Weather includes 21 indicators of weather, whose data is recorded every 10 min. Solar-Energy [9] (Solar) records the solar power production of 137 PV plants in 2006, which is sampled every 10 min. All the datasets are split into training, validation, and test sets in a 7:2:1 ratio.

Baselines. We compare DFNet with representative SOTA baselines including iTransformer, TSMixer, PatchTST, DLinear, TimesNet, FEDformer, and Autoformer.

Implementation Details. All experiments are implemented using Pytorch 2.5 and conducted on a single NVIDIA RTX 4080. We employ MSE as the loss function and present MAE and MSE results as the evaluation metrics. We use the Adam optimizer with an initial learning rate of 3×10^{-4}.

4.2 Main Results

As shown in Table 1, DFNet has almost provided the best predictive outcomes in all 7 datasets on average. Moreover, when compared to previous SOTA methods, DFNet has demonstrated significant advancements. For example, when the prediction window length is 96 on the ECL dataset, DFNet reduces the MSE error from 0.148 to 0.136, a significant reduction of about 8.11%. When the prediction window length is 96 on the Solar dataset, the MSE decreases from 0.203 to 0.192, a reduction of about 5.42%. These experimental results demonstrate the effectiveness of the DFNet model in multivariate time series prediction tasks and its ability to capture complex dependencies.

4.3 Model Analysis

Efficiency Analysis. Beyond predictive accuracy, we evaluate computational efficiency across four dimensions: parameter count, Multiply-Accumulate Operations (MACs), training time per epoch, and inference time per batch. We use the ECL dataset to compare the efficiency of DFNet with other methods with the same settings: lookback window length of 96 and prediction length of 720. The results are shown in Fig. 4. DFNet achieves a substantial reduction in the number of parameters and computational complexity compared to many heavy models. Specifically, compared to the lightest model, DFNet reduces MACs by

Table 1. Performance comparison across different prediction horizons with lookback window length $L = 96$. The best results are shown in **bold** and the second best are underlined.

Models		DFNet(ours)		iTransformer		TSMixer		PatchTST		DLinear		TimesNet	
Metrics		MSE	MAE	MSE	MAE	MSE	MAE	MSE	MAE	MSE	MAE	MSE	MAE
ETTh1	96	**0.374**	**0.395**	0.386	0.405	0.401	0.412	0.394	0.406	0.386	0.400	0.384	0.402
	192	**0.423**	**0.424**	0.441	0.436	0.452	0.442	0.440	0.435	0.437	0.432	0.436	0.429
	336	**0.469**	**0.445**	0.487	0.458	0.492	0.463	0.491	0.462	0.481	0.459	0.491	0.469
	720	**0.472**	**0.466**	0.503	0.491	0.507	0.490	0.487	0.479	0.519	0.516	0.521	0.500
ETTh2	96	0.295	0.346	0.297	0.349	0.319	0.361	**0.288**	**0.340**	0.333	0.387	0.340	0.374
	192	**0.372**	**0.394**	0.380	0.400	0.402	0.410	0.376	0.395	0.477	0.476	0.402	0.414
	336	**0.414**	**0.428**	0.428	0.432	0.444	0.446	0.440	0.451	0.594	0.541	0.452	0.452
	720	**0.426**	**0.445**	0.427	**0.445**	0.441	0.450	0.436	0.453	0.831	0.657	0.462	0.468
ETTm1	96	**0.316**	**0.357**	0.334	0.368	0.323	0.363	0.329	0.365	0.345	0.372	0.338	0.375
	192	**0.361**	**0.380**	0.377	0.391	0.376	0.392	0.380	0.394	0.380	0.389	0.374	0.387
	336	**0.397**	**0.405**	0.426	0.420	0.407	0.413	0.400	0.410	0.413	0.413	0.410	0.411
	720	**0.460**	**0.442**	0.491	0.459	0.485	0.459	0.475	0.453	0.474	0.453	0.478	0.450
ETTm2	96	**0.174**	**0.255**	0.180	0.264	0.182	0.266	0.184	0.264	0.193	0.292	0.187	0.267
	192	**0.239**	**0.297**	0.250	0.309	0.249	0.309	0.246	0.306	0.284	0.362	0.249	0.309
	336	**0.297**	**0.337**	0.311	0.348	0.309	0.347	0.308	0.346	0.369	0.427	0.321	0.351
	720	**0.397**	**0.396**	0.412	0.407	0.416	0.408	0.409	0.402	0.554	0.522	0.408	0.403
ECL	96	**0.136**	**0.230**	0.148	0.240	0.157	0.260	0.164	0.251	0.197	0.282	0.168	0.272
	192	**0.157**	**0.247**	0.162	0.253	0.173	0.274	0.173	0.262	0.196	0.285	0.184	0.289
	336	**0.175**	**0.265**	0.178	0.269	0.192	0.295	0.190	0.279	0.209	0.301	0.198	0.300
	720	**0.200**	**0.292**	0.225	0.317	0.223	0.318	0.230	0.313	0.245	0.333	0.220	0.320
Weather	96	**0.160**	**0.206**	0.180	0.222	0.166	0.210	0.176	0.217	0.196	0.255	0.172	0.220
	192	**0.208**	**0.250**	0.221	0.254	0.215	0.256	0.221	0.256	0.237	0.296	0.219	0.261
	336	**0.265**	**0.292**	0.278	0.296	0.287	0.300	0.275	0.296	0.283	0.335	0.280	0.306
	720	0.347	**0.344**	0.358	0.347	0.355	0.348	0.352	0.346	**0.345**	0.381	0.365	0.359
Solar	96	**0.192**	**0.230**	0.203	0.237	0.221	0.275	0.205	0.246	0.290	0.378	0.250	0.292
	192	**0.221**	**0.254**	0.233	0.261	0.268	0.306	0.237	0.267	0.320	0.398	0.296	0.318
	336	**0.237**	**0.267**	0.248	0.273	0.272	0.294	0.250	0.276	0.353	0.415	0.319	0.330
	720	**0.241**	**0.272**	0.249	0.275	0.281	0.313	0.252	0.275	0.356	0.413	0.338	0.337

over 68% compared to TimeMixer, while accelerating training by 58% and inference by 26% relative to iTransformer. These results further demonstrate that DFNet not only maintains strong predictive performance but also offers superior efficiency, making it well-suited for resource-constrained environments.

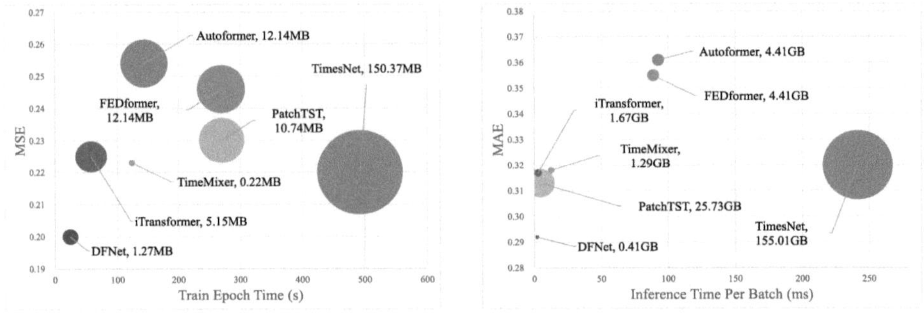

(a) (b)

Fig. 4. Efficiency comparison of DFNet and other SOTA models on the ECL dataset. MACs represent the computational cost for processing a single sample, and memory usage is measured with batch size of 16 for all models.

Visualization of Prediction. We present a prediction display on the Weather dataset with lookback window 96 and horizon 96, as shown in Fig. 5. We choose iTransformer and PatchTST as representative comparison methods. Compared to these different SOTA models, we can observe that DFNet provides better prediction of future sequence changes, which shows superior performance.

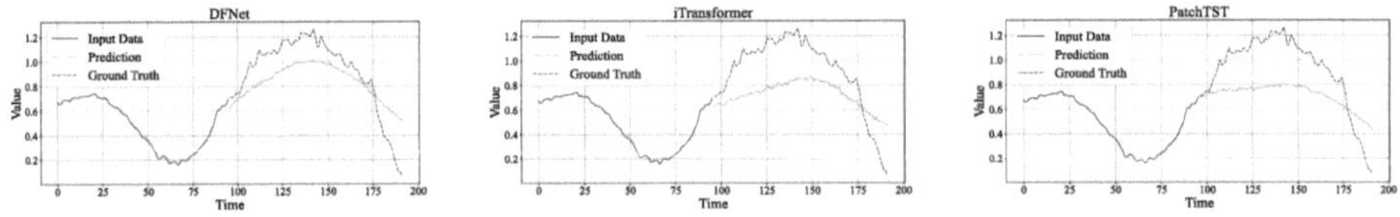

(a) (b) (c)

Fig. 5. Visualization of prediction on the Weather dataset.

Influence of Lookback Length. The lookback length determines the richness of historical information that the model can utilize. Generally, models are expected to perform better with longer input lengths if they possess robust long-term dependency modeling capabilities. We explore how varying the lengths L of these lookback windows impacts the forecasting performance for $L \in \{48, 96, 192, 336\}$ in all datasets with prediction length of 96. As shown in Fig. 6, DFNet could consistently improve its performance by effectively utilizing the enhanced data available from an extended lookback window. Also, DFNet performs consistently better than other models under different lookback window lengths.

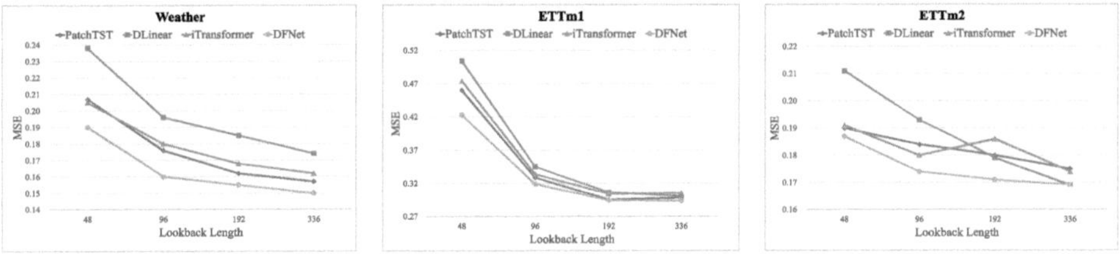

(a) (b) (c)

Fig. 6. Impact of lookback window length on forecasting performance (MSE).

4.4 Ablation Study

Comparison of Different Intermediate Channels. We ablate the intermediate channel size in the TLS module's pointwise to validate the effectiveness of the expand-and-compress design. As shown in Fig. 7, an expanded bottleneck generally improves performance over the "w/o expansions" baseline on both ETTh1 and ETTm1. Results show that a moderate expansion such as 21–35 channels yields optimal performance by enabling effective cross-channel interaction without overfitting, validating the efficacy of the expand-and-compress design for modeling LCCD.

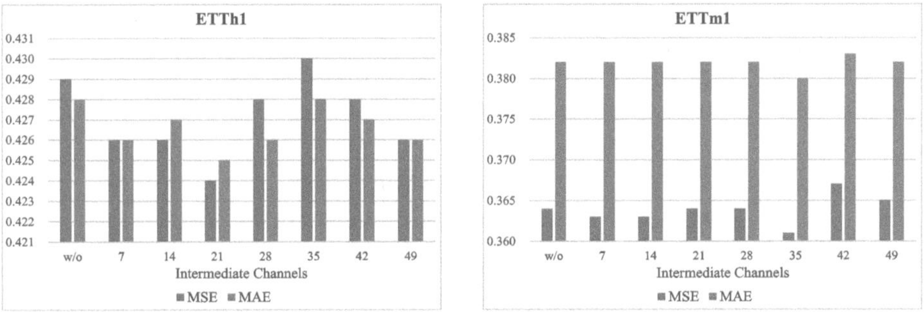

(a) (b)

Fig. 7. Ablation study on the number of intermediate channels in the TLS module's pointwise FFN. We report MAE and MSE on ETTh1 and ETTm1.

Effectiveness of the End-to-End TLS+FLS Architecture. We conduct an ablation study to validate the necessity of the unified TLS+FLS design. As shown in Table 2, the full model consistently achieves the best performance across all datasets and horizons. Three variants are evaluated: TLS+MLP replaces FLS with a standard MLP and yields 0.015 MSE degradation on Weather at 96-step horizon, highlighting the importance of frequency-domain modeling for long-range patterns; Linear+FLS substitutes TLS with a linear layer and incurs 0.024 MAE loss on ECL at 720-step horizon, confirming the necessity of nonlinear temporal modeling for LCCD; FLS-Only removes LCCD features and suffers 0.025 MSE decline on Solar at 192-step horizon, indicating that local temporal dynamics are crucial for representation learning. These results demonstrate that TLS and FLS provide complementary capabilities, and their end-to-end integration enables robust forecasting across diverse scenarios.

Table 2. Ablation study results for prediction horizon $H \in \{96, 192, 336, 720\}$ with lookback window length $L = 96$. The best results are shown in **bold** and the second best are <u>underlined</u>.

Models		TLS + FLS		TLS + MLP		Linear + FLS		FLS-Only	
Metrics		MSE	MAE	MSE	MAE	MSE	MAE	MSE	MAE
ETTm1	96	**0.316**	**0.357**	0.324	0.363	<u>0.320</u>	<u>0.360</u>	0.325	0.361
	192	**0.361**	**0.380**	<u>0.368</u>	0.386	0.370	<u>0.385</u>	0.375	0.389
	336	**0.397**	**0.405**	0.402	0.408	<u>0.400</u>	<u>0.407</u>	0.405	0.412
	720	**0.460**	**0.442**	<u>0.466</u>	<u>0.446</u>	0.470	0.447	<u>0.466</u>	0.447
Weather	96	**0.160**	**0.206**	0.175	0.217	<u>0.165</u>	<u>0.209</u>	0.166	0.210
	192	**0.208**	**0.250**	0.220	<u>0.252</u>	0.214	0.255	<u>0.210</u>	0.253
	336	**0.265**	**0.292**	0.276	0.295	0.269	<u>0.294</u>	<u>0.268</u>	0.295
	720	**0.347**	**0.344**	0.353	<u>0.346</u>	0.353	0.349	<u>0.350</u>	0.348
Solar	96	**0.192**	**0.230**	0.196	<u>0.233</u>	<u>0.195</u>	0.236	0.213	0.251
	192	**0.221**	**0.254**	<u>0.229</u>	<u>0.256</u>	0.232	0.264	0.246	0.275
	336	**0.237**	**0.267**	0.248	<u>0.270</u>	<u>0.240</u>	0.271	0.262	0.285
	720	**0.241**	**0.272**	0.252	<u>0.274</u>	<u>0.250</u>	0.281	0.263	0.287
ECL	96	**0.136**	**0.230**	<u>0.146</u>	<u>0.236</u>	0.147	0.242	0.150	0.240
	192	**0.157**	**0.247**	<u>0.161</u>	<u>0.251</u>	<u>0.161</u>	0.255	0.163	0.252
	336	**0.175**	**0.265**	<u>0.178</u>	<u>0.268</u>	0.179	0.274	0.183	0.272
	720	**0.200**	**0.292**	<u>0.209</u>	<u>0.296</u>	0.232	0.316	0.223	0.308

5 Conclusion

We propose DFNet, a novel framework that synergistically integrates time-domain and frequency-domain learning for multivariate time series forecasting. DFNet addresses the challenge of modeling heterogeneous temporal patterns where local aperiodic dynamics and global periodic structures coexist, through two complementary modules: TLS based on depthwise separable convolutions to efficiently capture local cross-channel dependencies with low computational cost, and FLS that leverages Fourier transforms with adaptive spectral gating to identify and emphasize informative frequency components while suppressing noise. The end-to-end collaboration between TLS and FLS enables joint modeling of fine-grained temporal details and long-range spectral regularities, effectively handling real-world sequences with mixed periodicity. Extensive experiments show that DFNet consistently outperforms state-of-the-art methods in both accuracy and efficiency. Our work demonstrates that unifying temporal granularity and spectral structure provides a powerful paradigm for complex time series forecasting and opens new directions for interpretable, frequency-aware architectures.

References

1. Ariyo, A.A., Adewumi, A.O., Ayo, C.K.: Stock price prediction using the Arima model. In: UKSIM, pp. 106–112. IEEE (2014)
2. Bai, S., Kolter, J.Z., Koltun, V.: An empirical evaluation of generic convolutional and recurrent networks for sequence modeling. arXiv preprint arXiv:1803.01271 (2018)
3. Chung, J., Gulcehre, C., Cho, K., Bengio, Y.: Empirical evaluation of gated recurrent neural networks on sequence modeling. arXiv preprint arXiv:1412.3555 (2014)
4. Han, L., Chen, X.Y., Ye, H.J., Zhan, D.C.: Softs: efficient multivariate time series forecasting with series-core fusion. In: NeurIPS, vol. 37, pp. 64145–64175 (2024)
5. Hochreiter, S., Schmidhuber, J.: Long short-term memory. Neural Comput. **9**(8), 1735–1780 (1997)
6. Ioffe, S., Szegedy, C.: Batch normalization: accelerating deep network training by reducing internal covariate shift. In: ICML, pp. 448–456. PMLR (2015)
7. Kadiyala, A., Kumar, A.: Multivariate time series models for prediction of air quality inside a public transportation bus using available software. Environ. Prog. Sustain. Energy **33**(2), 337–341 (2014)
8. Kim, T., Kim, J., Tae, Y., Park, C., Choi, J.H., Choo, J.: Reversible instance normalization for accurate time-series forecasting against distribution shift. In: ICLR (2021)
9. Lai, G., Chang, W.C., Yang, Y., Liu, H.: Modeling long-and short-term temporal patterns with deep neural networks. In: SIGIR, pp. 95–104 (2018)
10. Liu, Y., et al.: itransformer: inverted transformers are effective for time series forecasting. In: ICLR (2024)
11. Nie, Y., Nguyen, N.H., Sinthong, P., Kalagnanam, J.: A time series is worth 64 words: long-term forecasting with transformers. In: ICLR (2023)
12. Wu, H., Xu, J., Wang, J., Long, M.: Autoformer: decomposition transformers with auto-correlation for long-term series forecasting. In: NeurIPS, pp. 22419–22430 (2021)
13. Wu, H., Zhou, H., Long, M., Wang, J.: Interpretable weather forecasting for worldwide stations with a unified deep model. Nat. Mach. Intell. **5**(6), 602–611 (2023)
14. Xu, Z., Zeng, A., Xu, Q.: FITS: modeling time series with $10k$ parameters. In: ICLR (2024)
15. Yi, K., et al.: FilterNet: harnessing frequency filters for time series forecasting. In: NeurIPS, vol. 37, pp. 55115–55140 (2024)
16. Yi, K., et al.: Frequency-domain MLPs are more effective learners in time series forecasting. In: NeurIPS, vol. 36, pp. 76656–76679 (2023)

17. Yi, K., et al.: Deep coupling network for multivariate time series forecasting. ACM Trans. Inf. Syst. **42**(5), 1–28 (2024)
18. Zeng, A., Chen, M., Zhang, L., Xu, Q.: Are transformers effective for time series forecasting? In: AAAI, vol. 37, pp. 11121–11128 (2023)
19. Zhou, H., et al.: Informer: beyond efficient transformer for long sequence time-series forecasting. In: AAAI, vol. 35, pp. 11106–11115 (2021)
20. Zhou, T., Ma, Z., Wen, Q., Wang, X., Sun, L., Jin, R.: FedFormer: frequency enhanced decomposed transformer for long-term series forecasting. In: ICML, pp. 27268–27286. PMLR (2022)

On the Possibility of Enhancing Structured Pruning by Input Space Partitioning

Zoltán Domokos(✉) and Richárd D. Király

Department of Image Processing and Computer Graphics, University of Szeged, Árpád tér 2., Szeged 6720, Hungary
domokos@inf.u-szeged.hu

Abstract. In this article, the link between the efficiency of structured pruning and tiling of the input space is explored, assuming that the network as a predictor is applied separately over distinct tiles. Multi-layer neural networks with ReLU activations are considered with fully connected layers. However, our results are straightforward to extend to several other layer configurations. In order to demonstrate that our findings are applicable to naturally occurring data as well, we performed experiments on four grayscale datasets included in medMNIST.

Keywords: Structured pruning · Geometry of ReLU networks · Tilings

1 Introduction

The geometry of feedforward neural networks is well understood since the foundational work of Mont far et al. [1] has drawn significant attention to the field. Considering specifically ReLU activations, in [2] it is shown that every layer in such models implements a Max Affine Spline (MAS) operation, while the whole model is equivalent to the composition of a specific sequence of MAS operators. In a sequel [3] it is further analyzed, how a collection of multiple neurons residing in several layers subdivide the input space into power diagram regions, providing an elegant geometrical perspective.

Based on the aforementioned results, we investigate the possibilities of ReLU network pruning. The current work is focused on structured pruning. In contrast with former work that aims at practically applicable compactification [4–6], our goal is to construct a pruning algorithm that can be used as a mathematical tool to establish a set of sufficient conditions on the reducibility of network layer width, while guaranteeing equivalence with the initial model over a finite Piecewise Linear (PL) manifold. For this purpose, we borrow ideas from the literature of formal verification. However, instead of propagating star sets and zonotopes [7,8], or a union of polyhedra [9] through network layers in order to compute the intersection of reachable sets and unsafe regions, we propagate the input

Z. Zhang (Ed.): ISNN 2026, LNCS 16848, pp. 133–145, 2027.
https://doi.org/10.1007/978-981-92-5870-3_11

manifold's convex hull - see Algorithm 2 - to identify stable inactive neurons. Based on this approach, we apply formal reachability theory to construct a safe pruning method, which we employ as a mathematical tool to simplify our proofs.

The main contribution of this work is assessed in Theorem 2, which provides a mathematical foundation for a heuristic intuition, delineating the conditions under which the assumption remains valid. Given a convex tiling of the input space, it seems to be evident that pruning a ReLU network can be more efficient over the intersection of the input manifold and a specific tile than pruning over the whole manifold. We formalize conditions under which the hypothesis is indeed true, and demonstrate in an experiment with selected MedMNIST [10] datasets that pruning over tile-manifold intersections yields more compressed models.

2 Tiling and the Reducibility of Layer Width

A single fully connected neural network layer with ReLU activation is equivalent to a Max Affine Spline (MAS) operator. The sequence of layers in a feedforward multilayer network is a composition of single layer MAS operators [2], which we define as follows:

$$\mathcal{N}(\mathbf{x}) = \left(\mathcal{L}^{(L)} \circ \ldots \circ \mathcal{L}^{(l)} \circ \mathcal{L}^{(l-1)} \circ \ldots \circ \mathcal{L}^{(0)}\right)(\mathbf{x}), \tag{1}$$

where $\mathcal{L}^{(l)}$ is the MAS operator of the l-th layer, which is an affine transform, followed by the ReLU activation. Throughout this work, subsequences of the composition defined in (1) have special importance. For this reason, the notation $\mathcal{N}^{(l)}(\mathbf{x}) = \left(\mathcal{L}^{(l)} \circ \ldots \circ \mathcal{L}^{(0)}\right)(\mathbf{x})$ is introduced, while $x^{(l-1)}$ denotes $\mathcal{N}^{(l-1)}(\mathbf{x})$. The function corresponding to a given layer l can be written formally as:

$$\mathcal{L}^{(l)}\left(\mathbf{x}^{(l-1)}\right) := \sigma\left(\mathbf{W}^{(l)}\mathbf{x}^{(l-1)} - \mathbf{b}^{(l)}\right). \tag{2}$$

Here, $x^{(l-1)}$, the input of layer l, must be the output of layer $l-1$, due to the fact that we do not consider skip connections in the current work. Taking a closer look at (2), it can be seen that the k-th neuron of layer l defines a hyperplane of dimensions $dim\left(X^{(l-1)}\right) + 1$, as follows:

$$H_k^{(l)} := \left\{\left(h_k^{(l)}, \mathbf{x}^{(l-1)}\right) : h_k^{(l)} = \mathbf{w}_k^{(l)}\mathbf{x}^{(l-1)} - b_k^{(l)}\right\}. \tag{3}$$

Above, the gradient vector $\mathbf{w}_k^{(l)}$ is given as a row of $\mathbf{W}^{(l)}$, while the scalar $b_k^{(l)}$ is the bias parameter in machine learning terminology. After the affine transform, an activation function σ is applied. Let us notice that the hyperplanes $H_k^{(l)}$ define a positive halfspace $H_k^{(l)+}$, where $h_k^{(l)} > 0$ is satisfied, while $H_k^{(l)-}$ denotes the other halfpace, in which $h_k^{(l)} < 0$ must hold. Therefore, the intersection of $H_k^{(l)}$ with the input space $X^{(l-1)}$ is a zero level set of $h_k^{(l)}$, and can be viewed as the decision boundary for a single neuron with ReLU activation.

Taking any bounded subset from the input space of the l-th layer, assuming that the vertices $\mathcal{V}\left(C^{(l-1)}\right)$ of its convex hull $C^{(l-1)}$ are all mapped by $h_k^{(l)}$ to values that are not larger than zero, the neuron with index k must output zero for all the points contained by the aforementioned subset.

Lemma 1. *Let $C^{(l-1)}$ be a convex set in $X^{(l-1)}$. If all the vertices $\mathbf{c}^{(l-1)}$ in $\mathcal{V}\left(C^{(l-1)}\right)$ satisfy $h_k^{(l)}\left(\mathbf{c}^{(l-1)}\right) \leq 0$, then for all the points $\mathbf{x}^{(l-1)} \in C^{(l-1)}$ the inequality $h_k^{(l)}\left(\mathbf{x}^{(l-1)}\right) \leq 0$ must hold.*

Lemma 1 has an important consequence. If we want to decide whether a given neuron never fires - or in other words, always outputs zero - for a given set of points, it will be sufficient to verify that the neuron's output is zero for each vertex of the convex hull.

2.1 Structured Pruning with Equivalence Guarantees

Given the activation pattern P of a neural network model, we can iterate through all the $\mathcal{L}^{(l)}$ layers in the composition $\mathcal{N}$, and remove carefully selected rows and columns from the l-th layer's weight matrix $\mathbf{W}^{(l)}$, while some specific components from the bias $\mathbf{b}^{(l)}$ are needed to be discarded as well. From now on, we assume strict inactivity. More sophisticated techniques, mainly applied to pruning channels and filters exist [5,6]. However, for the sake of clarity, a simplified algorithm, that performs parameter removal is presented here, exclusively considering fully connected layers.

Algorithm 1. Compress Model

Input : $\mathcal{N}, P$
Output : $\tilde{\mathcal{N}}$
for $\mathcal{L}^{(l)}$ **in** $\mathcal{N}$ **do**
  $\mathbf{W}^{(l)}, \mathbf{b}^{(l)} \leftarrow \mathcal{L}^{(l)}$
  $\tilde{\mathbf{W}}^{(l)} \leftarrow$ **Remove Columns** $\left(\mathbf{W}^{(l)}, P^{(l-1)}\right)$
  if $l < L$ **then**
    $\tilde{\mathbf{W}}^{(l)}, \tilde{\mathbf{b}}^{(l)} \leftarrow$ **Remove Rows** $\left(\tilde{\mathbf{W}}^{(l)}, \tilde{\mathbf{b}}^{(l)}, P^{(l)}\right)$
  end if
  $\tilde{\mathcal{L}}^{(l)} \leftarrow \tilde{\mathbf{W}}^{(l)}, \tilde{\mathbf{b}}^{(l)}$
end for
$\tilde{\mathcal{N}} \leftarrow \tilde{\mathcal{L}}^{(L)}, ..., \tilde{\mathcal{L}}^{(0)}$
return $\tilde{\mathcal{N}}$

In Algorithm 1, the submethod **Remove Columns** removes the k-th column from $\mathbf{W}^{(l)}$, if and only if the element $(l-1, k)$ is not included in P. Then, from the resulting matrix $\tilde{\mathbf{W}}^{(l)}$, the submethod **Remove Rows** deletes the k-th row if and only if the element (l, k) is not included in P. Additionally, the row reduction stage, the k-th component of $\mathbf{b}^{(l)}$ is also discarded. Note that in both cases, there might be several columns and rows eligible for removal. Row reduction is not applied to the last operator. The output layer's dimensionality is preserved.

Lemma 2. *Let $\mathcal{N} : X \to Y$ be a composition of MAS operators and P a set of ordered pairs. Let $\mathbf{x}$ satisfy $h_k^{(l)}\left(\mathbf{x}^{(l-1)}\right) \leq 0$ for all (l,k) pairs that are not included in P. In this case, $\mathcal{N}(\mathbf{x}) = \tilde{\mathcal{N}}(\mathbf{x})$, where $\tilde{\mathcal{N}}(\mathbf{x})$ is the mapping returned by Algorithm 1, given $\mathcal{N}$ and P as inputs.*

Now, turning our attention to the activation pattern, we apply reachability analysis principles to identify regions where some specific neurons are stable and inactive. Algorithm 2 takes the vertex representation of the convex hull of a PL manifold, the latter is denoted by M. In every iteration, each vertex of the current convex hull is passed through the network layer $\mathcal{L}^{(l)}$ and we include (l,k) in the set P if and only if neuron k activates at least for one vertex. In case (l,k) is not in P, then it is guaranteed that the k-th neuron of the l-th layer never activates for any point contained by the manifold.

A layer with ReLU activation generally does not preserve convexity. Therefore, the convex hull $C^{(l)}$ needs to be computed in each iteration. For this purpose, the finite family of sets $\mathcal{S}$ is introduced, which contains convex polytopes S_i, acquired by recursively intersecting all S_i in $\mathcal{S}$ by the halfspaces $H_k^{(l)+}$ and $H_k^{(l)-}$. A detailed description of this subdivision process can be found in [9]. Finally, to all polytopes in $\mathcal{S}$, the layer's operator $\mathcal{L}^{(l)}$ is applied, and the convex hull of the result yields $C^{(l)}$.

Algorithm 2. Convex Hull Activation Pattern

```
Input : 𝒩, 𝒱(M)
Output : P
P ← ∅
C^(0) ← Convex Hull (𝒱(M))
for ℒ^(l) in 𝒩 do
    𝒮 ← {C^(l−1)}
    W^(l), b^(l) ← ℒ^(l)
    for w_k^(l), b_k^(l) in W^(l), b^(l) do
        H_k^(l) ← W^(l), b^(l)
        for c^(l−1) in 𝒱(C^(l−1)) do
            if 0 < w_k^(l) c^(l−1) − b_k^(l) then
                P ← P ∪ {(l, k)}
            end if
        end for
        𝒮* ← ∅
        for S_i ∈ 𝒮 do
            𝒮* ← 𝒮* ∪ {S_i ∩ H_k^(l)+, S_i ∩ H_k^(l)−}
        end for
        𝒮 ← 𝒮*
    end for
    C^(l) ← Convex Hull (ℒ^(l)(𝒮))
end for
return P
```

Lemma 3. *Let $\mathcal{N} : X \rightarrow Y$ be a composition of MAS operators. Furthermore, let M be a bounded PL manifold in X. If the set P returned by the Algorithm 2 does not contain (l, k) for the input $\mathcal{N}, M$, then for all $\mathbf{x} \in M$ the inequality $h_k^{(l)}\left(\mathbf{x}^{(l-1)}\right) \leq 0$ must hold.*

Proof. The relation $M^{(l)} \subseteq C^{(l)}$ is a loop invariant, as it is always true at the beginning of each iteration, which we prove by induction. In the first iteration the relation trivially holds. Let us assume that $M^{(l-1)} \subseteq C^{(l-1)}$ is true, from which it follows that the relation:

$$\mathcal{L}^{(l)}\left(M^{(l-1)}\right) \subseteq \mathcal{L}^{(l)}\left(C^{(l-1)}\right) \tag{4}$$

must hold. We subdivide $C^{(l-1)}$ by all $H_k^{(l)+}$ and $H_k^{(l)-}$ halfspaces of layer l, and store the result in $\mathcal{S}$, leading to the fact that $\mathcal{L}^{(l)}\left(C^{(l-1)}\right)$ is contained by the convex hull of $\mathcal{L}^{(l)}(\mathcal{S})$. Notice that $M^{(l)}$ equals the left hand side of (4), while the convex hull of $\mathcal{L}^{(l)}(\mathcal{S})$ is assigned to $C^{(l)}$ at the end of every iteration. Therefore, $C^{(l)}$ contains the right hand side of (4), leading to $M^{(l)} \subseteq C^{(l)}$. Indeed, the latter is a loop invariant. The pair (l, k) is not included in P if and only if all vertices of $C^{(l-1)}$ are in $H_k^{(l)-}$. As of Lemma 1, all the internal points of $C^{(l-1)}$ must be in the negative halfspace in this case. Above we proved that $M^{(l-1)} \subseteq C^{(l-1)}$ at the beginning of each iteration, which concludes the proof. □

Algorithm 2 requires the computation of $\mathcal{S}$, which can be performed by e.g., the Double Description Method (DDM) [11]. However, it is impractical in higher dimensional layers, due to the possibility of exponential increase in the number of sets S_i included in $\mathcal{S}$. This step is important in the formal proof of Theorem 3, as it guarantees that $M^{(l)} \subseteq C^{(l)}$ is not violated. Fortunately, as we will see in Sect. 3, even if we do not calculate all the exact intersections, it is possible to achieve reasonably accurate results. Even when a manifold is embedded in a high dimensional space, we conjecture that a high sample density combined with low manifold dimensionality leads to a moderate violation of equivalence between the initial and the pruned models. As a more elaborate method to acquire an inner approximation of the propagated convex hulls, a Randomized Linear Programming (RLP) approach might be successful. For convex hull approximation, a solution based on directional methods [12] provides acceptable results.

2.2 Conditions on Reducibility

In this subsection, we provide sufficient conditions on the reducibility of a neural network model $\mathcal{N}$, with theoretical guarantees that the mapping $\tilde{\mathcal{N}}$, found by the pruning method results in a model that is equivalent with the initial network over the convex hull of a piecewise linear manifold. Furthermore, sufficient conditions on the existence of an even more efficient pruning will be proposed, assuming that the manifold M can be partitioned by a tiling of the input space.

Theorem 1. *Let $M \subset X$ be a bounded PL manifold. Additionally, let the mapping $\mathcal{N} : X \to Y$ be a sequence of MAS operators. If Algorithm 2 returns the set P, given the inputs $\mathcal{N}, M$, and Algorithm 1 returns $\tilde{\mathcal{N}}$ for inputs $\mathcal{N}, P$, then $\tilde{\mathcal{N}}(M) = \mathcal{N}(M)$ holds.*

Proof. Algorithm 2 returns the set P, given the inputs $\mathcal{N}, M$. As of Lemma 3, for all (l,k) not included in P and for all $\mathbf{x} \in M$ the inequality $h_k^{(l)}\left(\mathbf{x}^{(l-1)}\right) \leq 0$ must hold. Then, Lemma 2 can be applied to each element $\mathbf{x}$ in the manifold M, leading to $\mathcal{N}(\mathbf{x}) = \tilde{\mathcal{N}}(\mathbf{x})$. □

Let us introduce a tiling $\mathcal{X}$, a countable collection of closed convex subsets X_i of X. Therefore, $X := \bigcup_{X_i \in \mathcal{X}} X_i$, and any $X_i, X_j \in \mathcal{X}$ has disjunct interior. Note that the X_i cells of the tiling are generally not need to be bounded. We consider finitely many tiles.

Theorem 2. *Let $M \subset X$ be a PL manifold, $\mathcal{X}$ a finite convex tiling of X, and $\mathcal{N} : X \to Y$ a sequence of MAS operators. We assume that a tile $X_i \in \mathcal{X}$ exists, such that $M_i \subset M$, where $M_i = X_i \cap M$. Let $\tilde{\mathcal{N}}$ be the result of pruning by Algorithm 2 for inputs $\mathcal{N}, M$, and the mapping $\mathcal{N}_i$ for $\mathcal{N}, X_i$, and $\tilde{\mathcal{N}}_i$ for $\mathcal{N}, M_i$. If $h_k^{(l)}$ exists, such that for all vertices $\mathbf{c}_i^{(l-1)}$ in $\mathcal{V}\left(C_i^{(l-1)}\right)$, the inequality $h_k^{(l)}\left(\mathbf{c}_i^{(l-1)}\right) \leq 0$ holds, while $0 < h_k^{(l)}\left(\mathbf{c}^{(l-1)}\right)$ for at least one vertex $\mathbf{c}^{(l-1)}$ in $\mathcal{V}\left(C^{l-1}\right)$, then the matrix $\tilde{\mathbf{W}}_i^{(l)}$ has strictly less rows than $\tilde{\mathbf{W}}^{(l)}$.*

Proof. If all elements in $\mathcal{V}\left(C_i^{(l-1)}\right)$ satisfy $h_k^{(l)}\left(\mathbf{c}_i^{(l-1)}\right) \leq 0$, the fact that $M_i^{(l-1)} \subseteq X_i^{(l-1)} \subseteq C_i^{(l-1)}$ leads to $h_k^{(l)}\left(\mathbf{x}^{(l-1)}\right) \leq 0$ for all $\mathbf{x}^{(l-1)}$ in $M_i^{(l-1)}$. A given set always contains the vertices of its convex hull. Therefore, the inequality must hold for all vertices of the convex hull of $M_i^{(l-1)}$. Consequently, Algorithm 2 will not include (l,k), when given $\mathcal{N}, M_i$ as inputs, and Algorithm 1 will not keep the k-th row of $\mathbf{W}^{(l)}$ in $\tilde{\mathbf{W}}_i^{(l)}$. If $0 < h_k^{(l)}\left(\mathbf{c}^{(l-1)}\right)$ for at least one vertex in $\mathcal{V}\left(C^{l-1}\right)$, the Algorithm 2 will keep (l,k), when processing $\mathcal{N}, M$, leading to Algorithm 1 keeping the k-th row of $\mathbf{W}^{(l)}$ in $\tilde{\mathbf{W}}^{(l)}$. □

Theorem 2 suggests that as long as the input space X has a tiling that is sufficiently fine-grained to partition the data manifold M, it is possible to achieve a compression ratio for a ReLU network's parameters within the boundaries of a given tile that is higher or equal compared to the global compression ratio. A straightforward example is tiling by Voronoi cells, which can be acquired by K-Means clustering [13]. In Sect. 3, we will demonstrate that Voronoi Tiling followed by pruning a model over individual cells can lead to even more efficient compression.

3 Numerical Experiments

In this section, we demonstrate that the phenomenon described formally in Sect. 2.2 is also present in realistic settings. We carried out several experiments on four datasets from MedMNIST. The MedMNIST collection features preprocessed medical images of various modalities, which are scaled to the size of MNIST. Still, these datasets highly resemble real-world data, making them ideal to demonstrate conceptual correctness in early-stage research. From this rich variety of datasets, OrganAMNIST, OrganSMNIST, OrganCMNIST, and BreastMNIST were selected. Our goal with the numerical experiment was to empirically demonstrate that the conditions formalized in Theorem 2 are satisfied in realistic data distributions, processed in a common application use-case.

3.1 The Experimental Setting

The data flow of the experiment can be split into two distinct phases, where the first phase highly resembles a well-known use-case, where an autoencoder (AE) is applied to learn features, and its latent space output is forwarded to another algorithm that performs classification [14]. Note that the classifier usually ends in a layer with softmax activations, and we solely focused on ReLU activations. We will address this concern later on in this section.

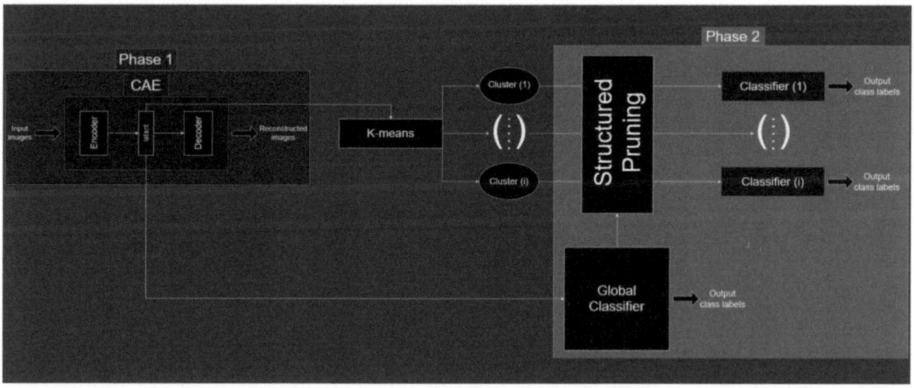

Fig. 1. Overview of the proposed experiment. In Phase 1, a simple CAE is trained for reconstruction. In the second phase, the flattened latent space output is forwarded to train a global classifier. Additionally, the flat latent data is clustered by K-means. Then, the global classifier is pruned, using all the available latent data. At the same time, the clusters identified by K-Means are used independently to acquire an array of independently pruned models. It is important to see that the setting is not end-to-end, training by backpropagation is restricted to the CAE and the global classifier, and even these two models are trained separately.

As a first step, a relatively simple Convolutional Autoencoder (CAE) is trained for image reconstruction. In the second phase, a shallow multi-layered fully-connected classifier model is trained for image classification over the latent layer data of the CAE. The resulting classifier will be referred to as the global classifier. After that, pruning by a relaxed variant of Algorithm 2 is performed over the global classifier. Finally, K-means clustering is applied, which tiles the CAE s latent space with Voronoi cells, and pruning of the initial global classifier is performed again for each tile-data intersection. Note that in the latter case we prune using only data that is enclosed by a given cell of the tiling induced by K-Means. The data-flow is visualized in Fig. 1.

The CAE model has six convolutional layers in both the encoder and decoder parts with ReLU activations and same padding. The number of filters is consistent through all layers except the latent layer, which has significantly less to avoid the curse of dimensionality effect, as it might be harmful for K-Means. We scale down the initial 28×28 image size by inserting max pooling layers, resulting in feature maps of size 7×7. Therefore, our latent space dimensionality is $7 \times 7 \times 2$. After training the CAE model, we apply flattening before K-means clustering. Our global classifier had to be relatively small. Consequently, it has only two fully-connected hidden layers with ReLU activations and an output layer with softmax. Note that the output layer neurons are not pruned, which justifies using arbitrary activations before the output. However, we have taken into account the consequences of softmax output. More details can be found in Subsect. 3.2.

Right after, we have trained the global classifier on the whole dataset, we apply structured pruning on it with the whole dataset s training set. Notice that pruning does not involve retraining in this process, neurons that are not removed must keep their parameter values acquired during training. Beside the pruned global model, from the initial global model we also derive as many local models as the number of clusters. In case of the local classifiers, pruning is performed exclusively on the subset of the whole training dataset which belongs to the corresponding cluster.

3.2 Measurement Results

We carried out measurements over all three OrganMNIST (A, S, C) datasets and BreastMNIST. Table 4 shows the results of measurements on the OrganMNIST sets with different number of clusters settings. In addition to the case where the number of clusters equals the number of classes (2 in BreastMNIST and 11 in the OrganMNIST variants), we measured different number of clusters settings as well. Decreasing the number of clusters seems to decrease the number of reducible neurons, caused by the possibly increasing size of clusters. The same is observable on BreastMNIST, which can be seen in Table 1.

All tests were repeated for different initial layer width values, varying between 256 and 16. We recorded the test accuracy of the global classifier, and the total number of neurons that can be removed from its hidden layers. The relatively high number of unnecessary neurons even in the low layer width cases - showing the symptoms of insufficient representational capacity - is in alignment with the observations of [15]. The next column shows the weighted average of neurons that can be omitted from the per-cluster reduced classifiers weighted by the number of training samples falling into that specific cluster. In the last column, the number of neurons is presented that can be removed from the most compressible local

Table 1. Comparison of accuracy and reduction metrics for the BreastMNIST dataset with the same number of clusters as their number of classes, and with 6, and 10. In the first column, N denotes the layer width of the initial global classifier. The Acc. column contains the test accuracy of the initial model, which in our measurements equaled to the accuracy of all pruned models. The Glob. column tells us how many neurons can be removed in case of global pruning, while Loc. is the weighted average of removable neuron quantity when pruning is applied per cluster. The fourth column provides information about the highest number of removable neurons for the most compressible local model.

N	clusters = 2				clusters = 6				clusters = 10			
	Acc.	Glob.	Loc.	Best	Acc.	Glob.	Loc.	Best	Acc.	Glob.	Loc.	Best
16	0.7564	4	4.95	7	0.7564	4	7.38	10	0.7564	4	7.9	10
32	0.8077	8	11.15	18	0.8077	8	14.68	21	0.8077	8	16.68	26
64	0.7628	11	15.78	24	0.7628	11	28.65	41	0.7628	11	33.79	54
128	0.7821	30	37.99	51	0.7821	31	53.41	77	0.7821	30	61.93	93
256	0.8013	54	69.49	101	0.8013	53	107.14	154	0.8013	54	129.27	206

Table 2. Comparison of the raw logit difference between the initial global model and pruned variants on the BreastMNIST dataset. The first column shows the MSE between the initial and pruned global model, while the second column shows the difference between the global model and per-cluster pruned local models.

N	Pruned Global					Locals				
	16	32	64	128	256	16	32	64	128	256
0	0	0	0	0	0	0	0	2.1e−4	1.7e−6	6.6e−6
1	0	0	4.2e−6	6.5e−6	2.2e−5	1.2e−5	8.9e−5	5.7e−5	6.5e−6	8.2e−5
2	0	0	0	0	0	0	0	0	0	2.6e−7
3	0	0	0	0	0	4.5e−4	7.4e−6	2.9e−5	0	2.0e−5
4	0	0	0	0	4.2e−5	0	0	0	0	0
5	0	0	1.4e−5	1.9e−5	4.4e−5	0	4.3e−3	3.8e−4	1.3e−4	9.4e−5

model. The average quantity of neurons that are removable from the local models is higher than the global value in all cases, which supports our theoretical results, even if the local models are only slightly more compressed in OrganMNIST. The phenomenon that we intended to demonstrate is more explicit in BreastMNIST.

Softmax activation was expected to suppress small deviations that are present between the raw-logit values of the pruned models and the initial model's raw-logit values. As it can be seen in Table 2 and Table 3, the Mean Squared Error (MSE) between the output of the pruned models and the initial classifier, measured per-cluster, right before softmax activation is reasonably low. Underestimation of the convex hulls introduces negligible or no error in the majority of cases in both the OrganMNIST and BreastMNIST sets. E.g., in BreastMNIST,

Table 3. Comparison of the raw logit difference between the initial global model and pruned variants across OrganMNIST datasets. The first group shows the MSE between the initial and pruned global model, while the second group shows the difference between the global model and per-cluster pruned local models.

K	Pruned Global					Locals				
	16	32	64	128	256	16	32	64	128	256
OrganAMNIST										
0	4.8e−6	1.1e−4	0	3.2e−6	7.6e−7	3.8e−1	1.1e−4	0	3.2e−6	6.2e−3
1	0	0	0	0	0	3.0e−4	0	0	3.5e−6	0
2	8.8e−5	2.2e−5	4.9e−6	2.4e−4	2.9e−6	8.8e−5	6.4e−6	4.9e−6	2.4e−4	2.9e−6
3	0	0	0	0	0	0	0	0	0	0
4	0	0	0	0	0	0	0	0	0	0
5	0	1.3e−5	0	5.1e−6	6.0e−8	0	1.3e−5	0	5.3e−6	1.2e−3
OrganSMNIST										
0	1.0e−5	1.2e−5	4.9e−6	6.1e−6	2.8e−7	1.0e−5	1.2e−5	4.9e−6	6.1e−6	2.8e−7
1	0	0	0	3.0e−8	0	0	0	0	3.0e−8	0
2	0	0	0	0	0	1.3e−5	0	0	0	4.4e−6
3	0	0	0	2.1e−7	2.0e−8	0	0	0	2.1e−7	5.0e−8
4	0	0	0	0	0	0	0	0	0	0
5	0	0	0	0	0	0	0	0	0	0
OrganCMNIST										
0	2.5e−5	3.6e−6	7.3e−5	6.1e−6	6.0e−6	2.5e−5	3.6e−6	8.5e−5	2.6e−5	6.0e−6
1	0	0	0	0	0	9.8e−3	6.4e−6	1.2e−2	0	6.0e−5
2	0	0	0	2.8e−7	8.0e−8	0	0	0	2.8e−7	8.2e−5
3	0	5.1e−6	0	7.1e−6	4.1e−6	0	5.1e−6	3.9e−6	1.4e−5	5.3e−6
4	0	0	0	0	0	2.2e−4	0	0	0	0
5	0	0	0	1.8e−5	2.6e−7	0	0	0	1.8e−5	4.2e−7

for layers with only 16 neurons, both the global pruned model and all the local models output exactly the same values as the initial model. By increasing layer width, the local models expose moderately higher MSE, while in two clusters equivalence is kept consistently. It is important to note that these differences do not affect the final prediction of softmax, which results in equivalent accuracy values throughout all measurements.

4 Summary

Sufficient conditions were proposed in Theorem 1, which guarantee the existence of a structurally pruned model that is equivalent to the initial model over a finite PL manifold, where the latter represents the support of a given data distribution. Furthermore, a rather interesting case, when the manifold is subdivided by a

Table 4. Comparison of accuracy and reduction metrics for OrganMNIST datasets with the same number of clusters as their number of classes, and with 6, and 4. In the first column, N denotes the layer width of the initial global classifier. The Acc. column contains the test accuracy of the initial model, which in our measurements equaled the accuracy of all pruned models. The Glob. column tells us how many neurons can be removed in case of global pruning, while Loc. is the weighted average of removable neuron quantity when pruning is applied per cluster. The fourth column provides information about the highest number of removable neurons for the most compressible local model.

N	clusters = 11				clusters = 6				clusters = 4			
	Acc.	Glob.	Loc.	Best	Acc.	Glob.	Loc.	Best	Acc.	Glob.	Loc.	Best
OrganAMNIST												
16	0.6681	8	9.68	11	0.6681	8	9.47	10	0.6681	8	8.82	10
32	0.7182	7	10.93	12	0.7182	6	10.2	11	0.7182	7	12.9	14
64	0.7498	12	13.73	15	0.7498	13	12.79	14	0.7498	13	13.73	15
128	0.7525	24	30.91	39	0.7525	23	27.81	33	0.7525	24	27.53	32
256	0.7744	65	78.10	86	0.7744	66	74.75	84	0.7744	66	73.62	82
OrganSMNIST												
16	0.5173	10	10.55	13	0.5173	6	6.06	7	0.5173	6	6.07	7
32	0.5694	9	9.47	12	0.5694	9	9.06	10	0.5694	9	9.07	10
64	0.5808	15	19.32	25	0.5808	15	17.92	23	0.5808	15	17.46	23
128	0.6011	31	33.74	38	0.6011	30	33.25	37	0.6011	31	32.82	37
256	0.6233	55	60.5	78	0.6233	54	58.71	73	0.6233	55	57.85	70
OrganCMNIST												
16	0.67	8	11	13	0.67	8	10.29	13	0.67	8	9.77	13
32	0.723	7	8.81	11	0.723	7	8.21	11	0.723	7	7.62	11
64	0.7535	13	17.57	21	0.7535	12	16.07	19	0.7535	13	15.21	19
128	0.77	32	37.35	45	0.77	31	34.95	38	0.77	32	34.25	38
256	0.7848	73	90.29	99	0.7848	73	86.69	97	0.7848	75	84.1	97

finite convex tiling was also considered. As long as a given tile contains a strict subset of the manifold and the additional conditions - formalized in Theorem 2 - are satisfied, another pruned model can be found, which is equivalent to the initial model over the corresponding submanifold, while it is even smaller than the globally pruned network. Algorithm 2 was proposed, which we applied as a mathematical tool to simplify the proof of Theorem 1 and Theorem 2. Even though, the mentioned algorithm serves merely formal purposes in this work, it can be implemented and provides highly accurate results when applied in low dimensions, where convex hulls and intersection vertices can be computed by the Double Description Method.

The hypothesis that a Voronoi tiling of the input space can enable us to find networks for given tiles, which are smaller than the model that is a result of global pruning, was tested in numerical experiments, carried out on some selected datasets of the MedMNIST collection. These measurements strengthen our theoretical findings. The importance of these measurements lies in demonstrating that the theoretical result assessed in Theorem 2 is not only a formal construction, but describes a phenomenon that is present in neural network models applied to realistic datasets. In the near future, we plan to explore what practical implications our results might have.

References

1. Montfar, G., Pascanu, R., Cho, K., Bengio, Y.: On the number of linear regions of deep neural networks. Adv. Neural Inf. Process. Syst. **27** (2014)
2. Balestriero, R., Baraniuk, R.: A spline theory of deep networks (extended version). stat **1050**, 14 (2018)
3. Balestriero, R., Cosentino, R., Aazhang, B., Baraniuk, R.: The geometry of deep networks: power diagram subdivision. Adv. Neural. Inf. Process. Syst. **32** (2019)
4. Hu, H., Peng, R., Tai, Y.W., Tang, C.K.: Network trimming: a data-driven neuron pruning approach towards efficient deep architectures. arXiv preprint arXiv:1607.03250 (2016)
5. Luo, J.H., Wu, J., Lin, W.: ThiNet: a filter level pruning method for deep neural network compression. In: Proceedings of the IEEE International Conference on Computer Vision, pp. 5058 C5066 (2017)
6. He, Y., Zhang, X., Sun, J.: Channel pruning for accelerating very deep neural networks. In: Proceedings of the IEEE International Conference on Computer Vision, pp. 1389–1397 (2017)
7. Bak, S., Tran, H.-D., Hobbs, K., Johnson, T.T.: Improved geometric path enumeration for verifying ReLU neural networks. In: Lahiri, S.K., Wang, C. (eds.) CAV 2020. LNCS, vol. 12224, pp. 66–96. Springer, Cham (2020). https://doi.org/10.1007/978-3-030-53288-8_4
8. Teuber, S., B uning, M.K., Kern, P., Sinz, C.: Geometric path enumeration for equivalence verification of neural networks. In: 2021 IEEE 33rd International Conference on Tools with Artificial Intelligence (ICTAI), pp. 200 C208. IEEE (2021)
9. Xiang, W., Tran, H.D., Johnson, T.T.: Reachable set computation and safety verification for neural networks with ReLu activations. arXiv preprint arXiv:1712.08163 (2017)
10. Yang, J., Shi, R., Ni, B.: MedMNIST classification decathlon: a lightweight AutoML benchmark for medical image analysis. In: 2021 IEEE 18th International Symposium on Biomedical Imaging (ISBI), pp. 191–195. IEEE (2021)
11. Fukuda, K., Prodon, A.: Double description method revisited. In: Deza, M., Euler, R., Manoussakis, I. (eds.) CCS 1995. LNCS, vol. 1120, pp. 91–111. Springer, Heidelberg (1996). https://doi.org/10.1007/3-540-61576-8_77
12. Nie, J., Ellingwood, B.R.: Directional methods for structural reliability analysis. Struct. Saf. **22**(3), 233–249 (2000)
13. Lloyd, S.: Least squares quantization in PCM. IEEE Trans. Inf. Theory **28**(2), 129–137 (1982)

14. Masci, J., Meier, U., Cireşan, D., Schmidhuber, J.: Stacked convolutional auto-encoders for hierarchical feature extraction. In: Honkela, T., Duch, W., Girolami, M., Kaski, S. (eds.) ICANN 2011. LNCS, vol. 6791, pp. 52–59. Springer, Heidelberg (2011). https://doi.org/10.1007/978-3-642-21735-7_7
15. Hanin, B., Rolnick, D.: Deep ReLu networks have surprisingly few activation patterns. Adv. Neural. Inf. Process. Syst. **32** (2019)

Quantum Wave Function Reconstruction via RNN

Shuang Cong(✉) and Limin Lin

University of Science and Technology of China, Hefei 230027, China
scong@ustc.edu.cn

Abstract. In this paper, a recurrent neural network (RNN) is proposed to reconstruct the probability distribution of a wave function in the ground state of a many-body quantum system described by the transverse field Ising model (TFIM). In the design of the RNN structure, the gated recurrent unit (GRU) is used as the core unit of the network. A second RNN is designed to generate training samples by leveraging autoregressive properties, eliminating the need for labeled inputs. The performance of the trained RNN is evaluated on the basis of the relative energy error and the energy variance per spin. In the numerical simulation reconstruction experiments, 4-qubit reconstruction experiments with different numbers of training samples and iterations are implemented first. The performances of different methods, including exact diagonalization (ED), the density matrix renormalization group (DMRG), and positive operator-valued measure (POVM), are subsequently compared with those of the proposed method, and the results are analyzed. The method proposed in this paper demonstrates advantages in terms of wave function reconstruction accuracy and inference speed across different qubit systems.

Keywords: Probability distribution · quantum state reconstruction · recurrent neural network · ground state

1 Introduction

A quantum state contains all the information of a quantum system, which can be represented by a wave function or density matrix in Hilbert space. However, traditional quantum tomography reconstruction methods necessitate measuring a set of 2^n complete measurement information, which increases exponentially with qubit n, introduce significant complexity in both theoretical and experimental analyses and face many difficulties for state reconstruction in multiqubit systems. In recent years, the rapid development of machine learning has driven people to pay attention to new reconstruction models in quantum physics [1], such as the restricted Boltzmann machines (RBMs) [2], feedforward neural networks (FNNs) [3], convolutional neural networks (CNNs) [4] and Transformers [5]. The transverse field Ising model (TFIM) is a condensed matter physics model used to describe the spin states of a one-dimensional spin chain or a two-dimensional square lattice [6]. In this model, each spin is affected by interactions with neighboring spins and an external transverse field. When the energy function represented

Z. Zhang (Ed.): ISNN 2026, LNCS 16848, pp. 146–156, 2027.
https://doi.org/10.1007/978-981-92-5870-3_12

by the Hamiltonian is minimized under given physical parameters, the wave function can be used to describe the ground state of quantum systems. Exact diagonalization (ED) [7] and the density matrix renormalization group (DMRG) [8] are common numerical methods for solving the ground state of condensed matter physics models, and their results are often used as references to evaluate the accuracy of other methods. In the ED method, the system's Hamiltonian is expressed as a matrix, and the eigenvalues and eigenstates are obtained by diagonalization. Although the ED method can theoretically provide an exact solution, its calculation time is constrained by the exponential growth in dimensionality when multiple qubit systems are being processed. The DMRG method is used to overcome these computational limitations by optimizing local low-rank tensor approximations for handling multiqubit systems [8], but its accuracy is sometimes limited.

Unlike the RBM [2], FNN [3], and CNN [4], the recurrent neural network (RNN) has a unique sequence-based architecture and provides an actionable likelihood estimation that can be used to estimate variational wave functions. The probability distribution of an n-qubit wave function can be modeled using the (n-1)-qubit distribution as a prior, which shows the Markov property of the probability distribution. Therefore, RNNs based on recursion have a natural advantage in modeling probability distributions. Therefore, RNNs based on recursion have a natural advantage in modeling probability distributions. The distribution can be decomposed into the product of conditional probabilities, which enables the learning of the normalized representation of quantum states. Unlike supervised learning methods such as FNN and CNN, RNN does not require additional labeled samples; and has a lower complexity than Transformer dose [5]. The autoregressive nature of the model provides independent samples for effective computation of the estimated value. Although some studies have attempted to use the RNN for quantum state reconstruction [9], they have not fully explored the accuracy and inference speed of the RNN in reconstructing the probability distributions of the ground state. Moreover, positive operator-valued measure (POVM) is a commonly used quantum measurement method to extract probability distributions from quantum states [10]. Comparing the probability distributions obtained by the RNN and POVM is also important for evaluating the practical performance of the RNN in terms of quantum state reconstruction.

In this paper, the probability distributions of ground-state wave functions are reconstructed by designing an appropriate RNN. The ground states of the quantum many-body system in the TFIM are pure states. The study of pure states can provide a theoretical foundation for understanding more complex quantum systems. In the design of the RNN structure, the gated recurrent unit (GRU) is used as a unit of the RNN. We also generate training samples from the designed RNN by means of its autoregressive characteristics. The expected energy of the system is estimated on the basis of the generated samples and network parameters. By minimizing the energy function, the RNN weights are optimized. The performances of the RNN, ED, DMRG, and POVM in reconstructing ground-state energy and probability distributions are compared and analyzed. We aim to design an RNN that approximates the lowest energy of the TFIM and enables the reconstruction of the ground state to be both time-efficient and accurate.

This paper is organized as follows: Sect. 2 presents the wave function and the transverse-field Ising model. Section 3 describes the design of the RNN structure used to reconstruct the probability distribution. Section 4 presents the training of the RNN parameters for reconstructing probability distributions. In Sect. 5, the numerical simulation of probability distribution reconstruction experiments and performance comparison analysis are studied. Finally, conclusions are given in Sect. 6

2 Wave Function and Transverse Field Ising Model

In the quantum domain, the wave function $|\Psi\rangle$ is typically represented in complex-valued form: $|\Psi\rangle = \sum_x e^{i\Phi(x)}\sqrt{P(\boldsymbol{x})}\,|\boldsymbol{x}\rangle$, where $|\boldsymbol{x}\rangle$ represents the basis states, and $P(\boldsymbol{x})$ and $\Phi(\boldsymbol{x})$ denote the probability distribution and phase, respectively. However, the probability distribution of the quantum system is often more focused rather than the phase information. The study of wave functions with only the positive real $P(\boldsymbol{x})$ can provide a more intuitive understanding of the system's core characteristics.

A real-valued wave function $|\Psi\rangle$ describes the probability of a quantum state collapsing onto a basis state $|\boldsymbol{x}\rangle$ according to the probability distribution $P(\boldsymbol{x})$. The wave function $|\Psi\rangle$ is expressed as a linear superposition of the basis states $|\boldsymbol{x}\rangle$, with the superposition coefficients related to $P(\boldsymbol{x})$ as follows:

$$|\Psi\rangle = \sum_{x} \psi(\boldsymbol{x})\,|\boldsymbol{x}\rangle = \sum_{x} \sqrt{P(\boldsymbol{x})}\,|\boldsymbol{x}\rangle \tag{1}$$

where $\psi(\boldsymbol{x})$ represents the wave function coefficient, and $P(\boldsymbol{x})$ satisfies the 1-norm condition $\|P(\boldsymbol{x})\|_1 = 1$.

The probability distribution $P(\boldsymbol{x})$ of an n-qubit wave function can be obtained by modeling the conditional probability using the probability distribution $P(x_1, \ldots, x_{n-1})$ with $(n-1)$ qubits: $P(\boldsymbol{x}) = P(x_1, \ldots, x_n) = P(x_n| x_{n-1}, \ldots, x_1)P(x_1, \ldots, x_{n-1})$. Then, $P(x_1, \ldots, x_{n-1})$ can be obtained from $P(x_1, \ldots, x_{n-2})$ in a similar manner. By analogy, according to the probability multiplication rule, the probability distribution $P(\boldsymbol{x})$ can be obtained by multiplying a series of conditional probabilities starting with $P(x_1)$:

$$P(\boldsymbol{x}) = P(x_1)P(x_2|x_1)\ldots P(x_n|x_{n-1},\ldots,x_1) \tag{2}$$

where $\boldsymbol{x} = (x_1, \ldots, x_n)$ is an n-bit binary number, and $x_i \in \{0, 1\}$; $P(x_i| x_{i-1}, \ldots, x_1) = P(x_i| x_{<i})$ represents the conditional probability of x_i given all x_j $(j < i, i \neq 1)$.

One can see from (1) that reconstructing the wave function $|\Psi\rangle$ is equivalent to reconstructing its probability distribution $P(\boldsymbol{x})$. Once the conditional probability $P(x_i| x_{<i})$ is known, one can obtain the complete characteristics of any possible probability distribution $P(\boldsymbol{x})$ through (2). Since the conditional probability $P(x_i| x_{<i})$ carries historical information and RNN models are inherently related to time series, using an RNN model to solve the probability distribution $P(\boldsymbol{x})$ is a reasonable and natural choice.

The one-dimensional Transverse Field Ising Model (TFIM) [6] is a condensed matter physics model that describes the spin states of a one-dimensional spin chain. Each spin is influenced by interactions with its neighboring spins and a transverse magnetic field. The Hamiltonian is expressed as follows:

$$H_{TFIM} = -J_z \sum_{<i,j>} \sigma_i^z \sigma_j^z - B_x \sum_{i} \sigma_i^x \tag{3}$$

where, J_z is the exchange coupling constant, B_x is the intensity of the external magnetic transverse field, and σ_i^x and σ_i^z are the x- and z-components of the Pauli matrices for the i-th spin, respectively:

$$\hat{\sigma}_i^x = \begin{bmatrix} 0 & 1 \\ 1 & 0 \end{bmatrix}, \hat{\sigma}_i^z = \begin{bmatrix} 1 & 0 \\ 0 & -1 \end{bmatrix} \tag{4}$$

The Hamiltonian operator H corresponds to the total energy of the system. The ground state is the state with the lowest energy in the system, which corresponds to the lowest eigenvalue of the Hamiltonian operator. It can be expressed using the steady-state Schrödinger equation as follows:

$$H\,|\Psi\rangle = E_0\,|\Psi\rangle \tag{5}$$

where, E_0 is the energy of the ground state; and $|\Psi\rangle$ is the ground-state wave function.

When J_z and B_x are fixed in (3), the Hamiltonian H is fully determined. The relationship between the ground-state energy E_0 and the wave function $|\Psi\rangle$ can be obtained from (5). Our task is to design an RNN to reconstruct the probability distribution of the wave function in the ground state. The probability distribution $P(\boldsymbol{x})$ at the minimum energy is considered the equilibrium point of the RNN. When the values of J_z and B_x are given, the RNN continuously optimizes its parameters to reach the equilibrium point. Once the equilibrium point is reached, the output of the RNN represents the estimated probability distribution of the minimum energy of the wave function.

3 Design of the RNN Structure for Reconstructing the Probability Distribution

The RNN structure used to reconstruct the probability distribution $\hat{P}(\boldsymbol{x})$ is designed as shown in Fig. 1a, in which x_i denotes the input of the i-th gated recurrent unit (GRU) [11] in the configuration $\boldsymbol{x}$, which is represented by the one-hot encoding of x_i. When x_i is 0 or 1, x_i takes the values (1, 0) or (0, 1), respectively. $\boldsymbol{h}_i$ represents the hidden state of the i-th GRU. Both $\boldsymbol{h}_0$ and $\boldsymbol{x}_0$ are initialized as zero vectors. The update of the GRU hidden state can be expressed as: $\boldsymbol{h}_t$= GRU ($\boldsymbol{x}_t$, $\boldsymbol{h}_{t-1}$), which means that when the hidden state $\boldsymbol{h}_t$ is updated, the GRU preserves the relevant past information $\boldsymbol{h}_{t-1}$ and incorporates the new input information x_t. The output of each GRU is passed through an independent column of Softmax operations. The results of these Softmax operations (y_1, y_2, ..., y_n) are designed to approximate the probability distribution. The RNN structure diagram as it folds over time is shown in Fig. 1b. We fully leverage the parameter-sharing capability of the GRU to efficiently model the probability distribution, using the same GRU parameters at each moment in one iteration.

The GRU is the core of the designed RNN structure for reconstructing the probability distribution. The GRU selectively ignores or updates portions of the input sequence to provide more flexible modeling capabilities and to better capture them.

The structure of the GRU is shown in Fig. 2, where the input is the hidden state $\boldsymbol{h}_{i-1}$ from the previous time step and the current input $\boldsymbol{x}_{i-1}$, and the output is $\boldsymbol{h}_i$. Here,

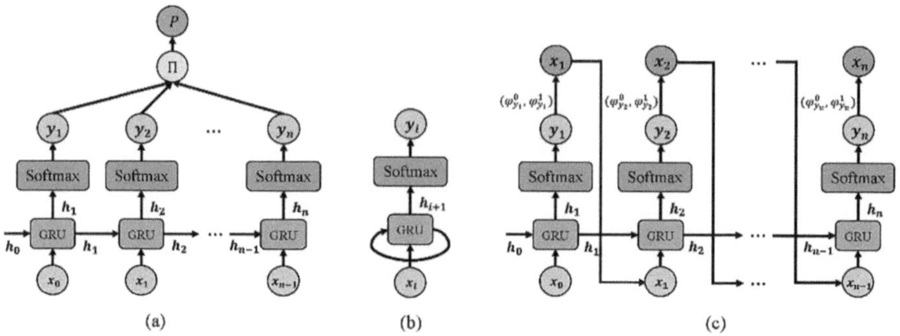

Fig. 1. Structural diagram of the QPP with quantum state $|0\rangle$ and $|1\rangle$ input: (a) The structure of the RNN for reconstructing the probability distribution unfolded in time; (b) The structure of the RNN for reconstructing the probability distribution folded in time; (c) The structure of the RNN for generating training samples

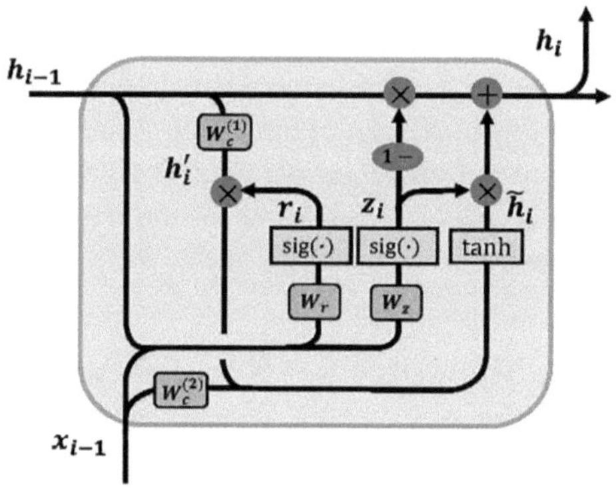

Fig. 2. The structure of the GRU

$\boldsymbol{x}_{i-1} \in \mathbb{R}^{1\times 2}$, and $\boldsymbol{h}_{i-1} \in \mathbb{R}^{1\times d_h}$, where d_h is the adjustable dimension of the hidden state.

The relationship between the input and output of the GRU can be specifically expressed as follows:

$$\boldsymbol{h}_i = (1 - z_i) * \boldsymbol{h}_{i-1} + z_i * \tanh\left(W_c^{(2)}\boldsymbol{x}_{i-1} + \boldsymbol{r}_i * \boldsymbol{h}_i' + \boldsymbol{b}_c^{(2)}\right)\Big) \tag{6}$$

where $\boldsymbol{r}_i = \mathrm{sig}(W_r[\boldsymbol{h}_{i-1}, \boldsymbol{x}_{i-1}] + \boldsymbol{b}_r)$, $z_i = \mathrm{sig}(W_z[\boldsymbol{h}_{i-1}, \boldsymbol{x}_{i-1}] + \boldsymbol{b}_z)$, and $\boldsymbol{h}_i' = W_c^{(1)}\boldsymbol{h}_{i-1} + \boldsymbol{b}_c^{(1)}$. The trainable parameters W_z and $W_r \in \mathbb{R}^{(d_h+2)\times d_h}$, $W_c^{(1)} \in \mathbb{R}^{d_h\times d_h}$, and $W_c^{(2)} \in \mathbb{R}^{2\times d_h}$ represent the weights; $\boldsymbol{b}_r$, $\boldsymbol{b}_z$, $\boldsymbol{b}_c^{(1)}$ and $\boldsymbol{b}_c^{(2)} \in \mathbb{R}^{1\times d_h}$ represent the biases. The symbol $[\bullet]$ denotes matrix concatenation, " $*$ " represents the Hadamard product, which is the elementwise multiplication of two matrices of the same order, and sig $(\bullet)$ represents the sigmoid normalization function.

Each output $\boldsymbol{h}_i$ of the GRU is fed through a fully connected layer with the Softmax function $S(\bullet)$, which results in $\boldsymbol{y}_i = \left(y_i^0, y_i^1\right)$. y_i^0 and y_i^1 represent the probabilities of x_i

being 0 or 1, respectively.

$$\boldsymbol{y}_i = S(W_s \boldsymbol{h}_i + b_s) \tag{7}$$

$$S(x_i) = \frac{e^{x_i}}{\sum_k e^{x_k}} \tag{8}$$

where $W_s \in \mathbb{R}^{2\times d_h}$ and $b_s \in \mathbb{R}^2$ are the weight and bias of the fully connected layer, respectively.

To obtain $\hat{P}(\boldsymbol{x})$, between y_i^0 and y_i^1, the component corresponding to $\boldsymbol{x}_i$ should be selected. Specifically, when $x_i = 0$ ($\boldsymbol{x}_i = (1, 0)$), y_i^0 is chosen to represent the conditional probability $P(x_i| x_{<i})$. Consequently, according to (2), the estimated probability distribution $\hat{P}(\boldsymbol{x})$, can be designed as the cumulative product of the dot product between $\boldsymbol{y}_i$ and the one-hot vector $\boldsymbol{x}_i$, as expressed by:

$$\hat{P}(\boldsymbol{x}) = \prod_{i=1}^{n} \boldsymbol{y}_i \bullet \boldsymbol{x}_i \tag{9}$$

where $\boldsymbol{y}_i \bullet \boldsymbol{x}_i$ is the inner product of the two, representing $P(x_i| x_{<i})$.

4 Training and Validation of the RNN for Reconstructing Probability Distributions

4.1 Generation of Training Samples

The RNN used to generate the training samples is shown in Fig. 1c, whose structure is similar to that in Fig. 1a, except that the output of each column serves as the input to the next GRU. To generate a sample $\boldsymbol{x} = (x_1, x_2, \ldots, x_n)$ of length n, the hidden state $\boldsymbol{h}_1$ is calculated on the basis of the initial values of $\boldsymbol{h}_0$ and $\boldsymbol{x}_0$, and the corresponding $\boldsymbol{y}_1 = (y_1^0, y_1^1)$ are obtained through the relevant formulas mentioned in (7, 8). A multinomial distribution is subsequently applied to $\boldsymbol{y}_1$, with the following formula:

$$\varphi_{y1}^0 = \frac{\log(y_1^0)}{\sum_{i=0}^{1} \log(y_1^i)},\ \varphi_{y1}^1 = \frac{\log(y_1^1)}{\sum_{i=0}^{1} \log(y_1^i)} \tag{10}$$

where φ_{y1}^0 and φ_{y1}^1 represent the probabilities of obtaining the values 0 and 1 according to the multinomial distribution applied to $\boldsymbol{y}_1$, respectively.

The output x_1 is generated as 0 or 1 with probabilities φ_{y1}^0 and φ_{y1}^1, and then encoded as $\boldsymbol{x}_1$, which is directly used as the input of the second GRU. This process repeated until $\boldsymbol{x}_n$ is generated. In this way, an n-qubit sample can be generated.

4.2 Variational Monte Carlo Optimization Strategy

Variational Monte Carlo (VMC) is a numerical optimization method that combines the variational principle with Monte Carlo sampling. In Sec. 4.1, we generate samples $\boldsymbol{x}$ on the basis of the probability distribution $\hat{P}(\boldsymbol{x})$ for Monte Carlo sampling. The basic idea of the variational principle is to find the ground state of a system by selecting a reasonable trial wave function with adjustable parameters and minimizing the expected energy of the system. We use the designed RNN to construct the parameterized trial wave function $|\hat{\Psi}_\lambda\rangle$, where λ represents the adjustable parameters. Substituting (3) into (5), one can obtain the energy expectation $\hat{E}(\lambda)$ as follows:

$$\hat{E}(\lambda) = \frac{\langle\hat{\Psi}_\lambda|H|\hat{\Psi}_\lambda\rangle}{\langle\hat{\Psi}_\lambda|\hat{\Psi}_\lambda\rangle} \tag{11}$$

where $|\hat{\Psi}_\lambda\rangle = \sum_{\boldsymbol{x}} \sqrt{\hat{P}(\boldsymbol{x})}\,|\boldsymbol{x}\rangle$, represents the RNN trial wave function, and $\langle\hat{\Psi}_\lambda|\hat{\Psi}_\lambda\rangle = 1$.

The objective of the VMC optimization strategy is to minimize $\hat{E}(\lambda)$ by taking the first-order partial derivative of (11) with respect to λ_k, as given by:

$$\frac{\partial \hat{E}(\lambda)}{\partial \lambda_k} = 2\,\mathrm{Re}\left(\left\langle\left(E_{\mathrm{loc}}(\boldsymbol{x}) - \langle E_{\mathrm{loc}}(\boldsymbol{x}')\rangle_{\hat{P}(\boldsymbol{x}')}\right)\frac{\partial \log\left(\hat{\psi}(\boldsymbol{x})\right)^*}{\partial \lambda_k}\right\rangle_{\hat{P}(\boldsymbol{x})}\right) \tag{12}$$

where λ_k denotes the model parameters at the k-th iteration, $E_{\mathrm{loc}}(\boldsymbol{x}) = \sum_{\boldsymbol{x}'} H(\boldsymbol{x},\boldsymbol{x}')\frac{\hat{\psi}(\boldsymbol{x}')}{\hat{\psi}(\boldsymbol{x})}$, $\boldsymbol{x}$ represents the current system state, and $\boldsymbol{x}'$ represents possible state transitions under the Hamiltonian. $\langle\bullet\rangle_{\hat{P}(\boldsymbol{x})}$ denotes the expectation over the distribution $\hat{P}(\boldsymbol{x})$. $\hat{\psi}(\boldsymbol{x})$ is the estimated wave function coefficient, and $\log\left(\hat{\psi}(\boldsymbol{x})\right)^*$ is the complex conjugate of $\log\left(\hat{\psi}(\boldsymbol{x})\right)$.

The process of optimizing the model parameters can be described as follows: Monte Carlo sampling is performed by using the initialized model, and the generated sample $\boldsymbol{x}$ is substituted into (9) to compute the energy gradient $g_k = \frac{\partial \hat{E}(\lambda)}{\partial \lambda_k}$. The parameters are updated using the adaptive moment estimation (Adam) optimization algorithm [12]:

$$\Delta\lambda_k = -\frac{\eta}{\sqrt{\hat{v}_k} + \epsilon}\hat{m}_k \tag{13}$$

where η is the learning rate; $\eta = 0.005$; $\hat{m}_k$ and $\hat{v}_k$ are the bias-corrected first and second moment estimates, respectively; $\hat{m}_k = \frac{m_k}{1-\beta_1^k}$; $\hat{v}_k = \frac{v_k}{1-\beta_2^k}$; $m_k = \beta_1 m_{k-1} + (1-\beta_1)g_k$; $v_k = \beta_2 v_{k-1} + (1-\beta_1)g_k^2$; β_1 and β_2 are the exponential decay rates for the moment estimates, and ϵ is a small constant to prevent division by zero.

The new model parameters λ_{k+1} are obtained from the update $\Delta\lambda_k$. The process repeated until the maximum number of iterations is reached. When the RNN network converges to the equilibrium point and the expected energy reaches its minimum value, the network output corresponds to the reconstructed value of the probability distribution.

4.3 Performance Indices for Verification

To validate the reconstruction performance of the RNN, exact diagonalization (ED) [7] and density matrix renormalization group (DMRG) [8] methods are used to generate accurate ground-state energy for comparison with the value estimated by the RNN. The

ED method is used for small systems, whereas the DMRG method is used for larger systems.

In this paper, we use the relative energy error ε to measure the similarity between the estimated and real values, which is defined as follows:

$$\varepsilon = \frac{\left|\hat{E}_0 - E_0\right|}{|E_0|} \times 100\% \tag{14}$$

where $\hat{E}_0$ represents the estimated ground state energy which is the minimum value of $\hat{E}(\lambda)$ in (11), and E_0 is the real ground state energy in (5).

On the other hand, the energy variance per spin σ^2 is used as another performance index to indicate whether the system has reached the state of minimum energy. It can be calculated as follows:

$$\sigma^2 = \frac{\left\langle H^2 \right\rangle - \langle H \rangle^2}{Q} = \mathrm{var}(\boldsymbol{E}_{loc}) \tag{15}$$

where var(•) represents the variance, and Q is the number of training samples.

5 Numerical Simulation Reconstruction Experiments and Performance Comparison Analysis

After training the RNN, the estimated probability distribution $\hat{P}(\boldsymbol{x})$ can be obtained, and the estimated wave function can be calculated by (1). In this section, we investigate the reconstruction performance of the designed RNN in different qubit cases. A reconstruction experiment with 4 qubits is conducted to analyse the impact of the number of samples on performance, and then the reconstructed probability distribution from the RNN is compared with the real results obtained from the POVM [10]. Finally, performance comparisons between the RNN and the real ground state energy methods ED and DMRG are conducted for 6, 10, and 12 qubit systems. In (3), both J_z and B_x were fixed at 1.0.

In the 4-qubit reconstruction experiments, we investigate the performance impact for different numbers of samples Q. With 4 qubits, $2^4 = 16$ basis states. Therefore, Q is chosen as 10, 30, 50, and 200, representing slightly less than 16, slightly greater than 16, greater than 16, and much greater than 16, respectively. The trend of the performance indices for different Q values as the number of iterations changes are shown in Fig. 3. When $Q = 10$, both curves oscillate significantly because of a lack of certain types of basis states to be fully trained. When $Q = 30$ or 50, the curve fluctuations gradually decrease, and the convergence rate increases. However, when $Q = 200$, although the accuracy improves, the improvement is not significant. The choice of Q requires a balance between accuracy requirements and computational resource consumption, and J_z and B_x were fixed at 1.0.

To achieve higher accuracy, with $Q = 200$, Table 1 presents the 4-qubit comparison result of $P(\boldsymbol{x})$ obtained from the POVM and RNN, which shows that: 1) nearly all accuracies exceed three decimal places; and 2) the accuracy even reaches four decimal places in some cases. The results in Table 1 are averaged over 10 independent training

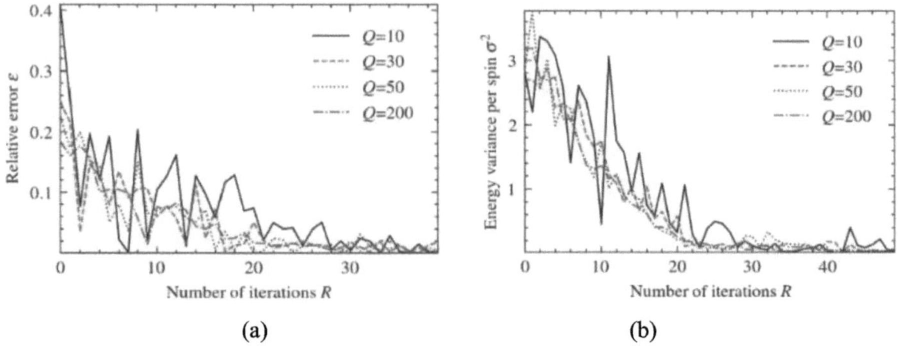

Fig. 3. Reconstruction performance versus iteration count for varying training sample sizes: (a) relative energy error for different numbers of training samples as the number of iterations changes; (b) energy variance for different numbers of training samples as the number of iterations changes

Table 1. Comparison results of $P(\boldsymbol{x})$ between POVM and RNN for a 4-qubit system

	$P(\boldsymbol{x})_{\text{POVM}}$	$P(\boldsymbol{x})_{\text{RNN}}$
$\|0000\rangle$	0.230430	0.23024155
$\|0001\rangle$	0.065040	0.06518517
$\|0010\rangle$	0.027528	0.02775929
$\|0011\rangle$	0.050098	0.05055941
$\|0100\rangle$	0.028144	0.02778910
$\|0101\rangle$	0.011750	0.01189571
$\|0110\rangle$	0.021412	0.02151079
$\|0111\rangle$	0.065518	0.06508549
$\|1000\rangle$	0.065612	0.06520832
$\|1001\rangle$	0.021498	0.02148216
$\|1010\rangle$	0.011628	0.01182633
$\|1011\rangle$	0.027624	0.02775346
$\|1100\rangle$	0.050410	0.05043275
$\|1101\rangle$	0.027330	0.02776419
$\|1110\rangle$	0.065224	0.06523533
$\|1111\rangle$	0.230754	0.23027090

runs with different random seeds to ensure statistical representativeness. The fidelity and KL divergence between the two distributions are 0.99999 and 1.53×10^{-5}, respectively.

To better illustrate the characteristics of the RNN, we compare the ground state energy E_0 in (5) obtained via the ED; and the DMRG, and reconstructed by the RNN for $n = 6$, 10, and 12. The results are shown in Table 2, from which one can see that the ED

Table 2. Comparison of the ED, DMRG and RNN for different qubits (500 samples)

Method	Qubit	Time/s	Ground state energy E_0
ED	6	0.20	−7.296230
	10	67.23	−12.381490
	12	1306.04	−14.925971
DMRG	6	0.90	−6.814083
	10	2.46	−12.082569
	12	2.68	−14.925971
RNN	6	33.88	−7.296189
	10	40.44	−12.381552
	12	68.99	−14.925936

method can achieve the most accurate results of −7.296, −12.381, and − 14.926, but it 1306.04 s are needed to compute the solution for $n = 12$. The DMRG method isfaster but yields somewhat unstable results: very accurate for $n = 12$, but inaccurate for $n = 6$ and 10 with values of −6.814 and − 12.083, respectively. In contrast, the values obtained by the RNN have very high accuracy for all qubits, with precisions of 3 decimal places, corresponding to a relative error energy below 1×10^{-5}, and further precision can be achieved by continuing the training.

6 Conclusion

We designed a recurrent neural network (RNN) in this paper to investigate the probability distribution reconstruction of the wave function in the lowest energy state of a many-body quantum system. The results of numerical reconstruction experiments revealed that as the number of qubits increased, the RNN-based wave function probability distribution reconstruction method proposed in this paper demonstrated superiority in terms of both time consumption and reconstruction accuracy. A study on phase reconstruction and its impact on modeling excited states or time-dependent systems can be found in [13]. A performance comparison between Transformer and an RNN can be found in [14].

Acknowledgments. This work was supported partly by the National Science Foundation of China under Grant No. 62473354.

References

1. A. Rocchetto, G. Edward, et al. Learning hard quantum distributions with variational autoencoders npj Quantum Inf. 4, 1 (June 2018), 28. (2018).
2. Carleo, G., Yusuke, N., Masatoshi, I.: Constructing exact representations of quantum many-body systems with deep neural networks. Nat. Commun. **9, 1**, 5322 (2018)

3. Choo, K., Carleo, G., Regnault, K.-N., et al.: Symmetries and many-body excitations with neural-network quantum states. Phys. Rev. Lett. **121, 16**, 167204 (2018)
4. Liang, X., Liu, W.-Y., Lin, P.-Z., et al.: Solving frustrated quantum many-particle models with convolutional neural networks. Phys. Rev. B. **98, 10**, 104426 (2018)
5. Zhang, Y.-H., Di Ventra, M.: Transformer quantum state: a multipurpose model for quantum many-body problems. Phys. Rev. B. **107, 7**, 075147 (2023)
6. Pfeuty, P.-P.: The one-dimensional Ising model with a transverse field. Ann. Phys. **57**(1) March 1970, 79–90 (1970)
7. Weiße, A., Fehske, H.: Exact diagonalization techniques. Comput. many part. phys. (2008)
8. Schollwöck, U.: The density-matrix renormalization group in the age of matrix product states. Ann. Phys. **326**(1), 96–192 (2011)
9. Wu, D., Rossi, R., Vicentini, F., et al.: From tensor-network quantum states to tensorial recurrent neural networks. Phys. Rev. Res. **5**(3), L032001 (2023)
10. S. A. Holevo, Probabilistic and Statistical Aspects of Quantum Theory. Pisa: Scuola Normale Superiore, chapter 1. (2011)
11. J. Chung, C. Gulcehre, K. Cho and Y. Bengio. Empirical evaluation of gated recurrent neural networks on sequence modeling. https://arxiv.org/abs/1412.3555. (2014)
12. D-P. Kingma and J. Ba. Adam: A method for stochastic optimization. https://arxiv.org/abs/1412.6980. (2014)
13. Cong, S., Lin, L.: A general RNN model for reconstructing arbitrary wave functions. In: The 17th International Conference on Advanced Computational Intelligence (ICACI2025), pp. 166–174, Bath, United Kingdom (2025)
14. Cong, S., Lin, L.: Generalized wave function reconstruction of high-qubit transverse-field Ising model. Pattern Recognit. Artif. Intell. **37**(12), 1083–1093. (in Chinese) (2024)

An Intelligent Fault Detection Method for Coke Ovens Oriented to Concept Drift

Gong Jiaqi[1] and Shan Yuan[1,2](✉)

[1] ACRE Coking & Refractory Engineering Consulting Corporation (Dalian), MCC, Dalian 116085, Liaoning, China
shanyuan@mail.dlut.edu.cn

[2] School of Energy and Power Engineering, Dalian University of Technology, Dalian 116024, Liaoning, China

Abstract. This paper addresses the performance degradation of fault detection models in coke oven production caused by concept drift and proposes a fault detection method based on concept drift. By employing sliding window technology to calculate the error rate of data streams in real time, combined with an embedded neural network for key feature extraction and a multilayer perceptron for drift point prediction, the operational status of the coke oven is determined through step transition stream analysis. Experiments were conducted on a JNX3–70-ID type compound heating top-charging coke oven, using data such as total flue gas composition, branch flue temperature, and heating wall temperature. The results demonstrate the effectiveness of the proposed method in both concept drift detection and fault identification tasks. Specifically, the detection accuracy for flue data reached 87.0%, a 2.5% improvement over baseline methods, while fault detection accuracy in the heating wall scenario increased by 5.5%. This study provides a novel solution for fault detection in coke oven production, offering significant industrial application value.

Keywords: Concept drift · Coke oven fault detection · Sliding window · Embedded neural network · Multilayer perceptron

In recent years, with the in-depth exploration of applying deep learning techniques to fault detection in the chemical industry, numerous advanced and practical research outcomes have been achieved.[1] Reference [2] proposed a minor fault detection method that integrates probabilistic information with sparse canonical variate analysis, enabling early warning of tiny faults. Reference [3] presented a chemical process fault diagnosis method based on an optimized convolutional neural network, which realizes timely and effective monitoring of multiple fault modes through flexible modelling strategies and efficient acquisition functions. Reference [4] proposed a neural-network-based fault diagnosis and classification model, achieving high accuracy and sensitivity in intelligent fault diagnosis and detection of chemical equipment. However, most existing studies are still limited to laboratory-scale testing and often ignore the model instability caused by concept drift.

Specifically, concept drift refers to a phenomenon in which the statistical properties of the prediction target change over time in an unpredictable manner, and such changes

Z. Zhang (Ed.): ISNN 2026, LNCS 16848, pp. 157–173, 2027.
https://doi.org/10.1007/978-981-92-5870-3_13

constitute a non-ignorable factor in fault detection tasks.[1] It is generally acknowledged that there are four types of concept drift: abrupt drift, gradual drift, incremental drift, and recurrent drift.[6] To address the issue of concept drift, an increasing number of concept drift detection mechanisms have been proposed. Generally, these methods follow a common framework comprising four stages[5–7]: Stage 1, data retrieval, which aims to extract data blocks from the data stream; Stage 2, data modelling, which focuses on modelling the retrieved data and extracting key features; Stage 3, test statistic calculation, which aims to quantify the degree of discrepancy between historical and newly arrived data; and Stage 4, hypothesis testing, which evaluates the statistical significance of changes between old and new data using hypothesis testing.

However, since the accuracy of drift detection is determined by hypothesis testing, this framework suffers from two limitations. First, as data streams do not provide all data at once, it is difficult to select the most appropriate hypothesis, thereby affecting the accuracy of concept drift detection.[8] Second, although concept drift can be categorized into four types, existing methods are generally limited to identifying whether concept drift has occurred, rather than determining the specific type of drift.[9] This is because, while changes in data distribution can be represented by test statistics such as the average error rate, these statistics are incapable of characterizing the evolution patterns of data distribution, i.e., they fail to capture the relationships between concept drifts at adjacent time stamps.[10] Therefore, this study designs a concept-drift-oriented fault detection method for coke oven production data streams. Based on a lazy learning strategy, sliding window technology is employed to compute the error rate of the data stream in real time, an embedded neural network is constructed for feature extraction, a multilayer perceptron architecture is adopted to predict drift points, and finally, the operating status of the coke oven is determined through step transition stream analysis.

1 Method

To address the performance degradation of fault detection models in coke oven production caused by concept drift, this paper proposes an online adaptive detection method based on a lazy learning strategy. As illustrated in Fig. 1, the core idea of the proposed method is to dynamically perceive variations in data streams and adjust the model in real time. The specific implementation consists of the following key steps. First, sliding window technology is employed to segment the real-time data stream, and the average error rate within each window is calculated to capture the statistical characteristics of local data distributions. Second, an embedded neural-network-based feature extraction mechanism is designed, and then nonlinear mappings are used to transform the original error features into a high-dimensional embedding space, thereby enhancing sensitivity to concept drift. Furthermore, a multilayer perceptron is utilized to construct a drift point prediction model, enabling accurate localization of drift positions through supervised learning. Finally, based on step transition stream analysis, the operating status of the coke oven is comprehensively diagnosed by integrating historical states with real-time prediction results (Fig. 2 and 4).

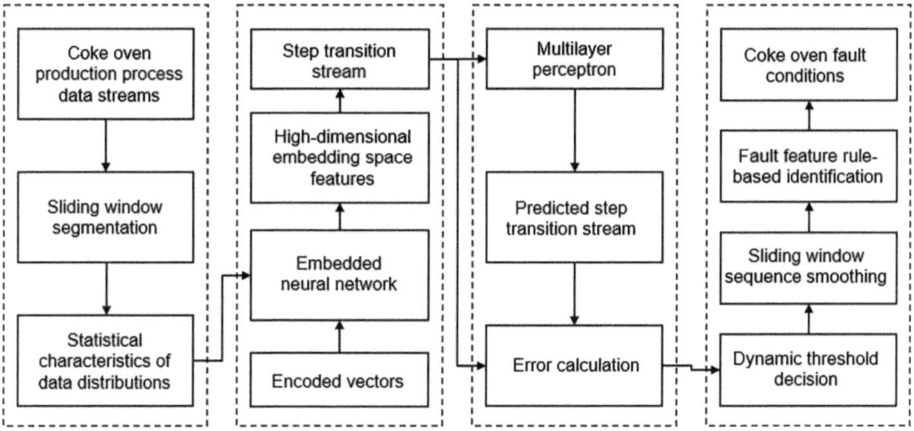

Fig. 1. Concept drift-based fault detection for coke ovens.

1.1 Drift Point Localization Method

The core of this method is capturing the variation patterns of error features in data streams through a dynamic window mechanism. Specifically, the prediction model is firstly applied to the data stream in real time to obtain the instantaneous error rate e_t at each time stamp t. To eliminate noise interference caused by instantaneous fluctuations, a sliding window W_i with a length of n is employed to smooth the error rate sequence. The selection of the window size requires a trade-off between detection sensitivity and noise robustness: a smaller window enables rapid response to concept drift but is more susceptible to noise, whereas a larger window can effectively smooth noise at the cost of delayed drift detection. Through experimental validation, n = 40 is ultimately determined as the optimal window size, which effectively suppresses random fluctuations while maintaining real-time performance. The average error rate $\hat{e}_t$ under the sliding window W_i can be expressed as:

$$\hat{e}_t = \frac{\sum_{t=1}^{n} e_t}{n} \tag{1}$$

During the window-sliding process, the system computes the arithmetic mean of the error rates within the window in real time, denoted as $\hat{e}_t$, which reflects the overall prediction performance over the current time interval. It should be noted that relying solely on the error rate of a single window is insufficient to effectively identify concept drift. Therefore, the difference between the error rates of adjacent windows, denoted as ap_i, is further introduced as a key feature for drift detection. This difference-based feature can capture abrupt changes in prediction performance and is particularly effective for identifying sudden concept drift.

The calculation formula for the average error rate difference Gap_i between two data streams is defined as follows:

$$Gap_i = \hat{e}_{t+1} - \hat{e}_t \tag{2}$$

Assume that each data stream contains $l \times n$ samples, where l denotes the number of sliding windows. Accordingly, each data stream yields $l - 1$ values of ap_i. During the pre-training stage, each sample can be represented as $\{X_i, y_i\} \in R^l$, where the feature vector $X_i = \{Gap_1, \ldots, Gap_{l-1}\}$, and the label $y_i \in \{1, \ldots, 4\}$ corresponds to the four types of concept drift. Through this representation, the distributional characteristics of the original data streams are effectively mapped into a supervised learning sample space $\{X_i, y_i\}$.

Based on the extracted error-feature dataset $G : \{(X_1, y_1), \ldots, (X_L, y_L), L > 0\}$, the system constructs a two-layer feature mapping framework to achieve accurate drift detection. In the first mapping layer, a learnable embedded neural network t_θ is employed to transform the original error features X_i into a low-dimensional embedding space. This network adopts a fully connected architecture with ReLU activation functions, enabling the effective extraction of deep patterns in the error features through nonlinear transformations, yielding the embedded representation $t_\theta(X_i)$.

In the second mapping layer, a prototype learning mechanism is innovatively introduced. For each concept drift type k, a corresponding prototype vector c_k is constructed by computing the mean of the hidden representations of samples in its support set S_k. These prototype vectors essentially serve as anchors of different concept drift types in the feature space, providing clear reference benchmarks for subsequent drift classification. The prototype vector is calculated as follows:

$$c_k = \frac{1}{|S_k|} \sum_{(X_i, y_i) \in S_k} t_\theta(X_i) \tag{3}$$

where c_k denotes the d-dimensional prototype vector of the k concept drift class, S_k represents the sample set of class k, $|S_k|$ is the cardinality of the support set, and $t_\theta(X_i)$ denotes the hidden feature representation of sample (X_i, y_i) after being mapped by the neural network.

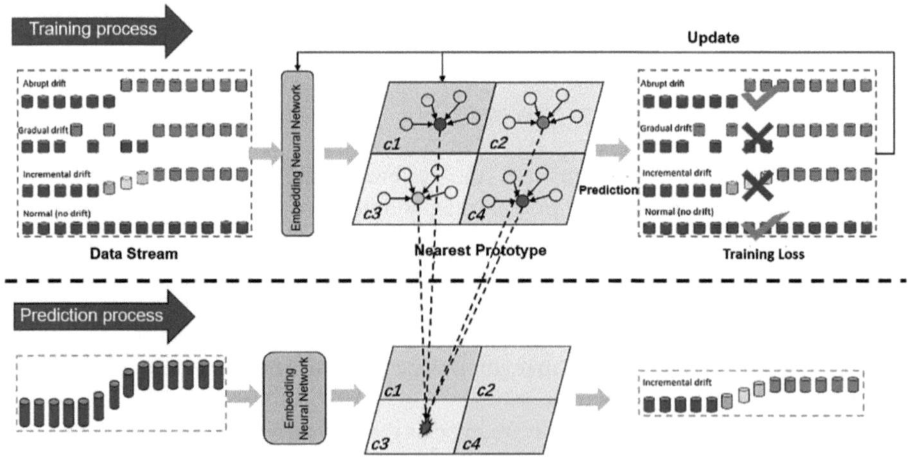

Fig. 2. Drift positioning method.

Furthermore, through feature concatenation and linear transformation, a joint feature representation $j_\theta(X_i)$ that integrates global class information and local temporal characteristics is generated. Based on this joint representation, a multilayer perceptron performs the final prediction, and its output $\hat{z}_\theta(X_i)$ directly corresponds to the potential drift point location in the data stream.

Assume L training data streams $D : \{(X_1, z_1), \ldots, (X_L, z_L) , \ L > 0\}$, where X_i denotes the error features associated with the corresponding drift point label z_i, and $z_i \in \{1, \ldots, l-1\}$.

An embedded neural network with learnable parameters θ is employed to map the error features X_i into an embedding space. Specifically, an embedding neural network v_θ is defined, where θ represents the learnable parameters. This network maps the error features X_i into the embedding space through a linear transformation followed by a *ReLU* activation function, which can be expressed as:

$$v_\theta(X_i) = \mathrm{ReLU}(\mathrm{W}_e X_i + \mathrm{b}_e)\# \tag{4}$$

Here, $v_\theta(X_i)$ denotes the embedding representation of the error features X_i, W_e represents the weight matrix of the embedding network with dimension $d \times d$, where d is the feature dimension, b_e denotes the bias term of the embedding network with dimension $d \times n$, where n is the number of training data streams, and ReLU denotes the rectified linear unit activation function.

By combining the prototype vector c_k and the encoded vector v_θ, a new feature vector is obtained. The prototype vector c_k and the encoded vector v_θ are concatenated and then transformed through a linear mapping followed by a *ReLU* activation function, which can be expressed as:

$$j_\theta(X_i) = ReLU(\mathrm{W}_c[c_k(X_i), v_\theta(X_i)] + b_c) \tag{5}$$

In this expression, $j_\theta(X_i)$ denotes the new feature vector of sample X_i, W_c represents the weight matrix of the linear transformation with dimension $d \times 2d$, where d is the feature dimension, b_c denotes the bias term with dimension $d \times N$, where N is the number of training data streams, and $[c_k(X_i), v_\theta(X_i)]$ indicates the concatenation of the prototype vector $c_k(X_i)$ and the encoded vector $v_\theta(X_i)$. ReLU denotes the rectified linear unit activation function.

A multilayer perceptron is then used for prediction. The joint embedding vector $j_\theta(X_i)$, which combines the prototype information and the input data, is fed into the multilayer perceptron, and the final prediction result is obtained through a linear transformation:

$$\hat{z}_\theta(X_i) = W_o j_\theta(X_i) + b_o \tag{6}$$

Here, $\hat{z}_\theta(X_i)$ denotes the predicted result for sample X_i, $j_\theta(X_i)$ represents the joint embedding vector, W_o denotes the weight matrix of the multilayer perceptron with dimension $1 \times d$, where d is the feature dimension, and b_o denotes the bias term with dimension $1 \times N$, where N is the number of training data streams.

During training, the model parameters are optimized by minimizing the mean squared error *MSE* between the predicted drift points and the true drift points. The *MSE* loss function is defined as:

$$L(\theta) = \frac{1}{N} \sum_{i=1}^{N} \|z_\theta(X_i) - \hat{z}_\theta(X_i)\|^2 \tag{7}$$

In this equation, $L(\theta)$ denotes the loss function, i.e., the *MSE*, N represents the number of training data streams, $z_\theta(X_i)$ denotes the true drift point in data stream X_i, and $\hat{z}_\theta(X_i)$ denotes the drift point predicted by the regression model for data stream X_i.

Finally, the drift point localizer takes the data stream X_i as input and outputs the locations at which drift occurs in the data stream, generating a step transition stream $S : \{(k_1, t_1), \ldots, (k_M, t_M)\ ,\ \ M \geq 0\}$, where each element indicates the type and time stamp of a detected drift event.

1.2 Fault Detection Method

Furthermore, a fault detection method based on step transition stream analysis is employed. The pseudocode of the proposed method is summarized as follows.

Algorithm 1: Coke Oven Fault Detection Method Based on Step Transition Stream

Input:

- Real-time data stream, X
- Trained recurrent neural network model, RNN_model
- Historical normal operating data, MSE'_max
- Dynamic threshold coefficient, λ
- Sliding window size, m

Output:

- Fault alarm signal, alarm
- Fault type diagnosis result, fault_type

```
1. // Step transition stream prediction
2. S_p = RNN_model.predict(X) // Generate the predict-
ed step transition stream {(k_1,t_1),...,(k_Mp,t_Mp)}

3. // Real-time step transition stream acquisition
4. S_r = get_rea_phase_stream() // Obtain the real
step transition stream{(k_1,t_1),...,(k_Mr,t_Mr)}

5. // Error calculation
6. MSE' = 0
7. for i = 1 to Mp do
8.     MSE' += (S_p.k_i - S_r.k_i)^2
9. end for
10. MSE' = MSE'/Mp

11. // Dynamic threshold decision
12. threshold = λ * MSE'_max
13. if MSE' > threshold then
```

```
    // Sliding window analysis
    window = get_last_m_samples(MSE')
    mean_MSE = average(window)
    std_MSE = standard_deviation(window)

    // Fault confirmation
    if mean_MSE > threshold AND std_MSE > σ_threshold then
        alarm = True

        // Fault type diagnosis
        if detect_sudden_jump(S_p.k_i) AND verify_sensor("flue") then
            fault_type = "Flue blockage or air leakage"
        else if check_correlation(S_p.k_i, temperature) then
            fault_type = "Abnormal combustion in the vertical flue "
        else if check_periodicity(S_p.k_i) then
            fault_type = "Abnormal coking in the carbonization chamber "
        else
            fault_type = " Unknown fault "
        end if
    end if
end if

return alarm, fault_type
```

A recurrent neural network is employed to receive the step transition stream S and generate the predicted step transition stream for the next reversing cycle, denoted as $S_p : \{(k_1, t_1), \ldots, (k_{M_p}, t_{M_p}), M_p > 0\}$, where S_p represents the predicted step transition stream, (k_i, t_i) denotes an element in the stream, k_i is the phase identifier, t_i is the corresponding time stamp, and M_p is the number of elements in the predicted stream.

During the subsequent reversing cycle, the real step transition stream is generated from the actual data stream and denoted as $S_r : \{(k_1, t_1), \ldots, (k_{M_r}, t_{M_r}), M_r = M_p\}$ where S_r represents the real step transition stream, and (k_i, t_i) is the number of elements in the real stream,, k_i is the phase identifier, t_i is the corresponding time stamp, and M_r is the number of elements in the predicted stream.

The mean squared error between S_p and S_r is calculated as:

$$MSE' = \frac{1}{M_p} \sum_{i=1}^{M_p} \left(k_{p_i} - k_{r_i}\right)^2 \tag{8}$$

Here, $MSE^{'}$ denotes the mean squared error, M_p is the number of elements in the step transition stream, k_{p_i} is the phase identifier of the i -th element in the predicted stream, and k_{r_i} is the phase identifier of the i -th element in the real stream.

A dynamic threshold decision mechanism is then adopted for fault determination. The core idea is to learn the fluctuation range of the step transition stream under normal operating conditions from historical data and to establish an adaptive fault decision criterion accordingly. Specifically, in the offline stage, the system first collects a sufficient amount of data under normal operating conditions and records, for each reversing cycle, the mean squared error $MSE^{'}$ between the predicted step transition stream S_p and the real step transition stream S_r. Through statistical analysis of these historical data, the maximum value MSE'_{max} is obtained and used as a reference baseline.

This is because MSE'_{max} represents the maximum deviation that may occur between the predicted step transition stream and the real step transition stream under normal operating conditions, and thus can be regarded as a conservative upper bound of normal fluctuations. However, in practical industrial scenarios, due to interference factors such as sensor noise and instability in coal quality, $MSE^{'}$ may slightly exceed the historical maximum in rare cases. To enhance system robustness and reduce the probability of false alarms, a safety coefficient λ is introduced to moderately relax this upper bound, thereby improving fault decision reliability. Therefore, the fault decision threshold is defined as $\lambda MSE'_{max}$.

During online operation, the system computes the current-cycle $MSE^{'}$ in real time and compares it with the dynamic threshold $\lambda MSE'_{max}$.

The parameter λ is empirically determined using a validation dataset collected under normal operating conditions. The selection criterion is to maintain sensitivity to fault-related deviations while minimizing the false alarm rate as much as possible.

To further enhance system robustness, a sliding window mechanism is introduced to smooth the $MSE^{'}$ sequence. The window size m determines the system's responsiveness to short-term fluctuations: a smaller window enables rapid detection of abrupt faults but is more sensitive to noise, whereas a larger window effectively suppresses noise interference at the cost of delayed fault detection. Therefore, comparative experiments with different values of m are required to comprehensively evaluate the trade-off between response speed and noise sensitivity, thereby determining the optimal window size. During the window-sliding process, the system not only computes the moving average $\overline{MSE'}$ within the window but also tracks its standard deviation σ in real time. This dual-indicator design allows the system to distinguish genuine fault signals from random noise. If $\overline{MSE'}$ persistently exceeds the dynamic threshold and σ increases significantly, it indicates a substantive change in the operating state, and triggering a fault alarm becomes more reliable. This adaptive strategy effectively overcomes the limitations of traditional fixed-threshold methods in complex industrial scenarios.

Subsequently, the fault type identification module performs precise classification of different fault types by analyzing the spatiotemporal distribution characteristics of abnormal phase identifiers k_i in the step transition stream, in conjunction with knowledge of coke oven operating mechanisms. In the temporal dimension, the system analyzes the occurrence frequency and duration of abnormal phase identifiers: abrupt faults are typically characterized by short-term, sharp jumps in k_i, whereas progressive faults manifest as gradual drifts in k_i. In the spatial dimension, the system monitors abnormal phase patterns across different equipment components, such as flues, vertical flues, and heating walls, and constructs a fault feature library to support accurate fault diagnosis.

The fault diagnosis strategy constructed in this study based on k_i is not a simple empirical assumption, but is derived from a comprehensive analysis of coke oven operating mechanisms and a systematic summary of historical fault records. Specifically, it is mainly based on domain knowledge provided by experienced personnel involved in the design, operation, and maintenance of compound heating top-charging coke ovens, as well as inspection reports, historical operational data, and archived fault logs of the equipment. Therefore, the variation patterns of abnormal phase identifiers k_i can, to a certain extent, reflect changes in the operating conditions of key coke oven components and provide a reliable basis for fault type identification.

The fault feature library established in this work is developed according to the structural characteristics and operating regime of the JNX3–70-1D type compound heating top-charging coke oven. Although the proposed step transition stream analysis framework has strong general applicability and can be extended to other coke oven types or similar industrial equipment, the detailed diagnostic rules and threshold settings may require recalibration and adjustment based on specific oven structures, sensor configurations, and operating conditions, in order to ensure stable and reliable diagnostic performance.

2 Experiments

2.1 Experimental Platform

In this study, a JNX3–70-1D type compound heating top-charging coke oven was selected as the experimental object. Based on this oven type, operational data were collected, including total flue gas composition, branch flue temperature and gas composition, as well as temperatures of the vertical flue and the heating wall. With respect to the experimental environment configuration, to ensure consistency in data processing and the reliability of experimental results, all computational tasks were performed on a unified hardware platform. Specifically, the server configuration consisted of an Intel Xeon Gold 6342 processor operating at 2.80 GHz, an NVIDIA A6000 GPU, and the Ubuntu 22.04.2 operating system.

2.2 Dataset

To simulate concept drift prediction, a total of 20,000 data streams were generated using the MOA tool. Each data stream contained 5,000 instances and included only a single type of drift. In this study, four types of drift were considered: abrupt drift, gradual drift, incremental drift, and normal (no drift). Subsequently, for each concept drift type, 4,000 data streams—amounting to a total of 16,000 data streams—were selected as the training set for drift feature extraction. The test set was constructed by mixing the remaining data streams, resulting in a training-to-testing ratio of 8:2.

The coke oven operational data were obtained from ACRE Coking & Refractory Engineering Consulting Corporation (Dalian), MCC. The dataset was compiled by experienced inspection personnel based on statistical analysis and historical production records. The faults mainly originated from the reversing valve, flue, and carbonization chamber. Reversing valve faults may be caused by loosening of the adjusting screw, incorrect switching angles, or lubrication and grinding issues, leading to air leakage. Flue-related faults may involve air leakage or cross leakage due to corrosion, abnormal temperature variations, or improper pressure regimes, as well as partial blockages caused by the accumulation of organic matter. Carbonization chamber faults may result in difficult coke pushing due to improper coking procedures, poor coal powder properties, or deformation of the oven wall or hearth bricks.

The organization of the coke oven fault detection dataset is summarized in Table 1, where the sampling interval between consecutive data groups is 1 min.

Table 1. Coke oven equipment fault detection dataset.

Labels	Flue		Vertical flue		Heating wall	
	Train	Test	Train	Test	Train	Test
Normal	8180	493	8259	505	8396	510
Switch cock abnormal	100	100	100	100	100	100
Flue abnormal	264	204	211	192	291	201
Heating wall abnormal	456	203	430	203	213	199

2.3 Experimental Design

The experiments are mainly divided into two parts. First, different prototype networks are employed to detect the location and type of concept drift on the generated data streams, to evaluate the performance of the proposed method under different prototype network configurations. Second, the fault detection performance of different methods on the same dataset is compared to assess their effectiveness in fault identification. Two mainstream strategies are considered for comparison: lazy and ensemble approaches. The lazy strategies include ADWIN,[11] DDM [12], EDDM,[13]> HDDM-A,[14] and KSWIN,[15] while the ensemble strategies include AAE,[16] BLAST,[17] and KUE.[18] To ensure fairness and comparability, all experiments adopt a unified learning model. Considering computational efficiency and partial fitting capability, the Gaussian Naive Bayes (GaussianNB) classifier from the sklearn[19] library is selected as the default classifier. Owing to its fast-processing speed and suitability for online learning, this model effectively supports the real-time analysis requirements of large-scale data streams. The experiments focus on the performance of different methods in concept drift detection and fault identification, without delving into the specific differences among learning models.

In addition, the dynamic threshold coefficient λ and the sliding window size m in the fault detection module are both determined through offline validation using data collected under normal operating conditions. Specifically, a grid search strategy is adopted to tune λ, while comparative experiments are conducted for different candidate values of m. By jointly considering the false alarm rate, detection accuracy, and alarm delay, the optimal parameter combination is finally selected, thereby ensuring the fairness and reproducibility of the experimental results.

The performance of the proposed method depends on the deep structures of the prototype networks and the error features extracted from the data streams. Therefore, during the experiments, we investigate how alternative deep prototype network architectures influence learning performance. The underlying idea is to generate multiple data streams with predefined forms of concept drift and then apply LDA-type methods to these streams using various prototype network architectures, while keeping the same error features, in order to observe the resulting performance. The objective is to verify whether the detector can correctly classify concept drift based on the error features in the data streams.

In the architecture of the prototype neural networks, four embedding functions are selected as nonlinear mappings: a full-attention mechanism neural network (FAN), a fully convolutional network (FCN), a recurrent neural network (RNN), and a four-layer fully connected neural network (FNN). For the parameters in the error features Xi, the error feature count l and the sampling frequency n are set to 50 and 25, respectively. During the training stage, five support data streams are randomly selected for each concept drift class to construct the prototypes. To evaluate the effectiveness of the model, 15 query data streams are selected, and during the testing stage, 20 support data streams are randomly chosen from the training dataset as the final prototypes. All experiments employ the ReLU activation function and use the Euclidean distance as the distance metric. All models are trained using stochastic gradient descent (SGD)[20] in conjunction with the Adam optimizer,[21] with the learning rate set to 0.01.

2.4 Experimental Results

2.4.1 Concept Drift Detection

The variation of the loss function during training provides important insights into the learning dynamics of the networks. As shown in Fig. 3, all four network architectures (FAN, FCN, RNN, and FNN) exhibit pronounced changes in loss values at the early stage of training, reflecting an initial exploration phase in which the models begin to capture concept drift features. As the number of training batches increases, the four architectures gradually demonstrate differentiated convergence behaviors.

Among them, FAN exhibits the most favorable learning curve characteristics. In the early training stage (0–10 batches), the loss value of FAN decreases rapidly, indicating its ability to efficiently capture key features associated with concept drift. During the intermediate training stage (10–30 batches), the loss shows a stable and approximately linear downward trend, suggesting that the network parameters are being finely adjusted. Finally, after 30 batches, the loss curve converges to a stable plateau, with a final loss value of 2.16, which is significantly lower than those of the other architectures. This result confirms the advantage of the attention mechanism in feature selection. Notably, FAN does not exhibit any obvious loss rebound throughout the training process, demonstrating strong training stability.

In contrast, the other network architectures exhibit varying degrees of limitation. Although FCN and RNN converge relatively quickly at the initial stage, they show noticeable loss fluctuations in the later training phase, indicating that convolutional and recurrent structures may be insufficient for handling temporally correlated concept drift. The FNN, on the other hand, displays a stable but slow convergence behavior, with a final loss value of 6.18, which is approximately three times that of FAN.

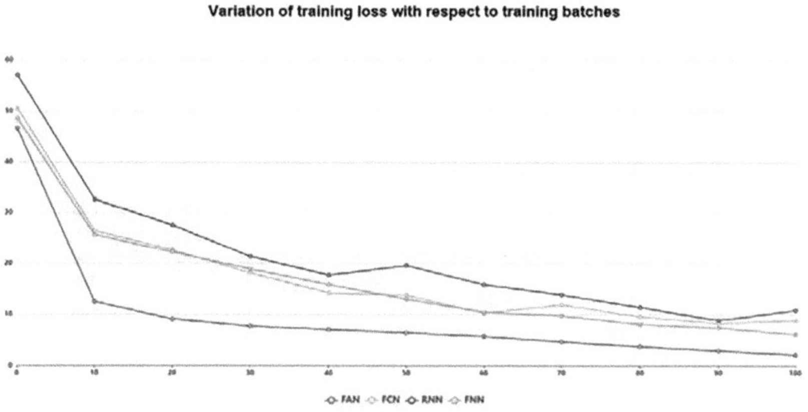

Fig. 3. Training loss of different deep structures of Prototypical Networks.

By analyzing the second-order derivatives of the loss function during training, deeper Insights into the robustness of different network architectures can be obtained. The FAN network exhibits a smooth second-order derivative curve, and in the later stages of training, its absolute values consistently remain at a low level. This indicates a stable parameter update process and low sensitivity of the network to variations in batch data. Such characteristics are particularly important in practical industrial applications, where production data inevitably contain noise and outliers.

As illustrated in Fig. 3, the LDA detector successfully identifies various types of concept drift in the simulated data streams. By comparing the detection performance across three typical drift types (abrupt, gradual, and incremental) as well as the normal (no-drift) scenario, significant differences in detection effectiveness are observed.

For the normal case, that is, determining whether concept drift has occurred, all detection methods exhibit excellent performance, with the FAN-based architecture achieving a detection accuracy of 95.8%. In contrast, incremental concept drift poses a considerably greater detection challenge. Experimental results show that FAN attains a detection accuracy of 79.2% in this scenario. The performance degradation is mainly attributed to the slow and subtle distributional changes associated with incremental drift, which are often accompanied by high-frequency noise. Notably, FAN still maintains an approximately 8.0% performance advantage over traditional methods, owing to its multi-scale attention mechanism, which can capture subtle patterns of feature evolution.

Overall, the results are consistent with expectations, with FAN demonstrating the best performance. This can be attributed to the ability of the attention mechanism to more effectively identify error features in data streams. Moreover, the results indicate that, even under identical settings, the detection performance varies significantly across different types of concept drift, with abrupt drift being the easiest to identify.

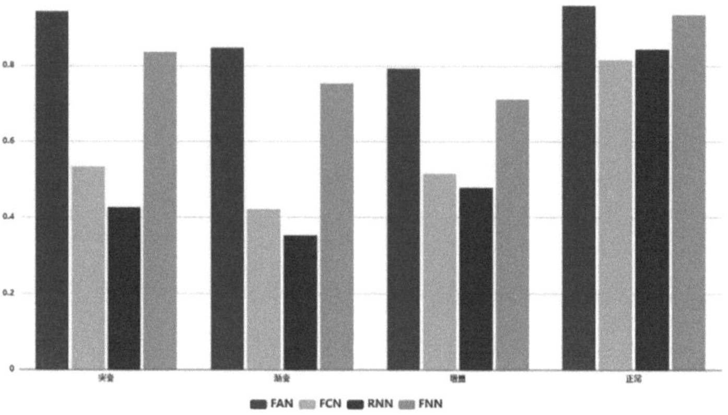

Fig. 4. Accuracy of different deep structures of prototypical networks.

2.4.2 Fault Detection

In this experiment, the performance of the proposed method is evaluated on a real-world dataset. Since real-world data streams contain unknown concept drift, it is not possible to compute the F1-score for concept drift detection. Consistent with most existing concept drift detection approaches [20and 21][22], predictive accuracy is therefore adopted as the primary metric to demonstrate the advantages of each algorithm. Meanwhile, fault detection is performed based on the step transition streams generated by different methods, and the effectiveness of fault detection is objectively evaluated.

Table 2. Coke oven equipment fault detection dataset

Models	Flue	Vertical flue	Heating wall
ADWIN	80.7	71.0	94.1
DDM	84.2	72.5	93.7
EDDM	84.3	72.5	94.1
HDDM-A	84.5	73.1	93.6
KSWIN	83.7	68.9	94.5
AAE	82.3	70.4	94.4
BLAST	82.4	72.3	94.1
KUE	83.5	71.3	94.0
Ours	**87.0**	**75.2**	**94.7**

Table 2 presents a comparison of classification accuracy achieved by different concept drift detection methods across three equipment-component datasets. Eight mainstream concept drift detection algorithms, including ADWIN, DDM, and EDDM, are selected as baseline methods. The experimental results indicate that all methods achieve a noticeable improvement in accuracy after detecting concept drift and retraining the classifier, thereby confirming the necessity of concept drift detection. The proposed method consistently attains the best performance across all three test scenarios. In particular, it achieves an accuracy of 87.0% on the flue dataset, representing an improvement of 2.5 percentage points over the second-best method. In addition, the detection accuracy for heating wall data is generally high (greater than 93%), which may be attributed to the relatively stable operating conditions of this component, whereas fault detection on vertical flue data remains the most challenging.

Table 3. Coke oven equipment fault detection dataset

Models	Flue	Vertical flue	Heating wall
ADWIN	77.5	66.9	88.0
DDM	80.0	68.0	90.7
EDDM	80.9	68.1	90.1
HDDM-A	81.7	68.7	90.9
KSWIN	78.9	68.5	90.6
AAE	78.7	66.7	89.9
BLAST	79.2	67.2	90.8
KUE	79.5	68.7	90.9
Ours	**85.4**	**74.2**	**93.5**

Table 3 provides a comparison of fault detection accuracy achieved by various concept drift detection techniques. The experimental results demonstrate that concept drift detection techniques can effectively enhance fault detection performance for coke oven equipment, thereby validating their value in industrial applications. The proposed method maintains a leading advantage in the fault detection task, particularly in the vertical flue scenario, where it achieves a detection accuracy of 74.2%, representing an improvement of 5.5 percentage points over the baseline methods.

3 Conclusion

This study proposes a concept-drift-based fault detection method for coke ovens. By integrating sliding window techniques with embedding neural networks and multilayer perceptions, the proposed approach effectively addresses the model instability caused by concept drift. Experimental results demonstrate that the proposed method significantly outperforms mainstream baseline approaches in both concept drift detection and fault identification tasks, showing particularly strong performance in complex scenarios such as flues and vertical flues. In addition, the Fully Attentive Network (FAN) exhibits the best feature extraction capability, further validating the robustness of the proposed framework.

This work not only provides a new technical pathway for coke oven fault detection, but also offers valuable insights for handling concept drift problems in other industrial scenarios. In future research, we will further explore multi-source data fusion by integrating additional sensor information, such as pressure and flow data, to construct more comprehensive fault detection models and enhance system adaptability in complex industrial environments. Meanwhile, to meet real-time requirements, we plan to investigate lightweight network architectures and edge computing technologies to improve computational efficiency and satisfy low-latency demands in industrial applications. Furthermore, improving adaptive mechanisms will be a key research direction, particularly the dynamic adjustment of sliding window sizes and thresholds, to better cope with varying concept drift characteristics under different operating conditions, thereby further enhancing the robustness and flexibility of the proposed method.

References

1. Yu, B.A.O., Shuo, C.H.E.N.G., Jingtao, W.A.N.G.: A review of deep learning-based fault detection and diagnosis in chemical processes. Chem. Ind. Eng. **39**(02), 9–22 (2022)
2. Xin, C.H.E.N., Donghui, P.A.N., Wenzhi, Y.A.N.G.: Chemical process minor fault detection by integrating probability information with sparse canonical variable analysis. Control Decis., 1–10. Accessed: 12 Mar 2025
3. Yupeng, H.U., Lijie, G.U.O., Zilong, Z.H.A.N.G., et al.: A fault diagnosis method for chemical processes based on optimized convolutional neural networks. J. Yanshan Univ. **48**(06), 550–560 (2024)
4. Yuan, H.U.A.N.G., Wentao, Z.H.O.U.: Research on distributed fault diagnosis Technology for Chemical Equipment Based on improved CNN. Chem. Equipment Pip. **60**(06), 8–17 (2023)
5. Agrahari, S., Singh, A.K.: Concept drift detection in data stream mining: a literature review. J. King Saud Uni. Comput. Inf. Sci. **34**(10), 9523–9540 (2022)
6. Lu, J., Liu, A., Dong, F., et al.: Learning under concept drift: a review. IEEE Trans. Knowl. Data Eng. **31**(12), 2346–2363 (2018)
7. Yu, H., Lu, J., Zhang, G.: Morstreaming: a multioutput regression system for streaming data. IEEE Trans Syst. Man Cybern. Syst. **52**(8), 4862–4874 (2021)
8. Elwell, R., Polikar, R.: Incremental learning of concept drift in nonstationary environments. IEEE Trans. Neural Netw. **22**(10), 1517–1531 (2011)
9. Gama, J., Žliobaitė, I., Bifet, A., et al.: A survey on concept drift adaptation. ACM comput. surv. (CSUR). **46**(4), 1–37 (2014)
10. Xu, S., Wang, J.: Dynamic extreme learning machine for data stream classification. Neurocomput. **238**, 433–449 (2017)
11. Bifet, A., Gavalda, R.: Learning from time-changing data with adaptive windowing. In: Proceedings of the 2007 SIAM International Conference on Data Mining. Society for Industrial and Applied Mathematics, pp. 443–448 (2007)
12. Gama, J., Medas, P., Castillo, G., et al.: Learning with drift detection advances in artificial intelligence–SBIA 2004. In: Proceedings 17th Brazilian Symposium on Artificial Intelligence, pp. 286–295. Springer Berlin Heidelberg, Sao Luis, Maranhao, Brazil., Sept 29-Oct. 1, 2004 (2004)
13. Baena-Garcıa, M., del Campo-Ávila, J., Fidalgo, R., et al.: Early drift detection method. In: Fourth International Workshop on Knowledge Discovery from Data Streams, vol. 6, pp. 77–86 (2006)
14. Frias-Blanco, I., del Campo-Ávila, J., Ramos-Jimenez, G., et al.: Online and non-parametric drift detection methods based on Hoeffding's bounds. IEEE Trans. Knowl. Data Eng. **27**(3), 810–823 (2014)
15. Raab, C., Heusinger, M., Schleif, F.M.: Reactive soft prototype computing for concept drift streams. Neurocomput. **416**, 340–351 (2020)

16. Krawczyk, B.: Active and adaptive ensemble learning for online activity recognition from data streams. Knowl.-Based Syst. **138**, 69–78 (2017)
17. van Rijn, J.N., Holmes, G., Pfahringer, B., et al.: The online performance estimation framework: heterogeneous ensemble learning for data streams. Mach. Learn. **107**, 149–176 (2018)
18. Cano, A., Krawczyk, B.: Kappa updated ensemble for drifting data stream mining. Mach. Learn. **109**(1), 175–218 (2020)
19. Pedregosa, F., Varoquaux, G., Gramfort, A., et al.: Scikit-learn: machine learning in python. J. mach. Learn. res. **12**, 2825–2830 (2011)
20. Saad, D.: Online algorithms and stochastic approximations. Online Learn. **5**(3), 6 (1998)
21. Kingma D P, Ba J. Adam: A method for stochastic optimization. arXiv preprint https://arxiv.org/abs/1412.6980, 2014.

DSLLM-Fin-7B: An Open-Source Chinese Financial Large Language Model

Kangping Zhang[1], Lijia Gao[1], Wangyang Ying[3], Wei Ju[1], Rong Xiao[1], Chenwei Tang[1,2(✉)], and Jiancheng Lv[1]

[1] College of Computer Science, Sichuan University,Chengdu610065, China
{zhangkangping,gao_lijia}@stu.scu.edu.cn,
{juwei,rxiao,tangchenwei,lvjiancheng}@scu.edu.cn
[2] National Key Laboratory of Fundamental Algorithms and Models for Engineering Simulation, Sichuan University, 610207 Chengdu, Sichuan, China
[3] School of Computing and Augmented Intelligence, Arizona State University,Tempe, AZ85281, USA
wangyang.ying@asu.edu

Abstract. Despite the growing promise of financial LLMs in investment decisions and knowledge acquisition, high-quality models remain mostly closed-source or English-centric, with limited Chinese support. To bridge this gap, we introduce **DSLLM-Fin-7B**, a Chinese financial LLM built upon the open-source *deepseek-llm-7b-chat* with strong reasoning capabilities. Specifically, we first meticulously curate a Chinese financial dataset **FinQA-Alpaca**, comprising 99,562 Chinese financial samples sourced from open-source projects via keyword retrieval and manual screening. FinQA-Alpaca features a unified format across four core financial reasoning tasks, i.e., financial report chain-of-thought, domain-specific Question Answering (Q&A), financial inference chains, and event causal analysis. Then, utilizing Low-Rank Adaptation (LoRA) for parameter-efficient fine-tuning on *deepseek-llm-7b-chat* with our collected FinQA-Alpaca dataset, we develop DSLLM-Fin-7B—a model supporting local deployment and exhibiting robust financial reasoning. Experiments on CFLEB, a Chinese financial benchmark dataset, demonstrate that the proposed DSLLM-Fin-7B outperforms its base model, confirming its superior performance in Chinese financial Q&A tasks.

Keywords: Chinese Financial Large Language Model · Open-Source · LoRA · Financial Reasoning and Question Answering

---

This work is supported by the National Major Scientific Instruments and Equipments Development Project of National Natural Science Foundation of China under Grant 62427820, the Sichuan Science and Technology Program under Grant 2025ZDZX0125, the Science Fund for Creative Research Groups of Sichuan Province Natural Science Foundation under Grant 2024NSFTD0035, the Ministry of Education Engineering Research Center Guiding Project for Machine Learning and Industrial Intelligence Applications under Grant SCU2024D013.

Z. Zhang (Ed.): ISNN 2026, LNCS 16848, pp. 174–185, 2027.
https://doi.org/10.1007/978-981-92-5870-3_14

1 Introduction

Large language models (LLMs) [4] embed human knowledge into high-dimensional representation spaces via self-supervised training, enabling strong language understanding and generation capabilities for financial tasks such as earnings analysis, policy interpretation, and market Q&A. However, financial language is domain-specific, structurally complex, and context-dependent, posing challenges for general LLMs in terminology comprehension, numerical reasoning, and factual consistency—especially in Chinese scenarios due to limited high-quality corpora and task-specific datasets.

Question	“资管新规”是什么意思？ *Translation: What does “New Asset Management Regulations ”means?*
llama2:7b	资管新规（Shareholder Value Maximization）是一种企业管理的核心理念，强调公司应该。。。 *Translation: New Asset Management Regulations (Shareholder Value Maximization) is a core principle of corporate management, emphasizing that companies should......*
DSLLM-Fin-7B	“资产管理业务新规”（Asset Management Industry New Regulations）是指中国政府为了规范和引导资产管理行业健康发展，于2018年4月发布的一系列规定。这些规定的目的是打破刚性兑付、去通道化、限制资金池运作等，以降低金融风险，促进金融服务实体经济。 *Translation: New Asset Management Regulations (Asset Management Industry New Regulations) refer to a series of regulatory measures issued by the Chinese government in April 2018 to standardize and guide the healthy development of the asset management industry. These regulations aim to break implicit guarantees, eliminate channel-based investments, and restrict the operation of capital pools, in order to reduce financial risks and promote the financial sector's support for the real economy.*

Fig. 1. Differences in understanding China-specific economic terms, between llama2:7b and our proposed DSLLM-Fin-7B.

Existing high-quality LLMs are primarily closed-source or English-centric. While BloombergGPT [18] achieves strong performance, it is not publicly available. Open-source alternatives such as LLaMA and FinBERT [8] support Chinese but often lack understanding of financial terminologies and policy-specific language styles. As illustrated in Fig. 1, they often lack sufficient understanding of Chinese financial terminologies and policy-specific language styles, e.g., terms like "New Asset Management Regulations". Meanwhile, financial Q&A datasets such as FinQA [2] and ConvFinQA [3] are English-dominant and fail to support Chinese reasoning tasks. Additionally, financial data from social media and news is typically noisy and unstructured. These limitations highlight the need for bilingual (Chinese-English) financial Q&A datasets with explicit reasoning support for downstream financial LLM fine-tuning.

To address these challenges, we first construct the **FinQA-Alpaca** dataset by collecting and processing financial Q&A reasoning data. Using this dataset, we perform LoRA fine-tuning [7] on the *deepseek-llm-7b-chat* model, iteratively optimizing parameters for enhanced performance. This process yields **DSLLM-Fin-7B**, a 7-billion parameter Chinese financial LLM, specialized in financial

Q&A reasoning. As shown in Fig. 1, when queried about "New Asset Management Regulations", llama2:7b produces factual hallucinations [9], whereas **DSLLM-Fin-7B**, trained on bilingual financial data, generates accurate and comprehensive responses. The main contributions are as follows:

1. We construct the **FinQA-Alpaca** dataset, containing 99,562 curated financial reasoning Q&A pairs, and analyze its distribution through visualization to reveal key thematic coverage.
2. We develop **DSLLM-Fin-7B**, an open-source financial LLM with enhanced Chinese comprehension and on-device deployment support via LoRA fine-tuning of *deepseek-llm-7b-chat* using FinQA-Alpaca.
3. DSLLM-Fin-7B achieves substantial gains on Chinese financial NLP benchmarks such as CFLEB, improving BLEU-4 by 21, ROUGE-1 by 23, ROUGE-2 by 10, and ROUGE-L by 24 over the base model.

2 Related Work

2.1 Base Model: Deepseek-Llm-7b-Chat

As a representative of domestic LLMs, the DeepSeek series collects high-quality Chinese training data and is optimized specifically for the characteristics of the Chinese language. The *deepseek-llm-7b-chat* [1] model is derived from *deepseek-llm-7b-base* through instruction fine-tuning, with a focus on dialogue optimization. It is capable of understanding context and generating coherent multi-turn responses, making it well-suited for building chat assistants.

2.2 LoRA: Low-Rank Adaptation

LoRA [7] is a parameter-efficient fine-tuning (PEFT) method that adjusts LLMs by introducing low-rank updates to the original model weights. Compared to other PEFT [5] methods—such as adding adapter modules or modifying input prefix vectors (Prefix Tuning)—these alternatives often either introduce latency during inference or are difficult to train and can degrade model performance. LoRA, on the other hand, offers an efficient adaptation strategy that avoids inference latency, maintains input sequence length, and significantly reduces the number of trainable parameters while preserving high model quality.

3 Chinese Financial LLM: DSLLM-Fin-7B

In this section, we present the construction of the Chinese financial LLM DSLLM-Fin-7B, including the sources and processing pipeline of the financial fine-tuning dataset FinQA-Alpaca, as well as the training procedure and parameter settings for LoRA fine-tuning based on the *deepseek-llm-7b-chat* model.

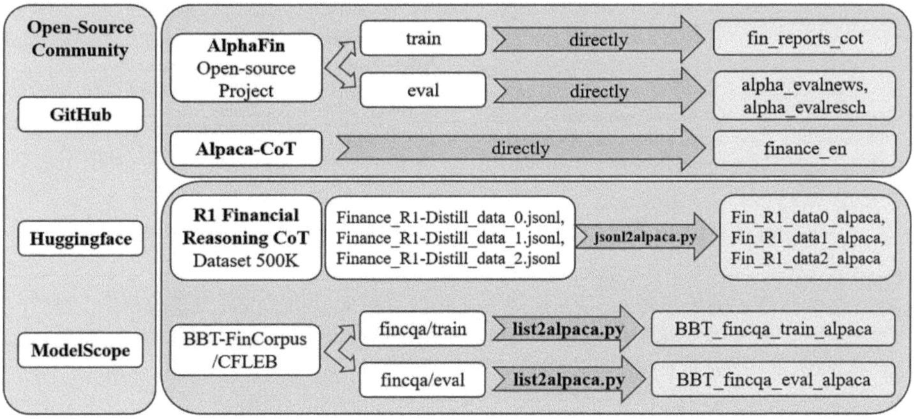

Fig. 2. Data sources and processing of the FinQA-Alpaca dataset. It is collected from various open-source communities and a portion of the data requires preprocessing using custom scripts (jsonl2alpaca.py or list2alpaca.py).

3.1 Financial Fine-Tuning Dataset: FinQA-Alpaca

In this paper, we construct a fine-tuning dataset in the Alpaca format for financial Q&A and reasoning, named FinQA-Alpaca, which is used for LoRA fine-tuning experiments on *deepseek-llm-7b-chat*. The construction process of FinQA-Alpaca is illustrated in Fig. 2.

As shown in Fig. 2, the initial data are sourced from the ModelScope community and several open-source projects. Specifically, financial keywords, e.g., finance and economics, are used to retrieve relevant content from ModelScope. After manual screening, high-quality financial Q&A reasoning data are retained, including Chain-of-Thought (CoT) and R1 deep reasoning data. Additionally, multiple open-source financial projects are investigated to extract useful financial statement data and causal analysis data of financial events. These are integrated to form the raw financial dataset. Specifically, the sources include the CoT dataset on financial report reasoning from the AlphaFin [12] open-source project (*fin_reports_cot*), financial Q&A instruction data from the Alpaca-CoT project [17], R1 financial reasoning CoT datasets [14] (*Finance_R1-Distill_data_0*, *Finance_R1-Distill_data_1*, and *Finance_R1-Distill_data_2*, totaling 500K samples), and financial event causal analysis Q&A datasets from the BBT-FinCUGE open-source project [15], specifically the CFLEB benchmark (*fincqa/train* and *fincqa/eval*). Additional evaluation datasets are obtained from AlphaFin project. Before performing LoRA fine-tuning, the original financial datasets need to be processed into the Alpaca format through keyword mapping and data unification, to prepare for subsequent Instruction Tuning (IT). The Alpaca format is a list of JSON objects, as illustrated in Fig. 3a. Among the key fields, instruction, input, and output are required, but the input field is optional and can be left empty. The processing of the original financial datasets is as follows.

```
[{
  "instruction": "User Instruction (required)",
  "input": "User Input (optional)",
  "output": "Model Response (required)",
  "system": "System Prompt (optional)",
  "history": ["First-round Instruction (optional)",
              "First-round Response (optional)"],
}]
```

(a) Alpaca Data Format

```
[{
  "id": "Unique Identifier ID",
  "user_input": "Original Financial Question",
  "reasoning_content": "LLM Reasoning Process",
  "answer_r1": "Final Answer Result",
  "created_by": "Ingenius_AI",
  "contact": "Ingenius AI Official Account"
}]
```

(b) R1 Financial Reasoning Chain

[["下文中有哪些因果事件?巴基斯坦对尿素的强烈需求拉动了国际尿素价格的上涨"]]
Translation: What are the causal events described in the following text? Pakistan s strong demand for urea has driven up international urea prices

[["需求增加导致市场价格提升"]]
Translation: The increase in demand has led to a rise in market prices

(c) Original Data Format of CFLEB dataset of BBT-FinCorpus

Fig. 3. Comparison between Alpaca format (a), original format of the R1 financial reasoning chain dataset (b), and the original data format of CFLEB dataset (c).

First, the datasets *fin_reports_cot.json*, *finance_en.json*, and the evaluation datasets from AlphaFin already conform to the Alpaca format, which is shown in Fig. 3a. Then, the datasets *Finance_R1-Distill_data_0.jsonl*, *Finance_R1-Distill_data_1.jsonl*, and *Finance_R1-Distill_data_2.jsonl* are structured as shown in Fig. 3b. Keyword mapping is used to convert the original fields: user_input is mapped to input in Alpaca format, answer_r1 is mapped to output, and other fields in the original format are discarded. The input field is set to empty. A custom Python script jsonl2alpaca.py is used to perform the conversion, resulting in the datasets *Fin_R1_data0_alpaca.json*, *Fin_R1_data1_alpaca.json*, and *Fin_R1_data2_alpaca.json*. Finally, the dataset *FinCUGE_Publish/fincqa/train.json* and its evaluation dataset are in a plain text list format without keywords, as shown in Fig. 3c, where each item can be seen as an array with two elements. An Alpaca dictionary object is constructed by mapping the first element to input and the second to output. Similarly, the input field is set to empty. This conversion is implemented using a custom script *list2alpaca.py*, resulting in the final datasets *BBT_fincqa_train_alpaca.json* and *BBT_fincqa_eval_alpaca.json*.

3.2 Fine-Tuning Process

This work performs LoRA fine-tuning, merged inference, and performance evaluation on the base model *deepseek-llm-7b-chat* using the open-source tool LLaMA-Factory [19]. The following section presents a detailed explanation of the LoRA fine-tuning process and parameter configuration.

Table 1. Key Parameter Settings for LoRA Fine-tuning.

Parameter Name	Value	Description
lora_rank	8	Rank of the LoRA matrix
lora_alpha	16	LoRA scaling factor
lora_target	all	Target modules to which the LoRA method is applied
lora_dropout	0	Dropout rate in LoRA fine-tuning
learning_rate	5.0e-05	Learning rate
gradient_accumulation_steps	8	Number of gradient accumulation steps
lr_scheduler_type	cosine	Learning rate scheduler type for dynamic adjustment
max_examples	1000	Maximum number of examples
cutoff_len	2048	Sequence length truncation
num_train_epochs	7.0	Number of training epochs
max_grad_norm	1.0	Maximum gradient norm for gradient clipping
bf16	true	Whether to use bf16 computation format
per_device_train_batch_size	2	Training batch size per GPU device
per_device_eval_batch_size	2	Evaluation batch size per GPU device
template	deepseek	Dataset template; must match the base model

LoRA fine-tuning under the current environment setup requires several essential third-party libraries, notably the PEFT library, which serves as a toolkit for efficient fine-tuning of pre-trained large language models (PLM). By introducing techniques such as LoRA and Adapter, PEFT enables effective fine-tuning by updating only a subset of model parameters. The 'LoraConfig' class defines LoRA-related hyperparameters, such as the rank of the low-rank matrix and the scaling factor. It helps users customize the fine-tuning process to improve training efficiency and effectiveness. The '*target_modules*' parameter is a key component in LoRA, specifying the model modules where the low-rank matrices are inserted for adaptation.

The core idea of LoRA is to insert low-rank matrices into specific layers of a PLM, allowing for parameter-efficient fine-tuning without modifying the original weights. In language models, modules with a significant impact on weight updates are typically selected, such as '*q_proj*' and '*k_proj*' (for query and key transformations), '*v_proj*' (for value transformation), and '*o_proj*' (for output projection). These modules are primarily located in self-attention and feedforward layers. Adjusting their weights through inserted low-rank matrices allows the model to adapt to new tasks while preserving its original capabilities, significantly reducing computational and memory costs. Table 1 summarizes several key parameter settings for training the LoRA adapter, as determined through empirical configuration. The template parameter refers to the structured format used to organize the training data, specifying how user inputs (questions) and model outputs (answers) are presented. Different foundation models or fine-tuning frameworks may require different template formats, and this value must be consistent with the underlying model.

4 Experiments

We investigate the following three questions in experiments: (1) What is the composition of the fine-tuning dataset **FinQA-Alpaca**? (2) Does the fine-tuned Chinese financial LLM, **DSLLM-Fin-7B**, demonstrate a significant improvement in answer quality for financial Q&A compared to the base model? and (3) Does **DSLLM-Fin-7B** retain its original generalization ability after fine-tuning?

4.1 Experiment Settings

The LoRA fine-tuning experiments based on the *deepseek-llm-7b-chat* are conducted within a Jupyter Notebook environment on the AutoDL cloud computing platform. The experiments are conducted on a Linux operating system using Python 3.10.16, with CUDA version 12.6. The hardware comprises a single NVIDIA RTX 4090D GPU (24GB memory), a 16 vCPU Intel Xeon Platinum 8481C processor, 80GB of RAM, and storage including a 30GB system disk and a 100GB data disk. Most LoRA training parameters are set based on general practice. Some hyperparameters, such as cutoff_len and num_train_epochs, are adjusted iteratively during training to minimize loss. Other parameters are determined empirically. Table 2 shows the mapping between different training hyperparameter configurations and the resulting LoRA adapter names, after filtering out experiments with high loss values.

Table 2. Different Training Process Hyperparameters and Trained LoRA Adapters.

LoRA_adapter_name	cutoff_len	num_train_epochs
train_fin1	1024	5
train_fin2	2048	5
train_fin3	2048	7

4.2 Analysis of FinQA-Alpaca Dataset

The processed datasets are consolidated into the financial fine-tuning dataset illustrated in Table 3, named **FinQA-Alpaca**. The table primarily presents the names, the corresponding number of data samples, size, the language used (Chinese or English), and a brief description of their core content of the sub-datasets within FinQA-Alpaca. To more intuitively reflect the key content of the FinQA-Alpaca dataset, we use the Counter class from Python's collections module to

Table 3. FinQA-Alpaca Financial Dataset Summary.

Name	Count	Size	Lang.	Description
fin_reports_cot	200	1.3M	CN	Financial Report Chain-of-Thought
finance_en	68912	40.1M	EN	Financial Q&A
Fin_R1_data0_alpaca	999	3.1M	EN	Complex Financial Reasoning Dataset
Fin_R1_data1_alpaca	992	3.1M	EN	Complex Financial Reasoning Dataset
Fin_R1_data2_alpaca	385	1.2M	EN	Complex Financial Reasoning Dataset
BBT_fincqa_train_alpaca	21965	8.5M	CN	Financial Event Causal QA Training Dataset
BBT_fincqa_eval_alpaca	2741	1.1M	CN	Financial Event Causal QA Evaluation Dataset
alpha_evalnews	1568	3.7M	CN	Financial News Evaluation Dataset
alpha_evalresch	1800	3.4M	EN	Financial Research Info Extraction & Summary Evaluation Dataset

perform word frequency analysis. As illustrated in Fig. 4, after filtering out function words, pronouns, and meaningless single-character words, we identify the top-6 frequent keywords in the dataset.

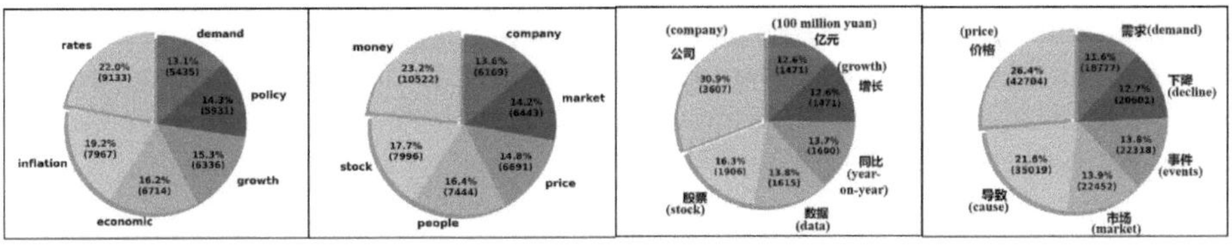

(a) Fin_R1_data* (b) finance_en (c) fin_reports_cot (d) fincqa_train

Fig. 4. Top 6 most frequent words pie charts for each dataset in FinQA-Alpaca, which illustrate their relative proportions and occurrence counts.

As shown in Fig. 4, based on the top-6 most frequent words, a more in-depth analysis can be conducted. The complex financial reasoning dataset (a combined analysis of the three *Fin_R1_data* subsets) focuses on macroeconomic topics and complex financial problem-solving, such as 'interest rates', 'inflation', 'economy', 'growth', 'policy', and 'demand'. The financial Q&A dataset *finance_en.json* centers on microeconomic and practical financial issues, including 'money', 'stocks', 'the public', 'prices', 'markets', and 'companies'. The financial report CoT dataset *fin_reports_cot.json* places greater emphasis on companies and the stock market, aiming to establish reasoning links between corporate financial reports and market behavior. In the causal analysis dataset *BBT_fincqa_train_alpaca.json*, the most frequently occurring words are 'price' and 'cause', indicating that the dataset primarily focuses on causal relationships among economic elements such as financial markets and price fluctuations.

4.3 Experiment Results

Different LoRA adapters are trained by adjusting parameters. Based on common text evaluation metrics, the best-performing experiment is selected to produce the financial LLM **DSLLM-Fin-7B**, which can be used for subsequent downstream tasks via Ollama deployment or further recursive training. And human-interactive Q&A is conducted to intuitively assess the training effectiveness.

Evaluation of Training Effectiveness: Batch Prediction with BLEU and ROUGE. The evaluation is conducted on the benchmark datasets **alpha_evalnews**, **alpha_evalresch**, and **BBT_fincqa_eval_alpaca** using commonly used text generation metrics BLEU [16] and ROUGE [13]. The evaluation results are shown in Table 4. For evaluating the similarity between generated text and reference text, higher values of BLEU-4, ROUGE-1, ROUGE-2, and ROUGE-L indicate greater similarity, with a maximum score of 100. In terms of model performance evaluation, predict_runtime refers to the total time (in seconds) required for the model to generate a batch of samples, while predict_samples_per_second denotes the number of samples the model can generate per second, typically used to assess inference speed. Considering both the similarity between generated and reference texts and the model's performance, **train_fin3** demonstrates the best training outcome. It not only produces outputs with higher similarity to the reference texts in the evaluation datasets but also performs well in terms of total generation time and inference speed. This adapter is merged with the original deepseek-llm-7b-chat to produce a complete model file named **DSLLM-Fin-7B**.

Table 4. Batch Prediction Text Generation Metrics on Evaluation Datasets.

Model Name	predict bleu-4 ↑	predict rouge-1 ↑	predict rouge-2 ↑	predict rouge-l ↑	predict runtime ↓	samples per second ↑
deepseek-llm-7b-chat	22.460	29.780	8.362	26.720	79.431	0.755
train_fin1	18.996	27.616	8.839	23.741	62.652	0.958
train_fin2	33.517	42.045	16.005	40.256	**22.255**	**2.696**
train_fin3	**43.128**	**53.578**	**18.628**	**50.879**	34.164	1.756

Table 5. Evaluation Results on General Capability Benchmarks.

Model Name	MMLU	CMMLU	CEval
DeepSeek-LLM-7B-Chat	49.4	49.7	47.0
train_fin1	49.44	49.72	46.29
train_fin2	49.96	49.54	46.21
DSLLM-Fin-7B	49.67	49.38	45.91

<table>
<tr><th colspan="2">问题：什么是边际效益递减？
Translation: Question: What is diminishing marginal efficiency?</th></tr>
<tr><th>deepseek-llm-7b-chat</th><th>DSLLM-Fin-7B</th></tr>
<tr><td>定义：边际效益递减是指在一定时间内，随着投入某项资源的增加，每单位新增资源所带来的收益(即边际效益)会逐渐减少的现象......
示例：例如，在农业生产中，......
总结：总之，......
Translation:
Definition: Diminishing Marginal Effectiveness is the phenomenon whereby the benefit per unit of additional resource (i.e., the marginal benefit) decreases over time as more resources are invested in a particular resource ......
Examples: For example, in agricultural production, ...
Conclusion: Overall, ...</td><td>定义：边际效率递减（DME）是指增加一个单位的投入所产生的产出增幅逐渐减小，从而降低整体经济效率的概念。当成本的增长速度快于产出或效益的增长速度时，就会出现这种情况，从而导致资源分配的次优化。下面是一个结构化的解释：
• 关键概念：......
• 示例：......
• 政策含义：......
• 数学公式：......
总结：为什么重要：......
Translation:
Definition: Diminishing Marginal Efficiency (DME) refers to the concept where an additional unit of input yields a progressively smaller increase in output, thereby reducing the overall economic efficiency. This occurs when costs rise faster than outputs or benefits, leading to suboptimal resource allocation. Here's a structured explanation:
• Key Concepts: ....
• Examples: ....
• Policy Implications: ....
• Mathematical Formulation: ....
Conclusion: Why It Matters: ....</td></tr>
</table>

(a) Basic financial knowledge Q&A

<table>
<tr><th colspan="2">下一个问题：为什么说内卷是上述现象的极致体现？
Translation: Next question: Why is it said that involution is the extreme manifestation of the above phenomenon?</th></tr>
<tr><th>deepseek-llm-7b-chat</th><th>DSLLM-Fin-7B</th></tr>
<tr><td>定义："内卷"一词通常用来形容一种社会现象，即......
联系：将"内卷"视为上述现象的极致体现，是因为......
总结：因此，"内卷"可以看作是上述现象的一种极端表现形式......
Translation:
Definition: The term "involution" is often used to describe a social phenomenon, i.e. ......
Connection: Involvement is seen as an extreme manifestation of the above phenomenon because ......
Conclusion: Thus, "involution" can be seen as an extreme manifestation of the above phenomenon ......</td><td>定义："内卷"一词源于社会学研究中的一个概念，指的是一种非理性的竞争行为，即......
联系：将"内卷"与边际效益递减相联系的关键在于:......
总结：因此，"内卷"可以被视为边际效益递减现象的极端体现：......
Translation:
Definition: The term "involution" is derived from a concept in sociological research that refers to an irrational competitive behavior, i.e. ......
Connection: The key to linking "involution" to diminishing marginal returns is that ......
Conclusion: To summarize: Thus, "involution" can be seen as an extreme manifestation of the phenomenon of diminishing marginal returns: ......</td></tr>
</table>

(b) Further financial analysis based on the above question

Fig. 5. Comparison of financial Q&A before and after fine-tuning. In (a), the base LLM and DSLLM-Fin-7B are separately evaluated on basic financial knowledge Q&A tasks. In (b), the inquiry is extended with more context-specific and in-depth follow-up questions based on the previous prompt. The differences in responses between the two models are highlighted in red text, illustrating their key performance gaps. (Color figure online)

Human Interactive Testing. The following section evaluates the training effectiveness through interactive Q&A on domain-specific financial questions, with the inference results shown in Fig. 5. The comparison reveals that, in answering questions related to fundamental financial knowledge, such as "What is diminishing marginal efficiency?", the base model typically follows a "definitionexampleconclusion" response structure. In contrast, the fine-tuned DSLLM-Fin-7B adopts a more professional and detailed approach, offering answers that follow a "definitionkey conceptsexamplepolicy implementationmathematical formulaconclusion" structure. In further financial analysis based on the above question, such as "Why is it said that involution is the extreme manifestation of **the above phenomenon**?", the base model fails to integrate the above text (diminishing marginal efficiency) well and only answers the current question. DSLLM-Fin-7B can clearly point out the connection between the current question and diminishing marginal returns and give a more professional and specific answer. These human-interaction tests demonstrate that, compared to the original base model, the fine-tuned DSLLM-Fin-7B shows a clear improvement in the quality of responses in the financial domain.

Mainstream LLM Benchmarks-MMLU, CMMLU, CEval. Although the goal of this work is to fine-tune a more domain-specialized Q&A assistant in the financial field using LoRA, an effective fine-tuning process should retain the original model's general capabilities while enhancing its domain-specific knowledge. To assess whether the generalization ability of the original model has degraded after fine-tuning, this study adopts three commonly used benchmark datasets for evaluating LLM generalization: MMLU [6], CMMLU [11], and CEval [10]. MMLU is a multi-task language understanding benchmark that covers a wide range of domains, primarily in English. C-MMLU is its Chinese counterpart, designed specifically to evaluate Chinese language models. The CEval dataset is mainly used to assess a LLM's knowledge and reasoning abilitiesspecifically, whether the model can comprehend broad world knowledge and perform reasoning and planning in a human-like manner. The evaluation results of the three different LoRA adapters used during the experiments are compared against the official baseline results of the original model, *DeepSeek-LLM-7B-Chat*. The results are shown in Table 5. As shown in Table 5, after LoRA fine-tuning on the financial dataset FinQA-Alpaca, the performance of the LLM on the MMLU, CMMLU, and CEval benchmarks differs only slightly from that of the base model. This indicates that the fine-tuned financial LLM has not only acquired domain-specific financial knowledge but has also largely retained its original general knowledge and generalization capabilities.

5 Conclusion

This work selected *deepseek-llm-7b-chat* as the base model and collected a high-quality financial dataset, which is processed into the fine-tuning dataset FinQA-Alpaca for financial Q&A and reasoning. The base model is then fine-tuned using LoRA on the FinQA-Alpaca dataset, with adjustments to certain parameters (such as cutoff length and number of training epochs) to identify the best training outcome. The result is the large financial reasoning and question-answering model DSLLM-Fin-7B, which possesses more specialized financial domain knowledge and demonstrates a solid understanding of Chinese.

Acknowledgments. This work was supported by the Sichuan Province Science and Technology Program under Grants 2025ZDZX0125.

References

1. Bi, X., et al.: DeepSeek LLM: scaling open-source language models with longtermism. arXiv preprint arXiv:2401.02954 (2024)
2. Chen, Z., et al.: FinQA: a dataset of numerical reasoning over financial data. arXiv preprint arXiv:2109.00122 (2021)
3. Chen, Z., Li, S., Smiley, C., Ma, Z., Shah, S., Wang, W.Y.: ConvFinQA: exploring the chain of numerical reasoning in conversational finance question answering (2022). arXiv:2210.03849 arXiv preprint
4. Guo, D., et al.: DeepSeek-R1: incentivizing reasoning capability in LLMs via reinforcement learning. arXiv preprint arXiv:2501.12948 (2025)
5. Han, Z., Gao, C., Liu, J., Zhang, J., Zhang, S.Q.: Parameter-efficient fine-tuning for large models: a comprehensive survey. arXiv preprint arXiv:2403.14608 (2024)
6. Hendrycks, D., Burns, C., Basart, S., Zou, A., Mazeika, M., Song, D., Steinhardt, J.: Measuring massive multitask language understanding (2020). arXiv:2009.03300 arXiv preprint
7. Hu, E.J., Shen, Y., Wallis, P., Allen-Zhu, Z., Li, Y., Wang, S., Wang, L., Chen, W., et al.: LoRA: low-rank adaptation of large language models. ICLR **1**(2), 3 (2022)
8. Huang, A.H., Wang, H., Yang, Y.: FinBERT: a large language model for extracting information from financial text. Contemp. Account. Res. **40**(2), 806–841 (2023)
9. Huang, L., et al.: A survey on hallucination in large language models: principles, taxonomy, challenges, and open questions. ACM Trans. Info. Syst. **43**(2), 1–55 (2025)
10. Huang, Y., et al.: C-Eval: a multi-level multi-discipline Chinese evaluation suite for foundation models. Adv. Neural. Inf. Process. Syst. **36**, 62991–63010 (2023)
11. Li, H., et al.: CMMLU: measuring massive multitask language understanding in Chinese (2023). arXiv:2306.09212 arXiv preprint
12. Li, X., et al.: AlphaFin: benchmarking financial analysis with retrieval-augmented stock-chain framework. arXiv preprint arXiv:2403.12582 (2024)
13. Lin, C.Y.: Rouge: a package for automatic evaluation of summaries. In: Text Summarization Branches Out, pp. 74–81 (2004)
14. Liu, C., et al.: The Chinese dataset distilled from deepseek-R1-671B (2025)
15. Lu, D., et al.: BBT-Fin: comprehensive construction of Chinese financial domain pre-trained language model, corpus and benchmark (2023). arXiv:2302.09432 arXiv preprint
16. Papineni, K., Roukos, S., Ward, T., Zhu, W.J.: Bleu: a method for automatic evaluation of machine translation. In: Proceedings of the 40th Annual Meeting of the Association for Computational Linguistics, pp. 311–318 (2002)
17. Si, Q., Wang, T., Lin, Z., Zhang, X., Cao, Y., Wang, W.: An empirical study of instruction-tuning large language models in Chinese. arXiv preprint arXiv:2310.07328 (2023)
18. Wu, S., et al.: BloomBergGPT: a large language model for finance (2023). arXiv:2303.17564 arXiv preprint
19. Zheng, Y., et al.: LlamaFactory: unified efficient fine-tuning of 100+ language models. arXiv preprint arXiv:2403.13372 (2024)

Energy Correlation for Fairness-Aware Bias Detection in Neural Recommendation Systems

Gam Hong Thi Le[1], Khoi Nguyen-Tan[2], and Hiep Xuan Huynh[3,4](✉)

[1] University of Economics Ho Chi Minh City, Ho Chi Minh City, Vietnam
gamlth@ueh.edu.vn
[2] University of Danang – University of Science and Technology, Da Nang, Vietnam
ntkhoi@dut.udn.vn
[3] Can Tho University, Can Tho, Vietnam
[4] CTU Leading Research Team on Automation, Artificial Intelligence, Information Technology and Digital Transformation (CTU-AIMED), Can Tho, Vietnam
hxhiep@ctu.edu.vn

Abstract. Neural recommendation systems influence decisions across e-commerce, streaming, and social media platforms, but their complex architectures may encode nonlinear demographic biases that are difficult to detect using conventional fairness metrics. This paper proposes Energy Correlation (dCor), a dependence measure from statistical energy theory, for fairness-aware bias detection in neural recommendation systems. Unlike linear correlation measures, dCor satisfies $\mathrm{dCor}(X, Y) = 0$ if and only if X and Y are statistically independent, enabling the detection of both linear and nonlinear bias patterns. Experiments on MovieLens-1M with MF, NCF, and LightGCN, together with a credit-scoring fairness audit on German Credit, show that NCF exhibits statistically significant gender bias ($\mathrm{dCor} = 0.051$, $p < 0.05$), while age bias is stronger than gender bias across all models ($\mathrm{dCor} = 0.102$ for NCF, $p < 0.01$). LightGCN shows the lowest bias levels, suggesting that graph-based propagation may reduce sensitive-attribute dependence. Compared to Pearson correlation, dCor also demonstrates superior measurement stability (CV: 26.8% vs. 73.9%), supporting its use in regulatory compliance audits.

Keywords: Neural Networks · Recommendation Systems · Fairness · Bias Detection · Energy Correlation · Distance Correlation

1 Introduction

Neural recommendation systems have become critical infrastructure shaping how billions of users discover content across e-commerce, streaming, and social media platforms [1,2]. While these systems deliver substantial value, they also learn implicit patterns from historical data that may encode societal prejudices. Unlike traditional rule-based systems where bias sources can be explicitly identified,

Z. Zhang (Ed.): ISNN 2026, LNCS 16848, pp. 186–197, 2027.
https://doi.org/10.1007/978-981-92-5870-3_15

neural networks operate as black boxes in which discrimination patterns may emerge from complex interactions among millions of parameters.

Fairness concerns have surfaced across multiple domains. Online platforms may expose users to different opportunities based on demographic attributes [3]; content recommenders can create filter bubbles that reinforce demographic preferences; and credit systems may perpetuate historical disparities. These issues have drawn regulatory attention worldwide, including mandatory fairness audits under the European Union's AI Act for high-stakes applications in employment, credit, and housing.

Traditional fairness metrics such as Disparate Impact [4] and Equal Opportunity [5] provide useful group-level criteria, but they may not fully capture complex nonlinear dependence patterns in deep architectures such as NCF [6] and LightGCN [7]. Pearson correlation and related linear measures face a more direct limitation: they can miss nonlinear bias that emerges through the transformation layers of neural networks. Since unfairness cannot be mitigated if it cannot be reliably detected, bias detection methods must match the expressive power of the architectures they audit.

This paper proposes Energy Correlation, also known as distance correlation (dCor), an energy-statistics-based dependence measure, for fairness-aware bias detection in neural recommendation systems. Distance correlation satisfies a key property: $\mathrm{dCor}(X,Y) = 0$ if and only if $X \perp Y$, enabling the detection of linear, nonlinear, and high-dimensional dependence patterns without parametric assumptions [8,9].

The main contributions of this paper are fourfold. We introduce an energy-statistics-based framework for fairness-aware bias detection in neural recommendation systems, formulate bias detection as a statistical independence testing problem, and evaluate the approach on MovieLens-1M with MF, NCF, and LightGCN as well as on German Credit for credit-scoring fairness analysis. The results show that dCor detects statistically significant bias patterns, including stronger age bias than gender bias, and provides more stable measurements than Pearson correlation for regulatory compliance audits.

The remainder of this paper is organized as follows. Section 2 reviews related work; Sect. 3 presents theoretical foundations; Sect. 4 describes the proposed algorithm; Sect. 5 reports experimental results; and Sect. 6 concludes the paper.

2 Related Works

Fairness in recommendation systems has been studied from several perspectives, including group fairness, individual fairness, causal fairness, and multi-stakeholder fairness. Group fairness methods aim to ensure comparable outcomes across demographic groups defined by sensitive attributes such as gender, age, race, or socioeconomic status. Common criteria include Disparate Impact [4], Equal Opportunity [5], and Statistical Parity Difference. In recommendation settings, Yao and Huang [10] proposed user-oriented fairness metrics that measure value, absolute, underestimation, and overestimation unfairness across groups. Burke [11] further emphasized multi-stakeholder fairness, where users, items, and providers must be considered simultaneously.

Individual fairness requires that similar individuals receive similar predictions [12], while counterfactual fairness uses causal reasoning to assess whether predictions would remain unchanged under counterfactual changes to sensitive attributes [13]. Although these approaches provide important conceptual foundations, they often require task-specific similarity metrics or strong causal assumptions, which are difficult to verify in high-dimensional recommendation environments.

Deep recommendation models have achieved strong predictive performance but also introduce new fairness challenges due to their complex latent representations. Matrix Factorization (MF) [2] models user–item preferences through low-rank latent factors. Neural Collaborative Filtering (NCF) [6] replaces the inner product with neural networks to capture nonlinear user–item interactions. LightGCN [7] propagates embeddings over the user–item bipartite graph, capturing higher-order collaborative signals through neighborhood aggregation. These architectures improve recommendation quality, but their nonlinear and high-dimensional representations may also encode demographic dependence in ways that are difficult to detect using conventional fairness metrics.

Existing fairness metrics remain limited for auditing such systems. Many metrics focus on group-level outcome disparities or scalar decision outputs, whereas recommendation systems produce high-dimensional score vectors. Linear measures such as Pearson correlation can miss nonlinear dependencies, and traditional fairness criteria often lack a direct statistical test for arbitrary dependence between recommendations and sensitive attributes. These limitations motivate the use of energy-statistics-based measures, particularly Energy Correlation, which can detect general dependence patterns and provides a theoretical characterization of statistical independence.

3 Theoretical Modeling

This work defines fairness through statistical independence. A recommendation system is considered unbiased if its outputs are statistically independent of sensitive attributes. Let $\mathcal{U} = \{u_1, \ldots, u_m\}$ be the set of users and $\mathcal{I} = \{i_1, \ldots, i_n\}$ be the set of items. A recommender produces a score matrix $R \in \mathbb{R}^{m\times n}$, where $r_{u,i}$ denotes the predicted preference of user u for item i. The recommendation vector for user u is $R_u = (r_{u,1}, \ldots, r_{u,n})^\top$, and each user has a sensitive attribute $s_u \in \mathcal{S}$, such as gender or age group. Fairness is therefore formulated as

$$P(R_u \mid s_u) = P(R_u) \quad \Longleftrightarrow \quad R \perp S. \tag{1}$$

Bias detection becomes a statistical independence testing problem, where the null hypothesis is $H_0 : R \perp S$ and the alternative is $H_1 : R \not\perp S$.

Energy Distance provides a statistical distance between probability distributions. For random vectors $X \sim P$ and $Y \sim Q$ in $\mathbb{R}^d$, it is defined as

$$\mathcal{E}(P, Q) = 2\mathbb{E}\|X - Y\| - \mathbb{E}\|X - X'\| - \mathbb{E}\|Y - Y'\|, \tag{2}$$

where $X, X' \sim P$ and $Y, Y' \sim Q$ are independent copies. This distance is non-negative and equals zero if and only if $P = Q$ [8,9].

Based on this idea, distance covariance measures dependence between random vectors. For $X \in \mathbb{R}^p$ and $Y \in \mathbb{R}^q$, distance covariance is defined as

$$\begin{aligned} \mathrm{dCov}^2(X, Y) = &\mathbb{E}[\|X - X'\|\|Y - Y'\|] + \mathbb{E}\|X - X'\|\mathbb{E}\|Y - Y'\| \\ &- 2\mathbb{E}[\|X - X'\|\|Y - Y''\|], \end{aligned} \tag{3}$$

where (X, Y), (X', Y'), and (X'', Y'') are independent copies from the joint distribution. The corresponding normalized measure, Energy Correlation or distance correlation, is

$$\mathrm{dCor}(X, Y) = \frac{\mathrm{dCov}(X, Y)}{\sqrt{\mathrm{dVar}(X)\mathrm{dVar}(Y)}}, \tag{4}$$

where $\mathrm{dVar}(X) = \mathrm{dCov}(X, X)$. A key property is that $\mathrm{dCor}(X, Y) = 0$ if and only if X and Y are statistically independent. This distinguishes dCor from Pearson correlation, which only captures linear dependence. As illustrated in Fig. 1, dCor can detect nonlinear quadratic and circular relationships even when Pearson correlation is close to zero.

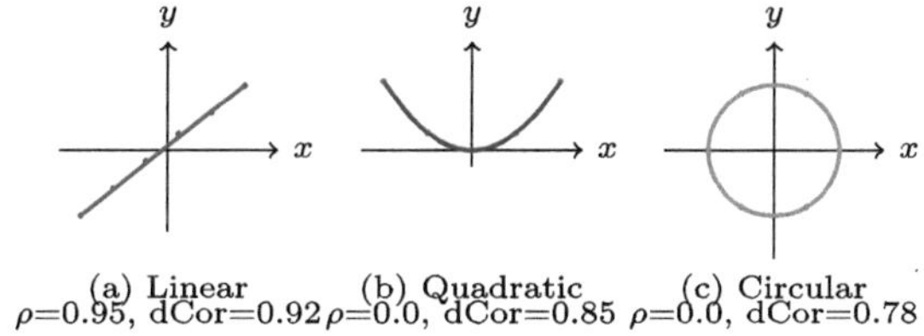

Fig. 1. Comparison of Pearson correlation (ρ) and Energy Correlation (dCor). Both metrics detect linear dependence, whereas dCor also detects nonlinear quadratic and circular dependence.

For paired observations $\{(x_i, y_i)\}_{i=1}^n$, sample distance matrices are computed as $a_{ij} = \|x_i - x_j\|$ and $b_{ij} = \|y_i - y_j\|$. After double centering,

$$\tilde{A}_{ij} = a_{ij} - \bar{a}_{i\cdot} - \bar{a}_{\cdot j} + \bar{a}_{\cdot\cdot}, \tag{5}$$

and analogously for $\tilde{B}$, the sample distance covariance is

$$\widehat{\mathrm{dCov}}_n^2 = \frac{1}{n^2} \sum_{i,j} \tilde{A}_{ij} \tilde{B}_{ij}. \tag{6}$$

The sample distance correlation is consistent, satisfying $\widehat{\mathrm{dCor}}_n \xrightarrow{a.s.} \mathrm{dCor}(X, Y)$ as $n \to \infty$.

To assess statistical significance, a permutation test is used under $H_0 : X \perp Y$. The observed statistic $T_{\mathrm{obs}} = n\widehat{\mathrm{dCov}}_n^2$ is first computed. The indices of Y are then randomly permuted N_p times, producing statistics $T^{(r)}$ for $r = 1, \ldots, N_p$. The p-value is calculated as

$$p = \frac{1}{N_p} \sum_{r=1}^{N_p} \mathbf{1}[T^{(r)} \geq T_{\text{obs}}]. \tag{7}$$

This nonparametric procedure controls Type I error under the null hypothesis and provides a practical significance test for fairness-aware bias detection.

4 Functional Modeling

4.1 System Architecture Overview

Four main processing stages are included in the bias detection framework. In Stage 1 (Data Preparation), the rating matrix $\mathcal{D}$ and demographic attributes S are loaded, and preprocessing including missing value handling, normalization, and stratified sampling is applied if needed for computational efficiency. In Stage 2 (Model Training), the recommendation model (MF, NCF, or LightGCN) is trained on the rating data, and user-level prediction vectors $\hat{R}_u$ representing the model's recommendations for each user are extracted. In Stage 3 (Distance Computation), pairwise distance matrices for both recommendations (A) and demographics (B) are computed, and double centering is applied to obtain $\tilde{A}$ and $\tilde{B}$. In Stage 4 (Statistical Analysis), Energy Correlation $\beta = \text{dCor}(\hat{R}, S)$ is computed, and the permutation test is performed to obtain the p-value for statistical significance (Fig. 2).

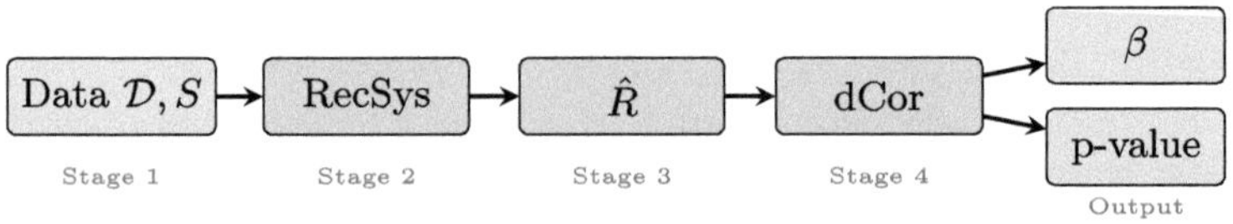

Fig. 2. System architecture for Energy Correlation-based bias detection. Four stages: data preparation, model training, prediction extraction, and dCor computation with permutation testing.

4.2 Main Algorithm

4.3 Key Functions and Complexity Analysis

The double centering transformation is applied by DoubleCenter(M) to remove marginal effects, computed as $\tilde{M}_{ij} = M_{ij} - \frac{1}{n}\sum_k M_{ik} - \frac{1}{n}\sum_k M_{kj} + \frac{1}{n^2}\sum_{k,l} M_{kl}$. Regarding complexity analysis, distance matrix computation requires $O(n^2 \cdot d)$ where n is the number of users and d is the embedding dimension, double centering and dCov computation each require $O(n^2)$, and the permutation test with N_p

Algorithm 1. Energy Correlation Bias Detection (EnergyBias)

Require: Rating data $\mathcal{D}$, Demographics S, Model type $\mathcal{M}$, Permutations N_p
Ensure: Bias score $\beta \in [0,1]$, p-value $p \in [0,1]$
// **Stage 1: Data Preparation**
$n \leftarrow |\mathcal{U}|$; $\mathcal{D}$ is preprocessed
// **Stage 2: Model Training**
if $\mathcal{M} = \text{MF}$ **then**
  $U, V \leftarrow \text{ALS}(\mathcal{D}, k, \lambda)$; $\hat{R} \leftarrow UV^\top$
else if $\mathcal{M} = \text{NCF}$ **then**
  $\theta \leftarrow \text{TrainNCF}(\mathcal{D})$; $\hat{R} \leftarrow \text{Predict}(\theta)$
else if $\mathcal{M} = \text{LightGCN}$ **then**
  $E \leftarrow \text{TrainGCN}(\mathcal{D})$; $\hat{R} \leftarrow E_U E_I^\top$
end if
// **Stage 3: Distance Computation**
for $i, j = 1$ to n **do**
  $A_{ij} \leftarrow \|\hat{R}_i - \hat{R}_j\|_2$; $B_{ij} \leftarrow |S_i - S_j|$
end for
$\tilde{A} \leftarrow \text{DoubleCenter}(A)$; $\tilde{B} \leftarrow \text{DoubleCenter}(B)$
// **Stage 4: Energy Correlation**
$\text{dCov}^2 \leftarrow \frac{1}{n^2} \sum_{i,j} \tilde{A}_{ij} \tilde{B}_{ij}$
$\beta \leftarrow \sqrt{|\text{dCov}^2| / \sqrt{\text{dVar}_A \cdot \text{dVar}_B}}$
// **Stage 5: Permutation Test**
$T_{\text{obs}} \leftarrow n \cdot \text{dCov}^2$; $count \leftarrow 0$
for $r = 1$ to N_p **do**
  $S^{(r)} \leftarrow \text{Permute}(S)$; Compute $T^{(r)}$
  if $T^{(r)} \geq T_{\text{obs}}$ **then**
    $count \leftarrow count + 1$
  end if
end for
$p \leftarrow count / N_p$
return β, p

permutations requires $O(N_p \cdot n^2)$. The total time complexity is $O(n^2 \cdot (d + N_p))$, and space complexity of $O(n^2)$ is required for storing distance matrices. For large-scale systems with millions of users, sampling-based approximations can reduce complexity to $O(s^2 \cdot N_p)$ where $s \ll n$ is the sample size.

5 Experiments

5.1 Datasets, Setup, and Evaluation Metrics

Experiments were conducted on MovieLens-1M [14] and German Credit [15]. MovieLens-1M contains 1,000,209 ratings from 6,040 users on 3,706 movies, with gender, age, and occupation information, and was used for recommendation fairness analysis. German Credit contains 1,000 credit applications with gender and age attributes, and was used for credit-scoring fairness analysis.

For MovieLens-1M, a stratified sample of 2,000 users was drawn to preserve the original gender distribution. Items with fewer than five ratings were removed, yielding 3,146 active items and 337,003 ratings. Five random seeds were used for variance estimation, and all dCor tests used 500 permutations with $\alpha = 0.05$.

We evaluated MF, NCF, and LightGCN. MF used 30 latent factors, regularization $\lambda = 0.1$, and 15 iterations; NCF used layers [64, 32, 30] with ReLU activation and learning rate 0.001; and LightGCN used three propagation layers with embedding dimension 30. dCor was computed using the `energy` package in R.

The primary bias metric is Energy Correlation (dCor), with Pearson correlation (ρ) used as a linear baseline. DI and SPD are reported as group fairness metrics, RMSE measures recommendation accuracy, and CV measures metric stability.

5.2 Scenario 1: E-Commerce Product Recommendation

E-commerce platforms such as Amazon, Shopee, and Lazada deploy recommendation systems that influence purchasing decisions for millions of users daily. A critical fairness concern in this context is whether product recommendations systematically differ based on user gender, potentially reinforcing stereotypes (e.g., showing technology products predominantly to male users while directing female users toward fashion and cosmetics). This scenario evaluates the ability of dCor to detect such gender-based recommendation disparities across three recommendation architectures. The results are presented in Table 1.

Table 1. Scenario 1: E-commerce gender fairness audit on MovieLens-1M (2,000 users). Significance: * $p < 0.05$.

Model	RMSE	dCor	Pearson	DI	SPD
MF	0.748	0.034	0.031	0.948	0.027
NCF	1.254	0.051*	0.051*	0.887	0.062
LightGCN	3.769	0.023	0.016	0.966	0.017

For e-commerce deployment, statistically significant gender bias was detected only in NCF (dCor = 0.051, $p = 0.048 < 0.05$), indicating that product recommendations generated by this neural architecture exhibit measurable dependence on user gender. This finding suggests that NCF-based e-commerce systems may require fairness interventions before deployment. Weak, non-significant bias was shown by MF (dCor = 0.034, $p = 0.25$), making it a safer choice for gender-sensitive applications. The lowest bias was exhibited by LightGCN (dCor = 0.023, $p = 0.62$), suggesting that graph-based collaborative filtering may naturally mitigate gender-specific recommendation patterns through neighborhood aggregation.

5.3 Scenario 2: Age Fairness in Recommendation

This scenario examines whether recommendation outputs depend on user age groups. Table 2 reports the results for MF, NCF, and LightGCN on MovieLens-1M.

Table 2. Scenario 2: Age fairness audit on MovieLens-1M. Significance: ** $p < 0.01$, *** $p < 0.001$.

Model	dCor	p-value	Pearson	p-value
MF	0.078**	0.002	0.079***	0.0004
NCF	**0.102****	0.002	0.071**	0.001
LightGCN	0.017	1.000	0.005	0.830

The results show that age dependence is stronger than gender dependence across the evaluated models. In NCF, dCor increases from 0.051 for gender to 0.102 for age, indicating more pronounced age-related dependence in the learned recommendation outputs. MF also shows statistically significant age dependence (dCor = 0.078, $p = 0.002$), whereas LightGCN exhibits minimal and non-significant dependence (dCor = 0.017, $p = 1.0$). Figure 3 summarizes the gender and age bias levels across models. The consistently low dCor values of LightGCN suggest that graph-based propagation may smooth sensitive-attribute signals, while nonlinear transformations in NCF may amplify demographic dependence.

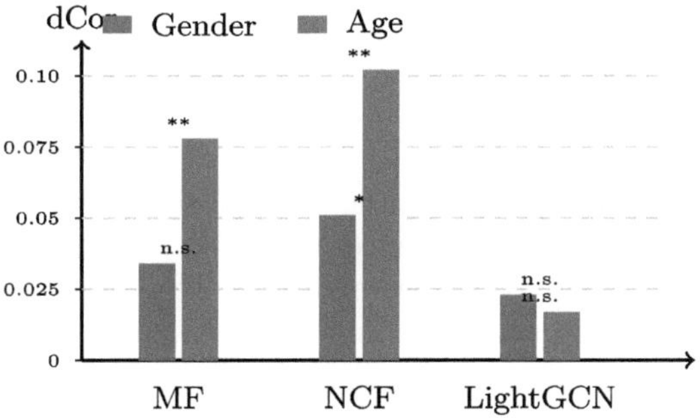

Fig. 3. Bias levels (dCor) for gender and age across recommendation models. NCF exhibits highest bias; LightGCN shows minimal bias. Significance: * $p < 0.05$, ** $p < 0.01$, n.s. = not significant.

5.4 Scenario 3: Credit-Scoring Fairness and Metric Stability

This scenario evaluates the applicability of Energy Correlation in a high-stakes credit-scoring setting and examines its stability as a fairness auditing metric.

Table 3. Scenario 3a: Credit-scoring fairness on German Credit.

Attribute	dCor	p-value	Pearson	DI
Gender	0.089**	0.004	0.075*	0.896
Age	0.112***	<0.001	0.094**	0.847

Table 3 shows statistically significant dependence for both gender and age. Age exhibits stronger dependence than gender (dCor = 0.112 vs. 0.089), consistent with the findings on MovieLens-1M. Notably, the observed gender Disparate Impact of 0.896 satisfies the four-fifths rule, yet dCor still detects statistically significant dependence, indicating that dependence-based auditing can reveal patterns not fully captured by group-level fairness criteria.

Table 4. Scenario 3b: Metric stability across 5 random seeds (MovieLens-1M, MF model).

Seed	Users	dCor	p-value	Pearson	DI
42	1,500	0.029	0.567	0.026	0.949
123	1,500	0.033	0.418	0.005	0.962
456	1,500	0.022	0.846	0.008	0.993
789	1,500	0.045	0.174	0.045	0.955
1024	1,500	0.040	0.229	0.032	0.981
Mean	—	0.034	—	0.023	0.968
SD	—	0.009	—	0.017	0.018
CV	—	26.8%	—	73.9%	1.9%

Table 4 shows that dCor is substantially more stable than Pearson correlation across random samples. The coefficient of variation is 26.8% for dCor, compared with 73.9% for Pearson. Figure 4 illustrates this difference, suggesting that dCor provides more consistent measurements for fairness auditing.

5.5 Discussion

The experimental results provide three main insights. First, NCF shows stronger demographic dependence than MF and LightGCN, with statistically significant gender bias in Scenario 1 and the highest age dependence in Scenario 2. This suggests that nonlinear neural transformations may amplify demographic patterns present in historical interaction data. Second, age dependence is consistently stronger than gender dependence across both recommendation and credit-scoring settings, indicating that fairness audits should consider multiple sensitive attributes rather than focusing only on gender. Third, LightGCN exhibits the

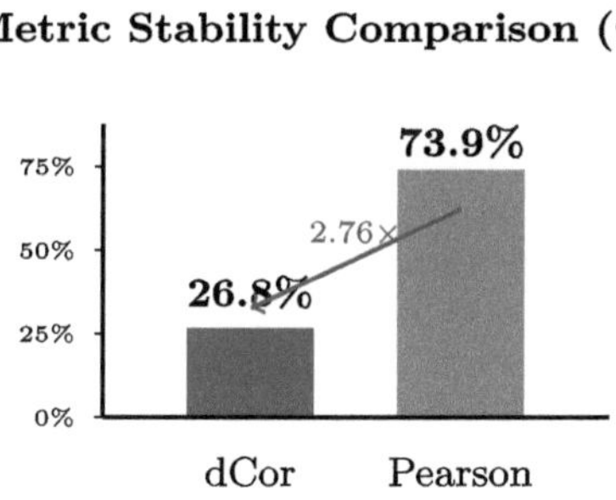

Fig. 4. Stability comparison: dCor exhibits 2.76× lower coefficient of variation than Pearson correlation across five random seeds.

lowest bias levels for both gender and age in MovieLens-1M. A possible explanation is that graph-based neighborhood aggregation smooths user representations and may reduce sensitive-attribute-specific signals during propagation. However, this effect should be interpreted cautiously, since graph propagation may also amplify bias in datasets with strong demographic homophily.

The German Credit results further show that dCor can detect statistically significant dependence even when Disparate Impact satisfies the four-fifths rule. In addition, dCor is more stable than Pearson correlation across random samples, with a lower coefficient of variation (26.8% vs. 73.9%). These findings support the use of Energy Correlation as a complementary fairness auditing metric for both recommendation and high-stakes decision-making systems.

This study has several limitations. The MovieLens-1M experiments rely on stratified subsampling for computational tractability, which may miss patterns visible only at full scale. The German Credit dataset is small and may not reflect the complexity of modern credit-scoring systems. Future work should validate the proposed approach on larger and more diverse datasets, and investigate whether LightGCN's lower bias is architecture-driven or dataset-dependent.

6 Conclusion

This paper introduced Energy Correlation (dCor) as a principled framework for fairness-aware bias detection in recommendation and decision-making systems. By relying on the independence characterization $\mathrm{dCor}(X, Y) = 0$ if and only if $X \perp Y$, the proposed approach can detect linear, nonlinear, and high-dimensional dependence between system outputs and sensitive attributes. This makes dCor a suitable complement to conventional group fairness metrics and linear correlation measures.

Experiments on MovieLens-1M and German Credit demonstrated the practical value of the proposed framework. NCF exhibited statistically significant gender bias, age dependence was stronger than gender dependence across scenarios, and LightGCN consistently showed lower bias levels, suggesting that graph-based propagation may reduce sensitive-attribute dependence. In the credit-scoring setting, dCor detected significant dependence even when Disparate Impact satisfied the four-fifths rule. Moreover, dCor showed greater measurement stability than Pearson correlation across random samples, supporting its use as a reproducible fairness auditing metric.

Future work will extend this framework to larger and more diverse datasets, develop differentiable dCor-based regularization for end-to-end fair learning, and investigate scalable approximations for large-scale recommendation systems.

Acknowledgments. This research is partly funded by University of Economics Ho Chi Minh City, Vietnam.

Disclosure of Interests. The authors have no competing interests to declare that are relevant to the content of this article.

References

1. Smith, B., Linden, G.: Two decades of recommender systems at Amazon.com. IEEE Int. Comput. **21**(3), 12–18 (2017)
2. Koren, Y., Bell, R., Volinsky, C.: Matrix factorization techniques for recommender systems. Computer **42**(8), 30–37 (2009)
3. Datta, A., Tschantz, M.C., Datta, A.: Automated experiments on ad privacy settings: a tale of opacity, choice, and discrimination. Proc. Priv. Enhan. Technol. **2015**(1), 92–112 (2015)
4. Zafar, M.B., Valera, I., Gomez Rodriguez, M., Gummadi, K.P.: Fairness beyond disparate treatment & disparate impact: Learning classification without disparate mistreatment. In: Proceedings of the 26th International Conference on World Wide Web, pp. 1171–1180. ACM, Perth (2017)
5. Hardt, M., Price, E., Srebro, N.: Equality of opportunity in supervised learning. In: Advances in Neural Information Processing Systems, vol. 30, pp. 3315–3323. Curran Associates, Barcelona (2016)
6. He, X., Liao, L., Zhang, H., Nie, L., Hu, X., Chua, T.S.: Neural collaborative filtering. In: Proceedings of the 26th International Conference on World Wide Web, pp. 173–182. ACM, Perth (2017)
7. He, X., Deng, K., Wang, X., Li, Y., Zhang, Y., Wang, M.: LightGCN: simplifying and powering graph convolution network for recommendation. In: Proceedings of the 43rd International ACM SIGIR Conference on Research and Development in Information Retrieval, pp. 639–648. ACM, Virtual Event (2020)
8. Székely, G.J., Rizzo, M.L., Bakirov, N.K.: Measuring and testing dependence by correlation of distances. Ann. Stat. **35**(6), 2769–2794 (2007)
9. Rizzo, M.L., Székely, G.J.: Energy distance. Wiley Interdiscip. Rev.: Comput. Stat. **8**(1), 27–38 (2016)
10. Yao, S., Huang, B.: Beyond parity: fairness objectives for collaborative filtering. In: Advances in Neural Information Processing Systems, vol. 30, pp. 2921–2930. Curran Associates, Long Beach (2017)
11. Burke, R.: Multisided fairness for recommendation. In: Proceedings of the Workshop on Fairness, Accountability and Transparency in Machine Learning, Halifax (2017)

12. Dwork, C., Hardt, M., Pitassi, T., Reingold, O., Zemel, R.: Fairness through awareness. In: Proceedings of the 3rd Innovations in Theoretical Computer Science Conference, pp. 214–226. ACM, Cambridge (2012)
13. Kusner, M.J., Loftus, J., Russell, C., Silva, R.: Counterfactual fairness. In: Advances in Neural Information Processing Systems, vol. 30, pp. 4066–4076. Curran Associates, Long Beach (2017)
14. Harper, F.M., Konstan, J.A.: The MovieLens datasets: history and context. ACM Trans. Int. Intell. Syst. **5**(4), 19:1–19:19 (2015)
15. Hofmann, H.: Statlog (German Credit Data). UCI Machine Learning Repository. https://archive.ics.uci.edu/dataset/144/statlog+german+credit+data (1994)

MC-JEPA: A Neural Network Framework with Multi-criteria Aggregation for Recommender Systems

Tri Minh Huynh[1], Van Hung Nang Nguyen[2], Hanh My Thi Le[2], and Hiep Xuan Huynh[3,4](✉)

[1] Kien Giang University, Rach Gia, Vietnam
hmtri@vnkgu.edu.vn
[2] The University of Danang - University of Science and Technology, Da Nang, Vietnam
{nnhvan,ltmhanh}@dut.udn.vn
[3] Can Tho University, Can Tho, Vietnam
[4] CTU Leading Research Team on Automation, Artificial Intelligence, Information Technology and Digital Transformation (CTU-AIMED), Can Tho, Vietnam
hxhiep@ctu.edu.vn

Abstract. Multi-Criteria Recommender Systems (MCRS) leverage multiple rating dimensions to provide more nuanced recommendations than single-rating approaches. This paper proposes MC-JEPA (Multi-Criteria Joint-Embedding Predictive Architecture), a neural network framework integrating self-supervised representation learning with interpretable multi-criteria aggregation operators. The architecture employs context and target encoders with exponential moving average updates, combined with Ordered Weighted Averaging (OWA) operators, Choquet integrals, and Shapley values. Four variants are presented: MC-JEPA-Base, MC-JEPA-OWA, MC-JEPA-Choquet, and MC-JEPA-Hybrid. Evaluation on three real-world datasets demonstrates consistent improvements over both classical methods (Matrix Factorization, SVD++) and modern deep learning baselines (NCF, VAE-CF): Yahoo Movies (RMSE=0.8123, NDCG@10=0.3056), TripAdvisor Hotels (RMSE=0.6123, AUC=0.8534), and BeerAdvocate (RMSE=0.3678, AUC=0.8856). The learned Shapley values reveal meaningful criteria importance: Location (21.3%) dominates hotel satisfaction while Taste (44.1%) dominates beer reviews.

Keywords: Neural Networks · Self-Supervised Learning · JEPA · Multi-Criteria Recommender Systems · Choquet Integral · OWA Operators · Shapley Value

1 Introduction

Deep neural networks have revolutionized recommender systems through learned representations [12]. However, traditional collaborative filtering relies on single

Z. Zhang (Ed.): ISNN 2026, LNCS 16848, pp. 198–209, 2027.
https://doi.org/10.1007/978-981-92-5870-3_16

overall ratings, failing to capture the multidimensional nature of user preferences [1]. Multi-Criteria Recommender Systems (MCRS) address this by considering ratings across multiple criteria, providing more nuanced recommendations.

MCRS have significant applications across diverse domains. In entertainment, movie platforms (Yahoo Movies, IMDb) allow ratings on story, acting, and visuals. In hospitality, platforms (TripAdvisor, Booking.com) collect multi-criteria reviews on cleanliness, location, and service. In e-commerce, platforms leverage ratings on quality, delivery, and price. Despite their potential, MCRS face challenges: exacerbated data sparsity (requiring ratings across multiple criteria), severe cold-start problems, and lack of interpretability in neural approaches [17].

The Joint-Embedding Predictive Architecture (JEPA) [12] offers promising directions by learning representations through predicting target embeddings from context embeddings in latent space, avoiding the reconstruction bottleneck while capturing semantic relationships. This paper proposes MC-JEPA, integrating JEPA's self-supervised learning with multi-criteria decision making methods derived from fuzzy set theory and cooperative game theory.

The contributions are fourfold: (1) a novel architecture combining context/target encoders with multi-criteria aggregation; (2) integration of OWA operators [16], Choquet integrals [6], and Shapley values [14]; (3) four model variants with different accuracy-interpretability trade-offs; (4) experimental evaluation demonstrating improvements over both classical (MF, SVD++) and deep learning baselines (NCF, VAE-CF) on both rating prediction (RMSE, MAE) and ranking metrics (HR@10, NDCG@10).

2 Related Work

Multi-criteria Recommender Systems. Adomavicius and Kwon [2] established MCRS foundations with three paradigms: aggregation-based approaches combining criteria ratings using functions ranging from simple weighted averaging to fuzzy logic operators; similarity-based approaches computing user/item similarities based on multi-criteria patterns; and model-based approaches learning joint predictive models including tensor decomposition. Lakiotaki et al. [11] applied multi-criteria decision analysis with UTA-Rec to infer user utility functions. Jannach et al. [9] found criteria contributions vary significantly across domains, highlighting personalized aggregation importance. Zheng [17] proposed criteria chains modeling sequential dependencies between criteria ratings.

Neural Recommender Systems. Deep learning approaches have achieved remarkable success in recommendation. Neural Collaborative Filtering (NCF) [8] combines matrix factorization with neural networks, using MLPs to learn non-linear user-item interactions. Variational Autoencoders (VAE-CF) [13] learn latent representations with principled uncertainty estimates through reconstruction of user-item interaction vectors. Graph Neural Networks model user-item interactions as bipartite graphs, leveraging structural information through message passing. Attention mechanisms enable focusing on relevant user history.

JEPA and Aggregation Operators. LeCun [12] proposed the Joint-Embedding Predictive Architecture (JEPA) with asymmetric context/target encoders and EMA updates, predicting in latent space rather than input space to avoid reconstruction bottleneck. I-JEPA [3] demonstrated strong image understanding, while V-JEPA [5] extended to video. VICReg [4] introduced variance-invariance-covariance regularization for stable training.

The OWA operator [16] provides parameterized aggregation with weights derived from linguistic quantifiers $Q(r) = r^\alpha$, where $\alpha \to 0$ represents "at least one", $\alpha = 1$ represents "average", and $\alpha \to \infty$ represents "all". The Choquet integral [6] incorporates criteria interactions through 2-additive fuzzy measures:

$$C_\mu(\mathbf{a}) = \sum_{k=1}^{K} \phi_k \cdot a_k + \sum_{k<l} I_{kl} \cdot \min(a_k, a_l) \tag{1}$$

where Shapley values ϕ_k [14] measure criterion importance satisfying efficiency, symmetry, null criterion, and additivity axioms, and interaction indices I_{kl} capture synergies between criteria.

3 Theoretical Modeling

3.1 Problem Formulation

Consider a set of M users $\mathcal{U} = \{u_1, \ldots, u_M\}$ and a set of N items $\mathcal{I} = \{i_1, \ldots, i_N\}$. In Multi-Criteria Recommender Systems, each user-item interaction involves ratings across K criteria, represented as $\mathbf{r}_{ui} = (r_{ui}^{(1)}, r_{ui}^{(2)}, \ldots, r_{ui}^{(K)})$, where $r_{ui}^{(k)} \in [1, 5]$ represents user u's rating for item i on criterion k. The primary goal is to predict the overall rating $\hat{r}_{ui}$ that captures user u's overall preference for item i, given partial or complete criteria ratings. Additionally, the proposed approach aims to provide interpretable criteria importance weights that explain the prediction.

The problem is formulated as a joint learning objective:

$$\min_{\theta,\phi,\psi} \mathcal{L}_{\text{pred}}(\theta, \phi) + \lambda_{\text{agg}} \mathcal{L}_{\text{agg}}(\psi) + \lambda_{\text{reg}} \mathcal{L}_{\text{reg}}(\theta, \phi) \tag{2}$$

where θ represents context encoder parameters, ϕ represents predictor parameters, $\psi = \{\alpha, \phi_k, I_{kl}, \gamma_1, \gamma_2, \gamma_3\}$ represents aggregation parameters (OWA quantifier α, Shapley values ϕ_k, interaction indices I_{kl}, hybrid weights γ), and λ_{agg} and λ_{reg} are regularization coefficients.

3.2 MC-JEPA Architecture

The MC-JEPA framework (Fig. 1) consists of five components: (1) **User/Item Embeddings** $\mathbf{E}^U \in \mathbb{R}^{M \times d}$, $\mathbf{E}^I \in \mathbb{R}^{N \times d}$; (2) **Criteria Encoder** transforming each rating $\mathbf{c}_k = \text{MLP}_k(r^{(k)})$ with learnable mask embeddings for missing ratings; (3) **Context Encoder** f_θ processing concatenated input $\mathbf{x}_c = [\mathbf{u}; \mathbf{i}; \mathbf{c}_1; \ldots; \mathbf{c}_K]$ via transformer layers; (4) **Target Encoder** $f_{\bar{\theta}}$ with EMA updates $\bar{\theta} \leftarrow \tau\bar{\theta} + (1-\tau)\theta$ where $\tau \in [0.99, 0.999]$; (5) **Predictor** g_ϕ mapping context to predicted target $\hat{\mathbf{z}}_t = g_\phi(\mathbf{z}_c)$.

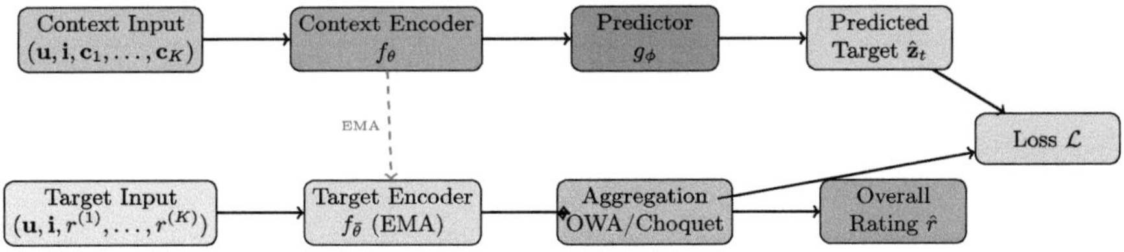

Fig. 1. MC-JEPA architecture with EMA-updated target encoder and multi-criteria aggregation.

3.3 Multi-Criteria Aggregation Module

Four aggregation variants are proposed, providing different trade-offs between prediction accuracy and interpretability:

MC-JEPA-Base: Simple averaging provides a strong reference point:

$$\hat{r}_{\text{Base}} = \frac{1}{K} \sum_{k=1}^{K} r^{(k)} \tag{3}$$

MC-JEPA-OWA: Ordered Weighted Averaging with learned quantifier parameter α:

$$\hat{r}_{\text{OWA}} = \sum_{j=1}^{K} w_j \cdot r^{(\sigma(j))}, \quad w_j = \left(\frac{j}{K}\right)^{\alpha} - \left(\frac{j-1}{K}\right)^{\alpha} \tag{4}$$

where σ orders criteria descending. Different α values correspond to different aggregation strategies: $\alpha < 1$ produces optimistic aggregation (emphasizing highest scores), $\alpha = 1$ produces balanced averaging, and $\alpha > 1$ produces pessimistic aggregation (emphasizing lowest scores).

MC-JEPA-Choquet: 2-additive Choquet integral with Shapley values ϕ_k and interaction indices I_{kl}:

$$\hat{r}_{\text{Choquet}} = \sum_{k=1}^{K} \phi_k \cdot r^{(k)} + \sum_{k<l} I_{kl} \cdot \min(r^{(k)}, r^{(l)}) \tag{5}$$

Positive I_{kl} indicates complementary criteria; negative I_{kl} indicates redundant criteria.

MC-JEPA-Hybrid: Ensemble combining all methods with learned weights:

$$\hat{r}_{\text{Hybrid}} = \gamma_1 \hat{r}_{\text{Base}} + \gamma_2 \hat{r}_{\text{OWA}} + \gamma_3 \hat{r}_{\text{Choquet}}, \quad \gamma_1 + \gamma_2 + \gamma_3 = 1 \tag{6}$$

Algorithm 1. MC-JEPA Training

```
Require: Training data 𝒟, learning rate η, EMA momentum τ
Initialize encoders f_θ, f_θ̄, predictor g_ϕ
for epoch = 1 to E do
    for batch ℬ ⊂ 𝒟 do
        Encode context: z_c = f_θ(x_c)
        Encode target: z_t = f_θ̄(x_t)
        Predict: ẑ_t = g_ϕ(z_c)
        Aggregate: r̂ = Agg(r)
        Compute ℒ, update θ, ϕ via gradient descent
        EMA: θ̄ ← τθ̄ + (1 − τ)θ
    end for
end for
```

3.4 Training Objective

The training loss follows VICReg [4] paradigm to prevent representation collapse:

$$\mathcal{L} = \mathcal{L}_{\text{JEPA}} + \lambda_{\text{rating}}\mathcal{L}_{\text{rating}} + \lambda_{\text{var}}\mathcal{L}_{\text{var}} + \lambda_{\text{cov}}\mathcal{L}_{\text{cov}} \tag{7}$$

The JEPA prediction loss encourages accurate target embedding prediction:

$$\mathcal{L}_{\text{JEPA}} = \|\hat{\mathbf{z}}_t - \text{sg}(\mathbf{z}_t)\|_2^2 \tag{8}$$

where $\text{sg}(\cdot)$ is the stop-gradient operation. The rating prediction loss is:

$$\mathcal{L}_{\text{rating}} = (\hat{r} - r)^2 \tag{9}$$

Variance regularization prevents collapse: $\mathcal{L}_{\text{var}} = \frac{1}{d}\sum_{j=1}^{d} \max(0, 1 - \sqrt{\text{Var}(z_j) + \epsilon})$. Covariance regularization decorrelates dimensions: $\mathcal{L}_{\text{cov}} = \frac{1}{d^2}\sum_{i \neq j}[\text{Cov}(z_i, z_j)]^2$. Coefficients are set as $\lambda_{\text{rating}} = 1.0$, $\lambda_{\text{var}} = 25.0$, $\lambda_{\text{cov}} = 1.0$.

4 Experiments

4.1 Datasets

Three real-world multi-criteria datasets are used (Table 1):

Yahoo Movies (YM): Standard MCRS benchmark from Yahoo Webscope with 4 criteria (Acting, Story, Direction, Visuals) on letter grade scale A+ to F (converted to 1–13). Contains 7,642 users, 11,915 movies, and 211,231 ratings with 99.77% sparsity.

TripAdvisor Hotels (TA): Hotel reviews with 7 criteria (Value, Rooms, Location, Cleanliness, Check-in, Service, Sleep Quality) on 1–5 star scale. Approximately 15,000 users, 3,000 hotels, and 20,000 reviews.

BeerAdvocate (BA): Beer reviews from Stanford SNAP with 4 criteria (Appearance, Aroma, Palate, Taste) on 0–5 scale. Subset of 500,000 reviews from ~33,000 users and ~66,000 items.

Table 1. Dataset characteristics.

Dataset	Users	Items	Ratings	K	Sparsity
Yahoo Movies	7,642	11,915	211,231	4	99.77%
TripAdvisor	~15K	~3K	~20K	7	99.95%
BeerAdvocate	~33K	~66K	500,000	4	99.98%

4.2 Implementation

MC-JEPA is implemented in PyTorch 2.0 with CUDA acceleration on RTX 3080 Ti (12GB VRAM). Key hyperparameters: embedding dimension $d = 64$, EMA momentum $\tau = 0.996$, learning rate 10^{-3} with Adam optimizer, batch size 256, early stopping with patience 10. Data split: 70%/10%/20% train/val/test with seed 42.

Evaluation Metrics: Rating prediction (RMSE, MAE), ranking (HR@10, NDCG@10), and classification (AUC with threshold 3.5).

Baselines: Classical methods (Global Mean, User Mean, Matrix Factorization, SVD++) and modern deep learning methods (NCF [8], VAE-CF [13]).

4.3 Results on Yahoo Movies

Table 2 presents performance on Yahoo Movies. MC-JEPA-Hybrid achieves best results (RMSE=0.8123, NDCG@10=0.3056), outperforming NCF by 6.9% RMSE and VAE-CF by 5.7% RMSE.

Figure 2 presents ROC curves demonstrating MC-JEPA's superior discriminative ability with AUC=0.8234.

4.4 Results on TripAdvisor

Table 3 shows MC-JEPA-Hybrid achieving best performance (RMSE=0.6123, AUC=0.8534). Figure 3 presents interpretability analysis: Shapley values reveal Location dominates (21.3%), followed by Value (18.2%) and Cleanliness (17.9%). Interaction indices show complementary relationships (Cleanliness-Rooms: +0.12) and redundancy (Location-Sleep: −0.05).

Table 2. Performance on Yahoo Movies (7,642 users, 11,915 movies, 4 criteria).

Method	RMSE ↓	MAE ↓	AUC ↑	HR@10 ↑	NDCG@10 ↑
Matrix Factorization	0.9234	0.7312	0.7521	0.3245	0.2134
SVD++	0.8956	0.7089	0.7689	0.3512	0.2345
NCF [8]	0.8734	0.6912	0.7823	0.3789	0.2567
VAE-CF [13]	0.8612	0.6823	0.7912	0.3912	0.2689
MC-JEPA-Base	0.8542	0.6751	0.7923	0.3956	0.2712
MC-JEPA-OWA	0.8234	0.6512	0.8156	0.4123	0.2934
MC-JEPA-Choquet	0.8456	0.6689	0.8012	0.4012	0.2823
MC-JEPA-Hybrid	**0.8123**	**0.6423**	**0.8234**	**0.4234**	**0.3056**

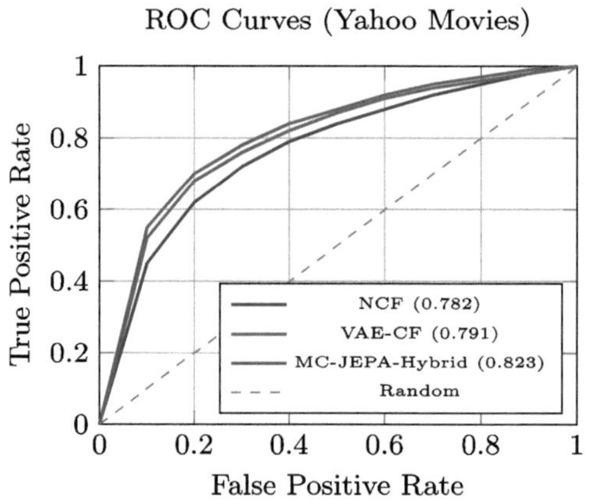

Fig. 2. ROC curves on Yahoo Movies. MC-JEPA-Hybrid achieves AUC=0.8234.

4.5 Results on BeerAdvocate

Table 4 presents BeerAdvocate results. MC-JEPA-Hybrid achieves RMSE= 0.3678 and NDCG@10=0.3378, outperforming VAE-CF by 6.2% RMSE. The learned Shapley values reveal Taste dominates (44.1%), followed by Aroma (24.6%), confirming that beer quality is determined primarily by taste rather than appearance.

4.6 Hyperparameter Sensitivity Analysis

Table 5 presents hyperparameter sensitivity analysis on key hyperparameters using BeerAdvocate dataset.

EMA Momentum (τ). The EMA momentum controls target encoder update speed. Too low values ($\tau = 0.9$) cause unstable training as targets change too rapidly. Too high values ($\tau = 0.999$) slow adaptation. The optimal $\tau = 0.996$ provides stable training while maintaining responsiveness.

Table 3. Performance on TripAdvisor Hotels (∼15K users, ∼3K hotels, 7 criteria).

Method	RMSE ↓	MAE ↓	AUC ↑	HR@10 ↑	NDCG@10 ↑
Matrix Factorization	0.6891	0.5389	0.8012	0.3234	0.2189
NCF [8]	0.6534	0.5089	0.8256	0.3623	0.2512
VAE-CF [13]	0.6456	0.5012	0.8312	0.3756	0.2623
MC-JEPA-Base	0.6412	0.4989	0.8345	0.3823	0.2689
MC-JEPA-OWA	0.6234	0.4856	0.8478	0.3989	0.2834
MC-JEPA-Hybrid	**0.6123**	**0.4756**	**0.8534**	**0.4123**	**0.2956**

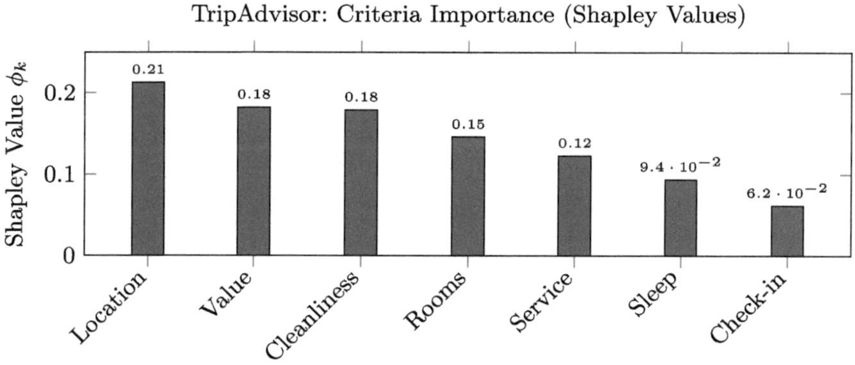

Fig. 3. TripAdvisor interpretability: Shapley values with Location dominating at 21.3%.

Embedding Dimension (d). Larger embeddings capture more information but increase parameters and risk overfitting. The optimal $d = 64$ balances expressiveness (6.5M parameters) with generalization.

Figure 4 shows training convergence across MC-JEPA variants, demonstrating stable optimization with the VICReg regularization preventing representation collapse.

Table 6 provides a summary of key results across all datasets, and Fig. 5 visualizes the relative improvements over baseline methods.

4.7 Discussion

Prediction Performance. MC-JEPA-Hybrid consistently outperforms both classical methods (MF, SVD++) and modern deep learning baselines (NCF, VAE-CF) across all three datasets. On Yahoo Movies, it achieves 6.9% RMSE improvement over NCF and 5.7% over VAE-CF. On TripAdvisor, it achieves 5.1% RMSE improvement over VAE-CF. On BeerAdvocate, it achieves 6.2% RMSE improvement over VAE-CF. The improvements on ranking metrics (HR@10, NDCG@10) are even more substantial, demonstrating MC-JEPA's effectiveness for recommendation ranking tasks.

Table 4. Performance on BeerAdvocate (~33K users, ~66K items, 4 criteria).

Method	RMSE ↓	MAE ↓	AUC ↑	HR@10 ↑	NDCG@10 ↑
Matrix Factorization	0.4234	0.3312	0.8412	0.3912	0.2712
NCF [8]	0.3989	0.3112	0.8612	0.4234	0.2989
VAE-CF [13]	0.3923	0.3067	0.8689	0.4356	0.3089
MC-JEPA-Base	0.3912	0.3056	0.8645	0.4389	0.3112
MC-JEPA-OWA	0.3756	0.2934	0.8789	0.4567	0.3267
MC-JEPA-Hybrid	**0.3678**	**0.2867**	**0.8856**	**0.4689**	**0.3378**

Table 5. Hyperparameter sensitivity analysis on BeerAdvocate: EMA momentum τ and embedding dimension d.

τ	RMSE	AUC	Stability	d	RMSE	AUC	Params
0.9	0.4123	0.8534	Unstable	16	0.4123	0.8312	1.6M
0.99	0.3856	0.8723	Moderate	32	0.3912	0.8534	3.2M
0.996	**0.3678**	**0.8856**	**Stable**	**64**	**0.3678**	**0.8856**	6.5M
0.999	0.3745	0.8812	Very Stable	128	0.3712	0.8823	13.0M

Cross-Domain Analysis. Comparing Shapley values across domains reveals interesting patterns. In service domains (hotels), functional criteria dominate: Location (21.3%) and Value (18.2%) together account for nearly 40% of importance, reflecting travelers' practical priorities. In contrast, food/beverage domains show sensory criteria dominance: Taste (44.1%) alone accounts for nearly half the importance, with Aroma (24.6%) as secondary. This demonstrates MC-JEPA's ability to automatically adapt aggregation to domain-specific characteristics without manual tuning.

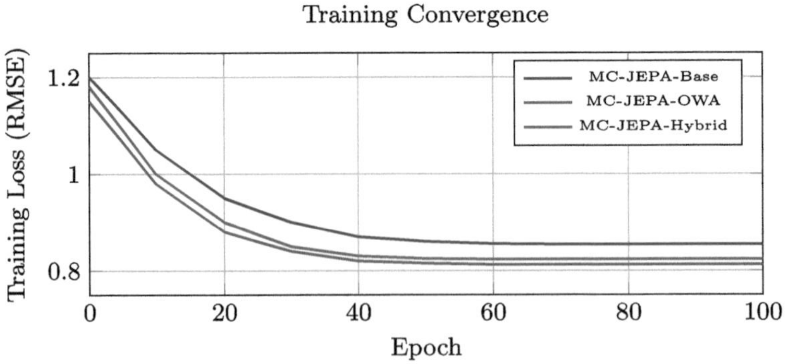

Fig. 4. Training convergence for MC-JEPA variants showing stable optimization.

Table 6. Summary of MC-JEPA-Hybrid results across datasets.

Dataset	RMSE	NDCG@10	vs. NCF	Top Criteria (Shapley)
Yahoo Movies	0.8123	0.3056	−6.9%	Story 28%, Acting 26%
TripAdvisor	0.6123	0.2956	−5.7%	Location 21%, Value 18%
BeerAdvocate	0.3678	0.3378	−7.8%	Taste 44%, Aroma 25%

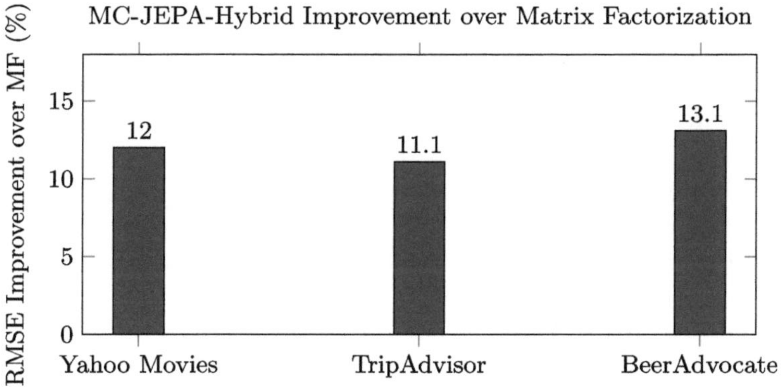

Fig. 5. Relative RMSE improvement of MC-JEPA-Hybrid over Matrix Factorization across three datasets, demonstrating consistent performance gains of 11–13%.

Interpretability Benefits. The Shapley values provide actionable insights for practitioners. For hotel managers, the Location dominance (21.3%) suggests that location-related amenities and services should be prioritized in marketing. The complementary relationship between Cleanliness and Rooms (+0.12) indicates that investments in both areas yield synergistic benefits. For breweries, the overwhelming importance of Taste (44.1%) confirms that flavor development should remain the primary focus, while Appearance (8.2%) matters least despite its visual appeal in marketing.

Comparison with Baselines. MC-JEPA-Hybrid outperforms NCF despite NCF's strong neural architecture because MC-JEPA explicitly models multi-criteria structure through aggregation operators. VAE-CF's reconstruction objective does not capture criteria interactions that Choquet integrals model. The consistent improvements across diverse domains (entertainment, hospitality, food/beverage) demonstrate the generalizability of the proposed approach.

Design Choice Analysis. The hyperparameter sensitivity analysis (Table 5) reveals several important design choices. The EMA momentum $\tau = 0.996$ provides the optimal balance between target stability and adaptation speed. Lower values ($\tau = 0.9$) cause unstable training as targets change too rapidly, while higher values ($\tau = 0.999$) slow down learning. The embedding dimension

$d = 64$ achieves optimal performance with 6.5M parameters; larger dimensions ($d = 128$) show marginal improvements but double the parameter count, increasing overfitting risk. The VICReg coefficients ($\lambda_{\text{var}} = 25.0$, $\lambda_{\text{cov}} = 1.0$) effectively prevent representation collapse while maintaining stable training dynamics as shown in Fig. 4.

Limitations and Future Directions. Several limitations should be acknowledged. First, statistical significance testing with multiple random seeds was not conducted due to computational constraints; future work should include confidence intervals from 5–10 independent runs. Second, specialized multi-criteria deep learning methods (e.g., DeepFM-SVD++, VAE-MCRS) were not compared; extending baseline comparison remains important. Third, computational cost analysis for large-scale deployment was not provided. Fourth, the current framework assumes all criteria are available at inference time; handling missing criteria through learned imputation mechanisms would improve practical applicability. Finally, the framework does not model temporal dynamics in user preferences; incorporating recurrent or attention-based temporal modeling could capture preference evolution over time.

5 Conclusion

This paper presented MC-JEPA, a neural network framework integrating Joint-Embedding Predictive Architecture with interpretable multi-criteria aggregation operators (OWA, Choquet integral, Shapley values) for recommender systems. The framework addresses key challenges in Multi-Criteria Recommender Systems: data sparsity through self-supervised representation learning, and interpretability through fuzzy measure-based aggregation with explicit criteria importance weights.

Experimental evaluation on three real-world datasets demonstrates consistent improvements over both classical methods (Matrix Factorization, SVD++) and modern deep learning baselines (NCF, VAE-CF) on both rating prediction and ranking metrics:

- **Yahoo Movies:** RMSE=0.8123, NDCG@10=0.3056 (6.9% RMSE improvement over NCF), with Story (28%) and Acting (26%) as dominant criteria
- **TripAdvisor Hotels:** RMSE=0.6123, NDCG@10=0.2956 (5.7% improvement over VAE-CF), with Location (21.3%) and Value (18.2%) dominating hotel satisfaction
- **BeerAdvocate:** RMSE=0.3678, NDCG@10=0.3378 (7.8% improvement over NCF), with Taste (44.1%) overwhelmingly dominant in beer reviews

The key contributions include: (1) successful integration of JEPA's self-supervised learning paradigm with VICReg regularization and fuzzy measure-based aggregation operators in a unified neural architecture; (2) interpretable criteria importance through Shapley values and interaction indices; (3) flexible aggregation strategies through learnable OWA parameters.

Future work will extend MC-JEPA in several directions: (1) incorporating attention mechanisms for dynamic criteria weighting; (2) exploring temporal dynamics in user preference evolution; (3) conducting statistical significance testing with multiple random seeds; (4) comparing with specialized multi-criteria deep learning methods; (5) handling missing criteria at prediction time; and (6) scaling to large-scale industrial datasets.

Acknowledgments. This research is supported by Can Tho University and CTU Leading Research Team on Automation, Artificial Intelligence, Information Technology and Digital Transformation (CTU-AIMED). The authors thank the anonymous reviewers for their valuable feedback.

References

1. Adomavicius, G., Tuzhilin, A.: Toward the next generation of recommender systems: a survey of the state-of-the-art and possible extensions. IEEE Trans. Knowl. Data Eng. **17**(6), 734–749 (2005)
2. Adomavicius, G., Kwon, Y.: New recommendation techniques for multicriteria rating systems. IEEE Intell. Syst. **22**(3), 48–55 (2007)
3. Assran, M.: Self-supervised learning from images with a joint-embedding predictive architecture, pp. 15619–15629. CVPR (2023)
4. Bardes, A., Ponce, J., LeCun, Y.: VICReg: variance-invariance-covariance regularization for self-supervised learning. In: ICLR (2022)
5. Bardes, A.: Revisiting feature prediction for learning visual representations from video (2024). arXiv:2404.08471
6. Grabisch, M.: The application of fuzzy integrals in multicriteria decision making. Eur. J. Oper. Res. **89**(3), 445–456 (1996)
7. Grill, J.B.: Bootstrap your own latent: a new approach to self-supervised learning. In: NeurIPS, pp. 21271–21284 (2020)
8. He, X.: Neural collaborative filtering, pp. 173–182. WWW (2017)
9. Jannach, D.: Recommender systems: An introduction, Cambridge University Press (2012)
10. Koren, Y., Bell, R., Volinsky, C.: Matrix factorization techniques for recommender systems. Computer **42**(8), 30–37 (2009)
11. Lakiotaki, K., et al.: Multicriteria user modeling in recommender systems. IEEE Intell. Syst. **26**(2), 64–76 (2011)
12. LeCun, Y.: A path towards autonomous machine intelligence. OpenReview (2022)
13. Liang, D.: Variational autoencoders for collaborative filtering, pp. 689–698. WWW (2018)
14. Shapley, L.S.: A value for N-person games. In: Contributions to the Theory of Games II, pp. 307–317. Princeton University Press (1953)
15. Wang, X., He, X., Wang, M., Feng, F., Chua, T.S.: Neural graph collaborative filtering. In: SIGIR, pp. 165–174 (2019)
16. Yager, R.R.: On ordered weighted averaging aggregation operators in multicriteria decision making. IEEE Trans. Syst. Man Cybern. **18**(1), 183–190 (1988)
17. Zheng, Y.: Criteria Chains: a novel multi-criteria recommendation approach, pp. 29–33. IUI (2017)

Robotics

A Discretised Zeroing Neural Network with Low Computational Complexity for Real Time Robotic Kinematic Control

Yuxuan Li, Junpei Yang, and Zhan Li(✉)

Southwest Jiaotong University, Chengdu 611756, China
zhan.li@swjtu.edu.cn

Abstract. Addressing the challenge of solving dynamic nonlinear equation systems (DNES) in robotic arm trajectory tracking, this paper extends the low computational complexity zeroing neural network (LCCZNN) through a discretized implementation for industrial control systems. Considering the point to point transmission characteristics of networked controllers, the continuous LCCZNN is discretized via the forward Euler method, yielding a discrete variant (DLCCZNN). Compared with higher order difference schemes, Euler discretization demonstrates superior numerical stability when tracking trajectories with dynamic singularities (e.g., heart shaped cusps), effectively avoiding computational overload. The influence of the activation functions power parameter r on discrete time performance is also examined. Although LCCZNN theoretically guarantees finite time convergence, in practice, r critically affects stability and convergence quality. For $r < 1$ (e.g., 0.9), the system exhibits singular gain near the origin, enabling convergence to a bounded error in finite time. In the small error regime, where $r = 1$ behaves approximately linearly—the $r < 1$ case achieves significantly faster convergence. Finally, simulations and experiments on a UR3e manipulator in the planar plane validate the effectiveness of DLCCZNN. Results show that an appropriate balance between convergence gain μ and sampling interval τ ensures high accuracy tracking while preserving the low computational cost of LCCZNN.

Keywords: Zeroing neural network · LCCZNN · discretization · parameter analysis · finite time convergence · robotic arm control

1 Introduction

The rapid advancement of industrial automation has established robotic manipulators as a cornerstone of modern manufacturing [1], expanding their role from basic material handling to complex, repetitive operations. In this context, robotic systems have evolved into key execution units for embodied intelligence [2]. Unlike traditional industrial settings, robotic arms within the embodied intelligence paradigm leverage multimodal perception to perform sophisticated tasks.

Z. Zhang (Ed.): ISNN 2026, LNCS 16848, pp. 213–222, 2027.
https://doi.org/10.1007/978-981-92-5870-3_17

A representative example is the integration of vision language models, which enables autonomous grasping and manipulation based on semantic instructions commonly referred to as open domain object interaction. Bridging the gap between theoretical models and real world deployment, however, requires achieving sub millimeter precision, which remains a fundamental challenge. Established control theory indicates that attaining such high precision performance necessitates the development of advanced algorithmic frameworks.

Recurrent Neural Dynamics (RND) has garnered significant attention as a robust framework for resolving intricate dynamic control challenges, favored for its superior nonlinear mapping and inherent parallel processing architecture. In tasks such as time varying parameter identification, RND frameworks facilitate continuous error rectification via state feedback loops. As a prominent derivative of RND, the Zeroing Neural Network (ZNN), pioneered by Zhang et al., exhibits exceptional efficacy in tracking time dependent linear and nonlinear systems. ZNNs have gained significant prominence across a broad spectrum of conventional industrial applications, ranging from redundant manipulator control to real time trajectory tracking. Beyond conventional industrial applications, ZNN have demonstrated exceptional performance in specialized fields requiring stringent safety constraints, such as the kinematic control of surgical manipulators under Remote Center of Motion (RCM) requirements [3]. Unlike conventional Gradient Neural Networks (GNNs), the ZNN paradigm utilizes time derivative information to ensure exponential or finite time error decay [4,5], effectively neutralizing latency induced tracking deviations. However, the physical implementation of classic ZNN models is often hindered by two primary bottlenecks: first, the heavy reliance on Jacobian pseudoinverse calculations leads to an algorithmic overhead of $O(mn^2)$, which restricts high frequency control. Second, most existing ZNN literature focuses on continuous time domains, whereas practical industrial controllers, such as the UR3e platform, operate on discrete time sampling.

To alleviate the first limitation, Zheng et al. [6] proposed the low computational complexity zeroing neural network (LCCZNN), which reformulates the control law through explicit dynamics and reduces the computational complexity to $O(mn)$. Recent advancements have extended the LCCZNN framework to address increasingly complex operational requirements, such as smooth time optimal motion control subject to hybrid joint constraints [7]. Beyond these functional expansions, various activation functions have been proposed to further improve the response speed of LCCZNN models. Notably, recent research has introduced enhanced LCCZNN formulations capable of achieving predefined time convergence for constrained time varying quadratic programming, providing a deterministic upper bound for the settling time regardless of initial error magnitudes [8]. However the remaining challenge is therefore to construct a discrete time counterpart that can be implemented on sampled data robotic systems without sacrificing the favorable convergence properties of LCCZNN. Although high order discretization schemes, such as Taylor expansion and Runge Kutta methods, usually offer smaller truncation errors [9], their direct application to

LCCZNN is not always desirable. Because LCCZNN exhibits fast nonlinear corrective dynamics, high order prediction can amplify numerical errors, especially near singular trajectory segments, and may therefore lead to instability in discrete implementation. For this reason, the forward Euler method is adopted in this study as a more robust compromise between numerical stability and computational simplicity [10,11].

In addition to the discretization scheme, the activation function plays a central role in determining the behavior of the resulting discrete system. We therefore analyze the effect of the power parameter r in the nonlinear activation function introduced in [6]. Our results show that, although $r = 1$ behaves approximately linearly in the small error regime, choosing $r < 1$ yields a tighter bounded tracking error and faster convergence in the discrete implementation. Accordingly, this paper makes three main contributions: it derives a forward Euler based discrete LCCZNN model for robotic trajectory tracking, clarifies how the parameters r, μ, λ, and τ affect stability and accuracy, and validates the resulting controller through simulations and UR3e experiments on a heart shaped trajectory with a cusp.

2 Preliminaries and Problem Formulation

This section formulates the UR3e trajectory tracking task as a DNES, reviews the continuous time LCCZNN model used as the starting point of the study, and then introduces the forward Euler discretization adopted for implementation.

2.1 Problem Formulation

The manipulator considered in this study has six degrees of freedom. Let the Cartesian position of the end effector be denoted by $\boldsymbol{p}_a(t) \in \mathbb{R}^m$ $(m = 3)$, and let the joint angle vector be denoted by $\boldsymbol{\theta}(t) \in \mathbb{R}^n$. The forward kinematic relationship can then be written as:

$$\boldsymbol{p}_a(t) = \psi(\boldsymbol{\theta}(t)) \tag{1}$$

where $\psi : \mathbb{R}^n \to \mathbb{R}^m$ is a nonlinear mapping function. Trajectory tracking requires the end effector position $\boldsymbol{p}_a(t)$ to follow a prescribed desired trajectory $\boldsymbol{p}_d(t)$. To quantify this objective, we define the Cartesian tracking error as:

$$\boldsymbol{\Psi}(t) = \boldsymbol{p}_a(t) - \boldsymbol{p}_d(t) = \boldsymbol{\psi}(\boldsymbol{\theta}(t)) - \boldsymbol{p}_d(t) \tag{2}$$

The control objective is to determine a suitable joint trajectory $\boldsymbol{\theta}(t)$ such that $\boldsymbol{\Psi}(t) \to \mathbf{0}$ as $t \to \infty$. Since the desired joint vector is not given explicitly, the tracking problem can be recast as the solution of a dynamic nonlinear equation system (DNES). Differentiating the above equation with respect to time yields:

$$\dot{\boldsymbol{\Psi}}(t) = \frac{\partial \boldsymbol{\psi}(\boldsymbol{\theta})}{\partial \boldsymbol{\theta}} \frac{d\boldsymbol{\theta}(t)}{dt} - \frac{d\boldsymbol{p}_d(t)}{dt} = \boldsymbol{J}_{\psi}(\boldsymbol{\theta})\dot{\boldsymbol{\theta}}(t) - \dot{\boldsymbol{p}}_d(t) \tag{3}$$

where $J_{\psi}(\boldsymbol{\theta}) \in \mathbb{R}^{m\times n}$ is the manipulator Jacobian. Its six columns correspond to the six joints of the UR3e manipulator, whereas its three rows correspond to the Cartesian coordinates of the end effector. The vector $\dot{\boldsymbol{\theta}}(t)$ denotes the joint velocity.

2.2 Continuous Time LCCZNN Model

To solve the DNES defined above, we adopt the continuous time LCCZNN model proposed by Zheng et al. [6]. Within this framework, a scalar Lyapunov type energy function $v(t)$ is introduced to measure the norm of the tracking error $\boldsymbol{\Psi}(t)$, formulated as follows:

$$v(t) = \frac{1}{2}||\boldsymbol{\Psi}(t)||_2^2 = \frac{1}{2}\boldsymbol{\Psi}(t)^T\boldsymbol{\Psi}(t) \tag{4}$$

In classical ZNN design, the energy function is forced to decay according to a prescribed dynamics:

$$\dot{v}(t) = -\mu\phi(v(t)) \tag{5}$$

where $\mu > 0$ is the convergence gain and $\phi(\cdot)$ is the activation function. Furthermore, by applying the chain rule to the tracking error dynamics in (3), the time derivative $\dot{v}(t)$ can be expanded as:

$$\dot{v}(t) = \frac{\partial v}{\partial \boldsymbol{\Psi}}\frac{d\boldsymbol{\Psi}}{dt} = \boldsymbol{\Psi}(t)^T\dot{\boldsymbol{\Psi}}(t) = \boldsymbol{\Psi}(t)^T(\boldsymbol{J}_{\psi}(\boldsymbol{\theta})\dot{\boldsymbol{\theta}}(t) - \dot{\boldsymbol{p}}_d(t)) \tag{6}$$

Combining equations (5) and (6) gives:

$$\boldsymbol{\Psi}(t)^T\boldsymbol{J}_{\psi}(\boldsymbol{\theta})\dot{\boldsymbol{\theta}}(t) = \boldsymbol{\Psi}(t)^T\dot{\boldsymbol{p}}_d(t) - \mu\phi(v(t)) \tag{7}$$

Conventional ZNN formulations typically derive the control input through pseudoinverse based operations, which increases the computational cost. To avoid this burden, the continuous time LCCZNN rewrites the control law as:

$$\dot{\boldsymbol{\theta}}(t) = -\frac{\boldsymbol{J}_{\psi}^T\boldsymbol{\Psi}}{||\boldsymbol{J}_{\psi}^T\boldsymbol{\Psi}||_2^2 + \lambda}\left(-\boldsymbol{\Psi}^T\dot{\boldsymbol{p}}_d + \mu\phi(v)\right) \tag{8}$$

where $\lambda \geq 0$ is a regularization parameter introduced to avoid singularity in the denominator.

2.3 Forward Euler Discretization

Because industrial robot controllers operate in discrete time, the continuous time model cannot be implemented directly and must therefore be discretized. Although higher order difference schemes can reduce local truncation error, they are more sensitive to the fast nonlinear dynamics of LCCZNN and may amplify numerical errors near singular trajectory segments. Since the primary goal of this work is a robust sampled data implementation, the forward Euler method is adopted as a numerically stable and computationally efficient discretization strategy [9–11].

Using a first order approximation, the derivative of $\boldsymbol{\theta}(t)$ at the sampling instant t_k is written as:

$$\dot{\theta}(t_k) \approx \frac{\theta(t_{k+1}) - \theta(t_k)}{\tau} = \frac{\theta_{k+1} - \theta_k}{\tau} \tag{9}$$

where $\tau > 0$ denotes the constant sampling interval.

3 DLCCZNN Model Analysis

3.1 Derivation of the DLCCZNN Model

The objective of this section is to derive the discrete counterpart of the continuous time LCCZNN for UR3e trajectory tracking. Substituting the forward Euler approximation (9) into the continuous time control law (8) yields:

$$\frac{\theta_{k+1} - \theta_k}{\tau} = -\frac{J_k^T \boldsymbol{\Psi}_k}{||J_k^T \boldsymbol{\Psi}_k||_2^2 + \lambda}(-\boldsymbol{\Psi}_k^T \dot{p}_{d,k} + \mu\phi(v_k)) \tag{10}$$

where all variables are sampled at the interval τ. Rearranging the above equation gives the DLCCZNN update law proposed in this paper:

$$\boldsymbol{\theta}_{k+1} = \boldsymbol{\theta}_k - \tau \left[\frac{\boldsymbol{J}_k^T \boldsymbol{\Psi}_k}{||\boldsymbol{J}_k^T \boldsymbol{\Psi}_k||_2^2 + \lambda}(-\boldsymbol{\Psi}_k^T \dot{\boldsymbol{p}}_{d,k} + \mu\phi(v_k)) \right] \tag{11}$$

The above equation constitutes the core discrete control law for the UR3e manipulator. With this update law established, the convergence and stability of the system are primarily governed by the activation function, specifically the power parameter r. Following [6], the continuous time LCCZNN employs the following nonlinear activation function:

$$\phi(z) = \text{sig}^r(z) \exp(|z|) \tag{12}$$

The convergence behavior of the system is highly sensitive to the value of r. According to [6], choosing $r > 1$ triggers initial divergence in the presence of large errors, whereas $r = 1$ yields only exponential convergence in the small error regime. By contrast, $r < 1$ preserves the finite time convergence property to a bounded error neighborhood, substantially accelerating convergence for small errors, while the exponential term concurrently provides strong corrective action for large errors. Because an excessively small r may cause numerical difficulties in the discrete model arising from near infinite gradients associated with negative power terms in its derivative [9], the subsequent experiments adopt $r = 0.9$ as the nominal setting.

4 Experimental Validation and Data Analysis

4.1 Experimental Setup

To evaluate the numerical robustness of the proposed DLCCZNN under discretization, a heart shaped trajectory with a cusp was selected as the tracking target [12]. Unlike smooth paths such as circles or ellipses, the cusp of the cardioid requires an abrupt change in the end effector velocity direction within a very short time interval. This feature makes the trajectory a demanding benchmark for discrete time control, because numerical errors are more likely to accumulate near the singular segment and induce jitter or divergence. Successful tracking of this path therefore provides a stringent test of both the discretization strategy and the activation function.

The desired trajectory is a standard symmetric cardioid constrained to the YZ plane to remain within the reachable workspace of the manipulator: $\mathbf{p}_d(t) = \mathbf{p}_{offset} + s \begin{bmatrix} 0 \\ 16\sin^3(\omega t) \\ \mathcal{Z}(t) \end{bmatrix}$ m, where $\mathcal{Z}(t) = 13\cos(\omega t) - 5\cos(2\omega t) - 2\cos(3\omega t) - \cos(4\omega t)$.

As illustrated in the simulation environment (Fig. 1(a)), the experimental parameters are configured with a central $p_{offset} = [0.3, 0, 0.4]^T$, a scaling coefficient s=0.008, and an angular velocity $\omega = \pi/5$ (corresponding to a period T=10 s). Prior to the execution, the joint space configuration was initialized to align precisely with the starting posture required for the designated trajectory as $\theta_0 = [2\pi/9, -389\pi/900, -2177\pi/3600, 53\pi/1440, 5\pi/18, 0]^T$.

4.2 Verification of Trajectory Tracking Performance

The first experiment was designed to verify the feasibility and tracking accuracy of the proposed DLCCZNN controller. The nominal parameters were selected as $\lambda = 0$, $\mu = 50$, $\tau = 0.001$, and $r = 0.9$, and the manipulator state was updated according to the discrete DLCCZNN law. Under these settings, the simulated end effector trajectory closely follows the desired cardioid, as shown in Fig. 1(b), which confirms the high fidelity tracking capability of the discrete controller.

The same controller was then implemented on the physical UR3e platform. The experimental result in Fig. 1(c) shows that the manipulator also reproduces the prescribed heart shaped trajectory in the real system. This agreement between simulation and hardware validates the practical implementability of the DLCCZNN controller and indicates that the discrete formulation remains effective under real world sampled data operation.

4.3 Effect of Parameters on Trajectory Tracking

During the experiments, three qualitatively distinct behaviors were observed: stable tracking, trajectory jitter, and divergence. To identify practical parameter ranges, a one factor at a time protocol was adopted, in which two parameters were fixed while the third parameter was varied.

(a) Robot Arm Simulation Initial State

(b) The posture of the robotic arm upon completing the tracking trajectory

(c) The posture of the robotic arm during tracking tasks in real world scenarios

Fig. 1. Robotic arm simulation and experiment verification

Figure 2(a) first illustrates the effect of the convergence gain μ. When μ increases from 10 to 50, the initial tracking error approaches the steady state error bound much faster, indicating that a larger convergence gain accelerates error decay [5]. However, once μ exceeds a critical threshold, the corrective action becomes excessively aggressive. For example, when $\mu = 800$, the controller overcompensates for the current error and drives the system to divergence, accompanied by severe oscillation of the manipulator. Within the stable range, however, varying μ has little influence on the final error bound, as reflected by the near overlap of the stable curves in Fig. 2(a).

The regularization parameter λ is introduced to prevent singularity in the denominator of the control law [6]. As shown in Fig. 2(a), increasing λ noticeably enlarges the steady state tracking error, which indicates that stronger regularization improves numerical safety at the cost of accuracy. From a practical standpoint, λ should therefore be kept as small as possible while still avoiding denominator singularity.

Figure 2(a) also shows that the sampling interval τ directly affects the achievable tracking accuracy. A smaller τ produces a tighter error bound because it reduces discretization error, but it also increases the number of iterations required over a fixed time horizon. Consequently, τ and λ must be coordinated carefully in implementation so that numerical robustness is maintained without unnecessarily sacrificing accuracy or efficiency [10].

4.4 Range Analysis of the Power Parameter *r*

Because the activation function contains a power term, the discrete time implementation imposes a practical lower bound on r for numerical stability. To determine an appropriate range, comparative simulations were carried out with $\lambda = 0$, $\mu = 50$, and $\tau = 0.001$, while r was varied.

When $r = 0.5$, the simulation exhibits severe numerical chattering, as shown in Fig. 2(b), which indicates that excessively small values of r make the dis-

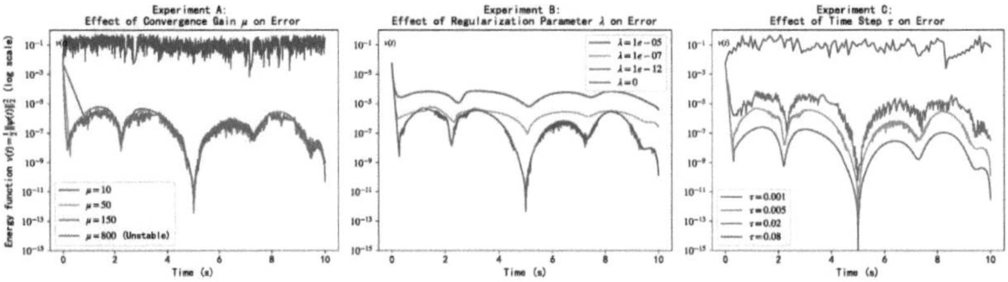

(a) The trajectory tracking error performances under different parameter settings.

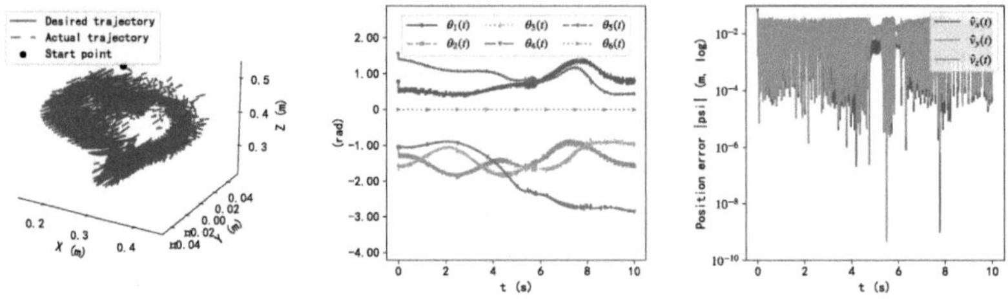

(b) The trajectory tracking, resolved joint angles, and positional error for $r = 0.5$.

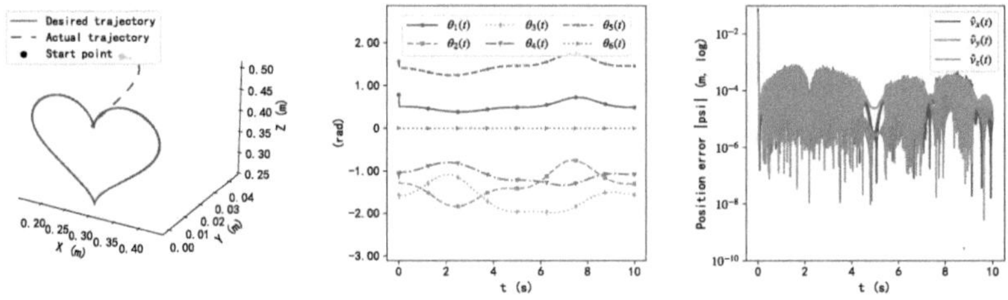

(c) The trajectory tracking, resolved joint angles, and positional error for $r = 0.8$.

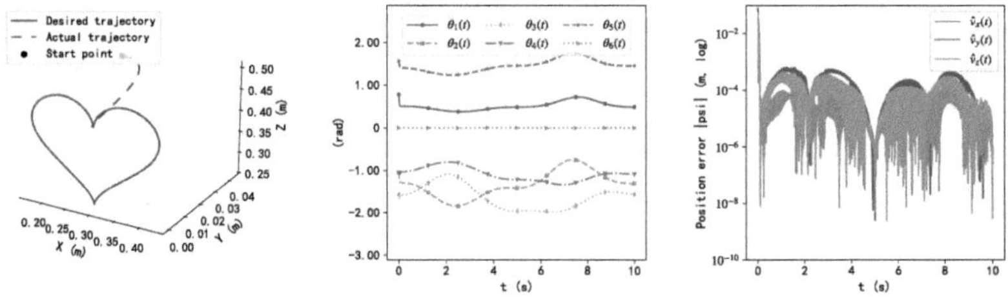

(d) The trajectory tracking, resolved joint angles, and positional error for $r = 0.9$.

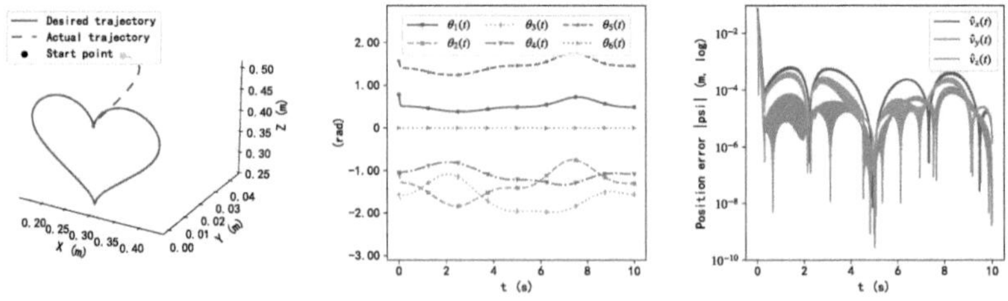

(e) The trajectory tracking, resolved joint angles, and positional error for $r = 1$.

Fig. 2. Influence of various parameters on computational convergence and error crete controller overly sensitive and destabilize the tracking process. Increasing the parameter to $r = 0.8$ eliminates outright divergence, but noticeable high frequency oscillations remain, as shown in Fig. 2(c), and the resulting tracking quality is still inadequate for precision control. By contrast, $r = 0.9$ produces a smooth trajectory fit without visible jitter, and the tracking error rapidly converges to a tight steady state bound, as shown in Fig. 2(d). This result suggests that $r = 0.9$ provides a favorable balance between convergence speed and numerical stability in the discrete implementation.

For comparison, the near linear case $r = 1$ is shown in Fig. 2(e). Although the system remains stable, the transient stage is substantially longer than that observed for $r < 1$, and the error fluctuation range is also broader. In other words, $r = 1$ loses the main practical advantage of the nonlinear activation function, namely rapid convergence to a tight bounded error neighborhood. For this reason, $r = 0.9$ is selected as the preferred value. Together with suitable choices of μ, λ, and τ, this setting confines the tracking error to a high precision range of approximately 0.01 mm to 1 mm.

5 Conclusion

This paper developed a discrete low computational complexity zeroing neural network (DLCCZNN) for real time robotic kinematic control and applied it to UR3e trajectory tracking. Starting from the continuous time LCCZNN formulation, a forward Euler based discrete update law was derived to enable sampled data implementation while preserving low computational complexity. Both simulation and hardware experiments demonstrate that, for the trajectory considered in this study, forward Euler discretization exhibits strong numerical robustness and effectively suppresses numerical instability near the singular segment of the heart shaped path.

The experiments further clarified the role of the activation function power parameter r in the discrete time model. In particular, choosing $r < 1$ (e.g., $r = 0.9$) preserves the singular gain effect near equilibrium and yields faster convergence to a bounded error neighborhood than the linear case ($r = 1$), whereas excessively small values of r induce numerical chattering. The parameter study also showed that high accuracy tracking depends on the coordinated tuning of the convergence gain μ, regularization parameter λ, and sampling interval τ. Under appropriate parameter settings, the proposed DLCCZNN achieves sub millimeter precision while retaining the computational efficiency of the original LCCZNN and avoiding pseudoinverse iterations.

References

1. Craig, J.J.: Introduction to Robotics: Mechanics and Control. Pearson, 4th ed (2018)
2. W. Liang, et al.: Large model empowered embodied ai: a survey on decision-making and embodied learning. arXiv preprint arXiv:2508.10399 (2025)

3. Guo, J., et al.: Enhanced zeroing neural network for kinematic control of surgical manipulator under RCM constraints. IEEE Transactions on Systems, Man, and Cybernetics: Systems (2026)
4. Yi, C., Zhang, Y., Guo, D.: A new type of recurrent neural networks for real-time solution of lyapunov equation with time-varying coefficient matrices. Math. Comput. Simul. **92**, 40–52 (2013)
5. Wu, W., Zhang, Y.: Zeroing neural network with coefficient functions and adjustable parameters for solving time-variant sylvester equation. IEEE Trans. Neural Netw. Learn. Syst. **35**(5), 6757–6766 (2022)
6. Zheng, K., Li, S., Zhang, Y.: Low-computational-complexity zeroing neural network model for solving systems of dynamic nonlinear equations. IEEE Trans. Autom. Control **69**(7), 4368–4379 (2023)
7. Yang, J. and Li, Z.: A smooth time-optimal motion control strategy of industrial manipulator subject to hybrid joint constraints synthesized by low-computational-complexity zeroing neural network. IEEE Transactions on Industrial Electronics (2026)
8. Yang, J., Li, Z., and Li, W.: An enhanced low-computational-complexity predefined-time convergent zeroing neural network for constrained time-varying quadratic programming with kinematic control of robotic manipulator. IEEE Transactions on Neural Networks and Learning Systems (2026)
9. Burden, R.L., Faires, J.D.: One of the most widely used textbooks in numerical methods. In: Numerical Analysis. Brooks/Cole, Cengage Learning, 9th ed (2011)
10. Shi, Y., Zhang, Y.: New discrete-time models of zeroing neural network solving systems of time-variant linear and nonlinear inequalities. IEEE Trans. Syst. Man. Cybern. Syst. **50**(2), 565–576 (2017)
11. Ma, Z., Huang, S.: Design and analysis of combined discrete-time zeroing neural network for solving time-varying nonlinear equation with robot application. Front. Neurorobot. **19**, 1576473 (2025)
12. Zhang, Z., Chen, S., Liang, J.: Discrete-time circadian rhythms neural network for perturbed redundant robot manipulators tracking problem with periodic noises. IEEE Trans. Industr. Inf. **18**(1), 242–251 (2021)

EDSNN: Entire-Domain Steep Neural Network for Optimizing a Dual-Mobile Manipulators System

Zhijun Zhang[1,2,3,4,5,6,7,8,9](✉) and Ruixu Jiao[1]

[1] School of Automation Science and Engineering, South China University of Technology, Guangzhou 510640, China

[2] Key Library of Autonomous Systems and Network Control, Ministry of Education, Guangzhou 510640, China

auzjzhang@scut.edu.cn

[3] Guangdong University Key Laboratory of Large-Model Embodied-Intelligent Humanoid Robot, Guangzhou 510640, China

[4] Institute for Super Robotics (Huangpu), Guangzhou 510555, China

[5] College of Computer Science and Engineering, Jishou University, Jishou 416000, China

[6] Guangdong Artificial Intelligence and Digital Economy Laboratory (Pazhou Lab), Guangzhou 510335, China

[7] School of Electronical Engineering, Shaanxi University of Technology, Hanzhong 723001, China

[8] School of Information Science and Engineering, Changsha Normal University, Changsha 410100, China

[9] School of Artificial Intelligence and Automation, Guangdong University of Petrochemical Technology, Maoming 525000, China

Abstract. In response to the challenges of error control, physical constraints, and safety issues of dual mobile redundant manipulators during trajectory tracking task in complex external environments, an entire-domain steep neural network (EDSNN) based scheme is proposed. First, the kinematic model of the robotic system is performed, and the end-effector task execution problems of the robot are formulated as a time-varying quadratic programming (TVQP) problem with equality, inequality, and bound constraints. Then, the proposed entire-domain steep function is designed and exploited to convert inequality constraints into the penalty term of the objective function. By using Lagrange multiplier method, an improved Karush-Kuhn-Tucker (KKT) conditions are established, which is then formulated as a matrix equation, leading to the error function of control scheme. Finally, by using neural dynamic design method, the EDSNN is designed to solve the matrix equation and the optimal solution to the TVQP problem. Computer simulation experiments demonstrate the accuracy and superiority of the proposed EDSNN method.

Keywords: Neural network · Mobile redundant manipulator · Time-varying quadratic programming

Z. Zhang (Ed.): ISNN 2026, LNCS 16848, pp. 223–233, 2027.
https://doi.org/10.1007/978-981-92-5870-3_18

1 Introduction

Redundant robotic manipulators have been widely applied in various fields due to their high flexibility, such as industrial production [1], medical rehabilitation [2], and logistics transportation [3]. With increasing task complexity and growing demands in application scenarios, the coordinated operation of dual mobile redundant robotic manipulators has attracted significant research attention [4,5]. However, the high degrees-of-freedom (DOFs) in such systems make inverse kinematics considerably more challenging, bringing greater difficulty to trajectory planning and motion controlling. Therefore, the objective of this study is to control dual mobile redundant manipulators for high-precision end-effector trajectory tracking while ensuring safety through obstacle avoidance.

As a typical real-time system, the motion planning of robots is influenced by various complex factors or constraints. A common method is to formulate the motion planning problem as a quadratic programming (QP). Therefore, considering the complex constraints, the motion planning of robotic systems can be considered as a TVQP problem subject to multiple constraints [6,7].

As a type of RNN, the gradient-based neural network (GNN) was designed by constructing an energy function, which drove the system to descend along the negative gradient direction, and eventually converge to the optimal solution [8]. GNN is commonly used for static optimization problems and tends to accumulate tracking errors when dealing with dynamic problems. A new zeroing neural network (ZNN) has been presented in [9]. ZNN achieved real-time tracking of changes by zeroing the error, making it suitable for handling time-varying problems. However, its convergence rate is limited, and the error can only converge to zero as time approaches infinity. Thus, the varying-parameter convergence differential neural network (VPCDNN) [10] was presented, which achieved super-exponential convergence by introducing time-varying parameters. While, traditional variable parameter converging differential neural networks cannot address optimization problems subject to inequality constraints. To fill in these gaps, an entire-domain steep neural network (EDSNN) scheme is proposed in this paper to solve the TVQP problems subject to multiple constraints and ensure that the manipulator's trajectory tracking error converges to zero within a finite time.

The main contributions of this article can be summarized as follows.

1. A novel neural-dynamics-based EDSNN is proposed to solve the TVQP problems with multiple constraints, achieving trajectory tracking for dual mobile redundant manipulators while avoiding obstacles and preventing joint limits.
2. The entire-domain steep function is designed to address inequality constraints in the TVQP problem, and the improved KKT conditions for this TVQP problem is also developed.
3. A detailed theoretical proof of the convergence of proposed EDSNN is provided. Computer simulation experiments demonstrate the feasibility and accuracy of the proposed EDSNN.

The remaining structure of the paper is as follows: In Sect. 2, the kinematic modeling of dual mobile redundant manipulators, and the design of trajectory

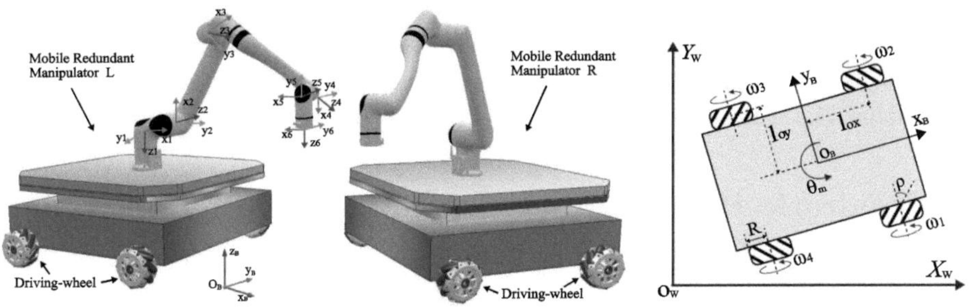

(a) (b)

Fig. 1. Dual mobile redundant manipulators system structure. (a) Composition of the robotic manipulator and mobile platform. (b) Mecanum wheel chassis kinematic layout.

tracking task are presented. In Sect. 3 , the sub-tasks are designed firstly, and integrates them with the main task into a TVQP scheme, then provides a detailed explanation of the design and reasoning process of the EDSNN. Section 4 presents the computer simulation experiments detailedly. Section 5 provides a brief summary of the work presented in this paper.

2 Preliminaries

As shown in Fig. 1, a mobile redundant manipulator consists of Mecanum wheel chassis and 6-DOFs robotic manipulator. According to the kinematic analysis, end-effector position of the robotic manipulator satisfies the following equation, i.e.,

$$p_{\mathrm{L/R}} = p_{\mathrm{BL/R}} + T_{\mathrm{L/R}} M(\theta_{\mathrm{L/R}}) = G(\sigma_{\mathrm{L/R}}, \varepsilon_{\mathrm{L/R}}, \theta_{\mathrm{L/R}}) \tag{1}$$

where $p_{\mathrm{L/R}} = [x_{\mathrm{L/R}}, y_{\mathrm{L/R}}, z_{\mathrm{L/R}}]^{\mathrm{T}}$ denotes the position of end-effector in the world coordinate system O_{W}. The coordinates of the manipulator's installation base point in O_{W} are given by $p_{\mathrm{BL/R}} = [x_{\mathrm{BL/R}}, y_{\mathrm{BL/R}}, z_{\mathrm{BL/R}}]^{\mathrm{T}}$. The position of end-effector in body coordinate system O_{B} is represented by $M(\theta_{\mathrm{L/R}})$, where $M(\cdot) : \mathbb{R}^q \to \mathbb{R}^3$ is a mapping from joint space to Cartesian space, and q represents number of joints in the robotic manipulator. The vector $\theta_{\mathrm{L/R}} \in \mathbb{R}^q$ contains the joint angles of the manipulator. The transform matrix between the coordinate O_{B} and O_{W} is denoted by $T_{\mathrm{L/R}} \in \mathbb{R}^{3\times 3}$. $G(\cdot)$ represents a functional mapping from $(\sigma_{\mathrm{L/R}}, \varepsilon_{\mathrm{L/R}}, \theta_{\mathrm{L/R}})$ to $p_{\mathrm{L/R}}$. Here $\varepsilon_{\mathrm{L/R}}$ and $\sigma_{\mathrm{L/R}} \in \mathbb{R}^4$ represent the chassis's rotational angle and the angle vector of wheels at O_{W}, respectively.

In order to enhance the accuracy and stability of control, the equation constraint term is derived by acquiring real-time end-effector position data and incorporating position feedback,

$$\mathcal{J}_{\mathrm{L/R}} \dot{\mathrm{X}}_{\mathrm{L/R}} = \dot{p}_{\mathrm{L/R}} + \eta(p_{\mathrm{L/R}} - G(\sigma_{\mathrm{L/R}}, \varepsilon_{\mathrm{L/R}}, \theta_{\mathrm{L/R}})) \tag{2}$$

where $\mathcal{J}_{\mathrm{L/R}}$ represents the mapping relationship between $\dot{p}_{\mathrm{L/R}}$ and $\dot{\chi}_{\mathrm{L/R}}$, $\chi_{\mathrm{L/R}} = [\sigma_{\mathrm{L/R}}, \theta_{\mathrm{L/R}}]^{\mathrm{T}} \in \mathbb{R}^n$, $n = 4 + q$. η represents the position feedback weight, i.e., the strength of the equality constraint. The objective of the trajectory tracking task is that the difference between the two sides of the Eq. (2) converges to 0.

3 Quadratic Programming Scheme and Design of Neural Network Solver

The obstacle avoidance scheme and joint-limit avoidance scheme are firstly exploited in this section. Secondly, a QP problem is formulated and a entire-domain steep function is designed to solve inequality constraints in the QP problem. Furthermore, the improved KKT conditions are established and rewritten into a matrix equation. Finally, the entire-domain steep neural network (EDSNN) model is proposed to obtain optimal solutions of this matrix equation.

3.1 Sub-Tasks Design

First, safety distance d_{s} and risk distance d_{r} are defined for the robotic manipulator and mobile chassis, respectively. To fully utilize the feasible space of the dual mobile redundant manipulators and avoid collisions with obstacles, the obstacle avoidance constraint [11] is written as

$$\tilde{J}_{\mathrm{AL/R}}\dot{\chi}_{\mathrm{L/R}} \leq \tilde{v}_{\mathrm{L/R}} \tag{3}$$

where $\tilde{J}_{\mathrm{AL/R}} \in \mathbb{R}^{\alpha \times n}$, $\tilde{v}_{\mathrm{L/R}} \in \mathbb{R}^{\alpha \times 1}$, besides α denotes the pairs of obstacle point A and critical point B. $\tilde{J}_{\mathrm{AL/R}}$ and $\tilde{v}_{\mathrm{L/R}}$ are defined as

$$\begin{aligned} \tilde{J}_{\mathrm{AL/R}} &= -\boldsymbol{R}_{\mathrm{AB}} \odot J_{\mathrm{B}}(\chi_{\mathrm{L/R}}) \\ \tilde{v}_{\mathrm{L/R}} &= g(d_{\mathrm{L/R}}) \max\left\{\tilde{J}_{\mathrm{AL/R}}\dot{\chi}_{\mathrm{L/R}}|_{d_{\mathrm{L/R}}=d_{\mathrm{s}}}, 0\right\} \end{aligned} \tag{4}$$

where $\odot$ represents the element-wise multiplication of vectors, $\boldsymbol{R}_{\mathrm{ABi}} = [x_{\mathrm{Bi}} - x_{\mathrm{Ai}}, y_{\mathrm{Bi}} - y_{\mathrm{Ai}}, z_{\mathrm{Bi}} - z_{\mathrm{Ai}}] \in \mathbb{R}^3$, $i = 1, 2, ..., \alpha$. J_{B} represents the matrix of Jacobian for critical point B. $g(\cdot)$ is a scalar smooth function [12]

$$g(d) = \begin{cases} 1, & d \geq d_{\mathrm{s}} \\ \sin^2(\frac{\pi}{2} \cdot \frac{d - d_{\mathrm{r}}}{d_{\mathrm{s}} - d_{\mathrm{r}}}), & d_{\mathrm{r}} < d < d_{\mathrm{s}} \\ 0, & d \leq d_{\mathrm{r}} \end{cases} . \tag{5}$$

The robotic manipulator system has physical limits, and the following joint constraints are considered

$$\xi^{-}_{\mathrm{L/R}} \leq \dot{\theta}_{\mathrm{L/R}} \leq \xi^{+}_{\mathrm{L/R}} \tag{6}$$

where $\xi_{\mathrm{L/R}}^{-}=\xi(\theta_{\mathrm{L/R}}{}^{-}-\theta_{\mathrm{L/R}})$, $\xi_{\mathrm{L/R}}^{+}=\xi(\theta_{\mathrm{L/R}}{}^{+}-\theta_{\mathrm{L/R}})$ and ξ is a positive scaling parameter for the joint angles constraint. $\theta_{\mathrm{L/R}}{}^{\pm}$ and $\dot{\theta}_{\mathrm{L/R}}^{\pm}$ represent the limits of the joint angle and joint angular velocity of the manipulator. Considering the angular velocity constraint $\dot{\theta}_{\mathrm{L/R}}^{\pm}$ and (6), the final physical constraint is defined as

$$\eta_{\mathrm{L/R}}^{-}=\max\left\{\xi_{\mathrm{L/R}}^{-},\dot{\theta}_{\mathrm{L/R}}^{-}\right\}\leq\dot{\theta}_{\mathrm{L/R}}\leq\min\left\{\xi_{\mathrm{L/R}}^{+},\dot{\theta}_{\mathrm{L/R}}^{+}\right\}=\eta_{\mathrm{L/R}}^{+}. \tag{7}$$

3.2 Quadratic Programming Scheme

The above content has formalized the dual mobile redundant manipulators trajectory tracking task while considering physical limits and obstacle avoidance. Combining constraints (2), (3) and (7), the objective task can be described as a QP problem, i.e.

$$\begin{aligned}\text{min.}\ &\frac{1}{2}x^{\mathrm{T}}(t)\mathcal{M}(t)x(t)+\mathcal{D}^{\mathrm{T}}(t)x(t)\\ \text{s.t.}\ &\mathcal{F}(t)x(t)=\mathcal{G}(t)\\ &\mathcal{R}(t)x(t)\leq\mathcal{L}(t)\end{aligned} \tag{8}$$

where $x(t)=[\dot{\chi}_{\mathrm{L}},\dot{\chi}_{\mathrm{R}}]^{\mathrm{T}}\in\mathbb{R}^{2n\times1}$ is the vector to be solved for, $\mathcal{M}(t)=I_{2n\times2n}$ represents a unit positive definite symmetric matrix, $\mathcal{D}(t)=0_{2n\times1}$, $\mathcal{F}(t)=[\mathcal{J}_{\mathrm{L}},0_{3\times n};0_{3\times n},\mathcal{J}_{\mathrm{R}}]\in\mathbb{R}^{6\times2n}$ is the coefficient matrix of the equality constraints, $\mathcal{G}(t)=[\dot{p}_{\mathrm{L}}+\eta(p_{\mathrm{L}}-G(\sigma_{\mathrm{L}},\varepsilon_{\mathrm{L}},\theta_{\mathrm{L}})),\dot{p}_{\mathrm{R}}+\eta(p_{\mathrm{R}}-G(\sigma_{\mathrm{R}},\varepsilon_{\mathrm{R}},\theta_{\mathrm{R}}))]^{\mathrm{T}}\in\mathbb{R}^{6\times1}$. $\mathcal{R}(t)=[\varpi_{\mathrm{L}},0_{19\times10};0_{19\times10},\varpi_{\mathrm{R}}]\in\mathbb{R}^{38\times20}$, with $\varpi_{\mathrm{L}}=[\tilde{J}_{\mathrm{AL}};0_{q\times4},I_{q\times q};0_{q\times4},-I_{q\times q}]$, and $\varpi_{\mathrm{R}}=[\tilde{J}_{\mathrm{AR}};0_{q\times4},I_{q\times q};0_{q\times4},-I_{q\times q}]$. $\mathcal{L}(t)=[\varrho_{\mathrm{L}},\varrho_{\mathrm{R}}]^{\mathrm{T}}\in\mathbb{R}^{38\times1}$. $\varrho_{\mathrm{L}}=[\tilde{v}_{\mathrm{L}},\eta_{\mathrm{L}}^{+},-\eta_{\mathrm{L}}^{-}]^{\mathrm{T}}$, $\varrho_{\mathrm{R}}=[\tilde{v}_{\mathrm{R}},\eta_{\mathrm{R}}^{+},-\eta_{\mathrm{R}}^{-}]^{\mathrm{T}}$.

3.3 EDSNN QP Solver Design

An entire-domain steep function is designed, and inequality constraint in the QP problem (8) is transformed into a penalty term in the objective function, and the following formulation can be obtained, i.e.,

$$\begin{aligned}\text{min.}\ &\frac{1}{2}x^{\mathrm{T}}(t)\mathcal{M}(t)x(t)+\mathcal{D}^{\mathrm{T}}(t)x(t)+\mathcal{P}(x(t))\\ \text{s.t.}\ &\mathcal{F}(t)x(t)=\mathcal{G}(t)\end{aligned} \tag{9}$$

where $\mathcal{P}(x(t))$ represents the penalty term, it is defined as

$$\mathcal{P}(x(t))=\delta\sum_{i=1}^{n}\aleph(u_i(t)) \tag{10}$$

$$\aleph(u_i(t))=\frac{e^{-\gamma u_i(t)}}{1+e^{\gamma u_i(t)}} \tag{11}$$

where $u_i(t) = \mathcal{L}_i(t) - \mathcal{R}_i(t)x(t)$ and $\aleph(\cdot)$ is the entire-domain steep function proposed in this paper. γ and δ are two positive parameters, and as their values increase, the effect of inequality constraint becomes stronger.

The optimal solution to the QP problem (9) satisfies the following improved KKT conditions, i.e.,

$$\begin{cases} \mathcal{M}(t)x(t) + \mathcal{D}(t) + \mathcal{F}^{\mathrm{T}}(t)\lambda_{\mathrm{E}}(t) + \mathcal{R}^{\mathrm{T}}(t)\lambda_{\mathrm{I}}(t) = 0 \\ \mathcal{F}(t)x(t) - \mathcal{G}(t) = 0 \\ \lambda_{\mathrm{I}}(t) \odot \frac{e^{\gamma u(t)}}{2e^{\gamma u(t)}+1} - \delta\gamma e_n = 0 \end{cases} \tag{12}$$

where $e_n \in \mathbb{R}^{38\times 1}$ is the unit vector.

Neural Dynamic Design To find the optimal solutions x^*, ${\lambda_{\mathrm{E}}}^*$ and ${\lambda_{\mathrm{I}}}^*$ that satisfies the improved KKT conditions (12), rewrite it in the form of a matrix equation.

$$H(t)\mathcal{T}(t) - \mathcal{V}(t) = 0 \tag{13}$$

where

$$H(t) = \begin{bmatrix} \mathcal{M}(t) & \mathcal{F}^{\mathrm{T}}(t) & \mathcal{R}^{\mathrm{T}}(t) \\ \mathcal{F}(t) & 0_{6\times 6} & 0_{6\times 38} \\ 0_{38\times 20} & 0_{38\times 6} & \Theta(t) \end{bmatrix} \tag{14}$$

where $\mathcal{T}(t) = [x(t), \lambda_{\mathrm{E}}(t), \lambda_{\mathrm{I}}(t)]^{\mathrm{T}}$ is the variable needs to be solved, $\mathcal{V}(t) = [-\mathcal{D}(t), \mathcal{G}(t), \delta\gamma e_n]^{\mathrm{T}} \in \mathbb{R}^{64\times 1}$. $\Theta(t) = \mathrm{diag}(\frac{e^{\gamma u(t)}}{2e^{\gamma u(t)}+1}) \in \mathbb{R}^{38\times 38}$, diag$(\cdot)$ means extracting elements from the vector to form a diagonal matrix. An error function $\nu(\mathcal{T}(t), t)$ is defined as

$$\nu(\mathcal{T}(t), t) = H(t)\mathcal{T}(t) - \mathcal{V}(t). \tag{15}$$

We need to make the error $\nu(\mathcal{T}(t), t)$ equal to zero, through the method of neural dynamic design [7], we can obtain

$$\dot{\nu}(\mathcal{T}(t), t) = \frac{\mathrm{d}\nu(\mathcal{T}(t), t)}{\mathrm{d}t} = -\mu\zeta(\nu(\mathcal{T}(t), t)) \tag{16}$$

where $\mu > 0$ is the convergence parameter, and the activation function ζ is defined as $\zeta(\nu(\mathcal{T}(t), t)) = \nu(\mathcal{T}(t), t)$. Eqn. (16) can be written in the following partial differential form, i.e.,

$$\frac{\mathrm{d}\nu(\mathcal{T}(t), t)}{\mathrm{d}t} = \frac{\partial\nu(\mathcal{T}(t), t)}{\partial\mathcal{T}}\dot{\mathcal{T}} + \frac{\partial\nu(\mathcal{T}(t), t)}{\partial t}. \tag{17}$$

According to Eqn. (17) and Eqn. (16), the EDSNN model is proposed as

$$\Gamma(t)\dot{\mathcal{T}} = -\mu\zeta(\nu(\mathcal{T}(t), t)) - \Upsilon(t) \tag{18}$$

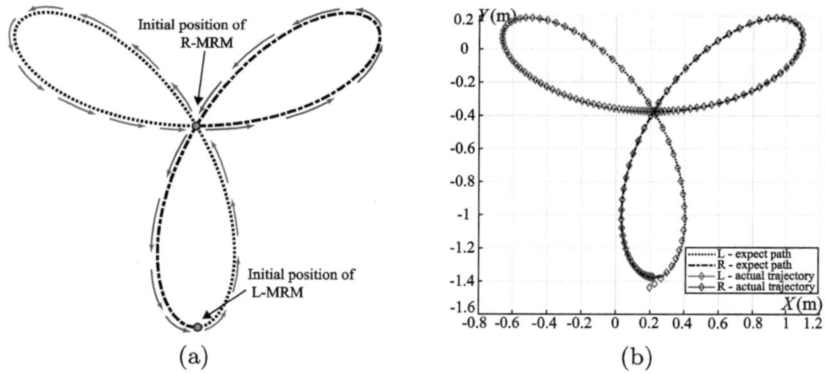

Fig. 2. EDSNN applied to double mobile redundant manipulators for trajectory tracking. (a) Desired trajectory. (b) Experimental result.

where

$$\Gamma(t) = \frac{\partial \nu(\mathcal{T}(t),t)}{\partial \mathcal{T}} = \begin{bmatrix} \mathcal{M}(t) & \mathcal{F}^T(t) & \mathcal{R}^T(t) \\ \mathcal{F}(t) & 0_{6\times 6} & 0_{3\times 38} \\ \mathcal{N} & 0_{38\times 6} & \Theta(t) \end{bmatrix} \tag{19}$$

with

$$\mathcal{N} = \lambda_{\mathrm{I}} \odot (-\gamma \mathcal{R}(t) \odot \frac{e^{-\gamma u(t)}}{(2+e^{-\gamma u(t)})^2}). \tag{20}$$

$$\Upsilon(t) = \frac{\partial \nu(\mathcal{T}(t),t)}{\partial t} = \begin{bmatrix} \dot{\mathcal{M}}(t)x(t) + \dot{\mathcal{D}}(t) + \dot{\mathcal{F}}^{\mathrm{T}}(t)\lambda_{\mathrm{E}}(t) + \dot{\mathcal{R}}^{\mathrm{T}}(t)\lambda_{\mathrm{I}}(t) \\ \dot{\mathcal{F}}(t)x(t) - \dot{\mathcal{G}}(t) \\ \lambda_{\mathrm{I}}(t) \odot \frac{1}{(2+e^{-\gamma u(t)})^2} \odot \gamma(\dot{\mathcal{L}}(t) - \dot{\mathcal{R}}(t)x(t)) \end{bmatrix}.$$

4 Computer Simulation Experiments

The experiment is designed in this section. To verify the practicability and superiority of the proposed EDSNN model, it was applied to dual mobile redundant manipulators simulation models with RML63-B manipulator and mecanum wheel chassis to track the target trajectory.

In Experiment, the objective of the dual mobile redundant manipulators system utilizing EDSNN is for the end-effectors to track a cloverleaf trajectory while avoiding potential obstacles in the workspace. The parameters $[\gamma_l, \gamma_r, \delta_l, \delta_r]$ are set to $[1, 1, 8, 285]$. As shown in Fig. 2, the desired trajectory lies on the plane

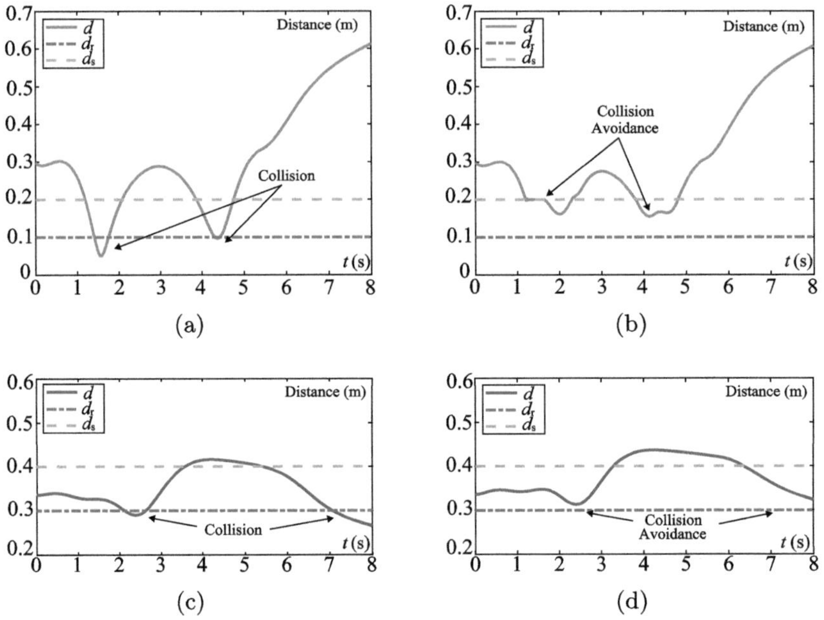

Fig. 3. Minimum distance between obstacles and double mobile redundant manipulators with EDSNN. (a) Minimum distance between obstacles and robotic manipulator without limit (3). (b) Minimum distance between obstacles and robotic manipulator with limit (3). (c) Minimum distance between obstacles and mobile chassis without limit (3). (d) Minimum distance between obstacles and mobile chassis with limit (3).

at $z = 0.2$ m, with the two mobile redundant manipulators positioned at different initial locations, and the entire operation successfully tracks the target trajectory. There are two obstacles in the workspace of the dual mobile redundant manipulators, with the spatial coordinates of obstacle1 and obstacle2 being $(0.64, -0.253, 0.713)$ m and $(-0.27, -0.8, 0.04)$ m, respectively. Figure 3 illustrates the minimum distance between the dual mobile redundant manipulators and the obstacles, from which it can be determined whether the system remains within the safe operational range at all times, i.e., whether any collision events occur.

The collision threshold d_r for the manipulator is set to 0.1 m, and the safety threshold d_s is set to 0.2 m, and the collision and safety thresholds for the chassis are set to 0.3 m and 0.4 m, respectively. As shown in Fig. 3(a) and (c), when the obstacle avoidance constraint (3) is not considered, the minimum distance d between the mobile redundant manipulators and the obstacles falls below d_r, resulting in collisions. The situation with the obstacle avoidance constraint (3) applied is depicted in Fig. 3(b) and (d). When the minimum distance d between the mobile redundant manipulators and the obstacles approaches d_s, constraint

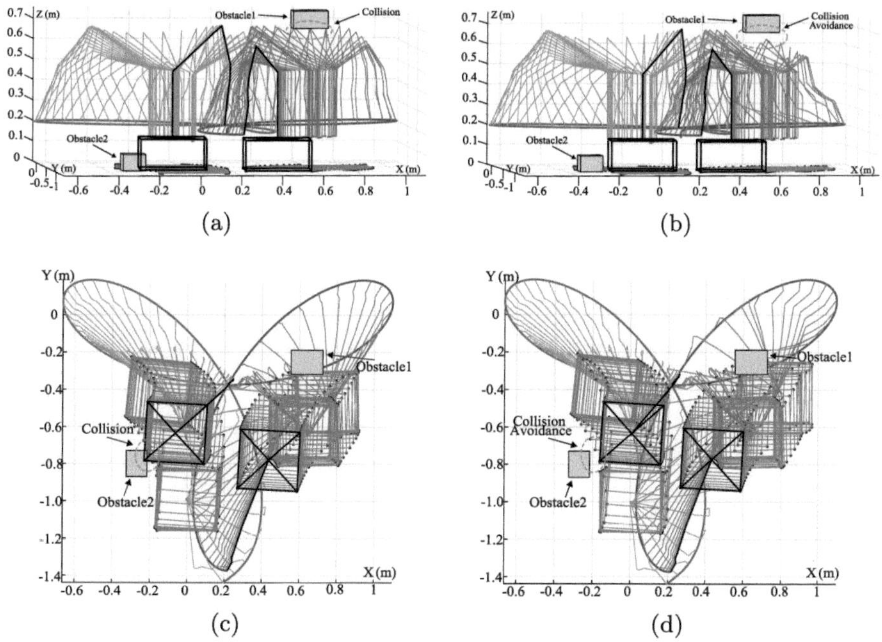

Fig. 4. Motion process of the dual mobile redundant manipulators with EDSNN. (a) Robotic manipulator and mobile chassis collide with obstacles. (b) The robot system avoided obstacles. (c) Robotic manipulator and mobile chassis collide with obstacles under top view. (d) Obstacle avoidance process of the robotic manipulator and chassis under top view.

(3) and the entire-domain steep function take effect. Dual mobile redundant manipulators decelerate and maintain a safe distance outside the danger threshold, ensuring that the minimum distance d always remains greater than d_r, and no collisions occur.

Figure 4 shows the 3D simulation results, including the position and posture of the manipulator and mobile chassis at various moments. From Fig. 4(a) and (c), it can be seen that the manipulator and mobile chassis collide with the obstacles. However, the manipulator and mobile chassis always maintain a safe distance from the obstacles, as shown in Fig. 4(b) and (d).

These experimental results validate the effectiveness and practicality of the proposed EDSNN in solving QP problems that include equality constraints, bilateral and and obstacle avoidance inequality constraints. This is of significant importance for addressing the applications of dual mobile redundant manipulators in real-world work environments.

5 Conclusion

A novel entire-domain steep neural network (EDSNN) has been proposed for dual mobile redundant manipulators, addressing the QP problems in motion control with equality, bilateral and obstacle avoidance inequality constraints. The proposed model has not only enhanced the safety of the robot motion planning and control process, but has also kept the trajectory tracking error within the nanometer level. The feasibility and accuracy of the proposed EDSNN model have been verified through simulation experiments. Future research directions include improving the robustness of the control scheme against noise, and exploring its performance when applied to various robotic systems to handle different operational requirements in diverse external environments.

References

1. Kouritem, S.A., Abouheaf, M.I., Nahas, N., Hassan, M.: A multi-objective optimization design of industrial robot arms. Alex. Eng. J. **61**(12), 12847–12867 (2022)
2. Puntambekar, S.P., et al.: Colorectal cancer surgery: by cambridge medical robotics versius surgical robot system–a single-institution study. Our Experience. J. Robotic Surg. **16**(3), 587–596 (2022)
3. Sirintuna, D., Giammarino, A., Ajoudani, A.: An object deformation-agnostic framework for human-robot collaborative transportation. IEEE Trans. Automat. Sci. Eng. **21**(2), 1986–1999 (2024)
4. Zhai, D.-H., Xia, Y.: Adaptive fuzzy control of multilateral asymmetric teleoperation for coordinated multiple mobile manipulators. IEEE Trans. Fuzzy Syst. **24**(1), 57–70 (2015)
5. Lee, H., Kim, H.J.: Constraint-based cooperative control of multiple aerial manipulators for handling an unknown payload. IEEE Trans. Ind. Inf. **13**(6), 2780–2790 (2017)
6. Zhang, Z., Chen, S., Liang, J.: Discrete-time circadian rhythms neural network for perturbed redundant robot manipulators tracking problem with periodic noises. IEEE Trans. Ind. Inf. **18**(1), 242–251 (2022)
7. Zhang, Z., Zheng, L., Chen, Z., Kong, L., Karimi, H.R.: Mutual-collision-avoidance scheme synthesized by neural networks for dual redundant robot manipulators executing cooperative tasks. IEEE Trans. Neural Netw. Learning Syst. **32**(3), 1052–1066 (2021)
8. Y. Zhang, K. Chen, X. Li, et al: Simulink modeling and comparison of zhang neural networks and gradient neural networks for time-varying lyapunov equation solving. In: 2008 Fourth International Conference on Natural Computation. vol. 2, pp. 521-525 (2008)
9. Guo, D., Zhang, Y.: Zhang neural network for online solution of time-varying linear matrix inequality aided with an equality conversion. IEEE Trans. Neural Netw. Learning Syst. **25**(2), 370–382 (2013)
10. Zhang, Z., Deng, X., Zheng, L.: A review on varying-parameter convergence differential neural network. Neurocomputing **490**, 54–65 (2022)

11. Zhang, Z., Yang, S., Chen, S., Luo, Y., Yang, H., Liu, Y.: A vector-based constrained obstacle avoidance scheme for wheeled mobile redundant robot manipulator. IEEE Trans. Cogn. Dev. Syst. **13**(3), 465–474 (2021)
12. Guo, D., Zhang, Y.: A New Inequality-Based Obstacle-Avoidance MVN Scheme and Its Application to Redundant Robot Manipulators. IEEE Trans. Syst. Man Cybern. Part C **42**(6), 1326–1340 (2012)

Distributed Synchronized Control of Multiple Robotic Arms: A Multi-objective Optimization via State-Coupled Neural Network

Qi Wang, Xingru Li, and Zhijun Zhang(✉)

South China University of Technology, Guangzhou 510640, China
auzjzhang@scut.edu.cn

Abstract. This paper proposes a State-Coupled Neural Network (SCNN) for the synchronized collaboration of Distributed Multiple Robotic Arms (DMRAs), ensuring synchronization in both the end-effector's Cartesian space and the joint velocity space. The problem is formulated as a quadratic programming framework that incorporates multiple objectives: minimum velocity two-norm, infinity-norm, and repetitive motion planning. A novel SCNN, designed based on the system's communication topology, is developed to solve this optimization, with its stability rigorously proven via Lyapunov theory. Simulation and experimental results validate the effectiveness and advantages of the proposed SCNN in solving the DMRAs collaboration problem.

Keywords: State-coupled neural network · multiple robotic arms · Multi-objective optimization · distributed system · synchronized collaboration

1 Introduction

In recent decades, multi-robot systems have garnered significant attention for executing complex tasks beyond the capability of a single robot [1–3]. Control architectures for such systems include decentralized, centralized, and distributed strategies. Among them, distributed control, relying on local communication, offers a balance between collaborative capability and system robustness [4–6].

For redundant manipulators, a fundamental challenge is inverse kinematics. Existing work on multi-arm collaboration either focuses on end-effector coordination in task space [7,8] or on joint synchronization without inter-manipulator collaboration [9,10]. A unified method achieving both simultaneous task-space collaboration and joint-space synchronization is currently lacking.

The substantial computational demand inherent in controlling multiple redundant manipulators necessitates an efficient, real-time solution. Recurrent Neural Networks (RNNs), with their proven real-time processing capability in

Z. Zhang (Ed.): ISNN 2026, LNCS 16848, pp. 234–245, 2027.
https://doi.org/10.1007/978-981-92-5870-3_19

robot control [12] (e.g., for dual-arm collision avoidance [13]), are a compelling candidate for this application.

(1) A State-Coupled Neural Network (SCNN) within a multi-objective optimization framework is proposed for multi-robot arm synchronous collaboration. It ensures both end-effector collaboration and joint-space synchronization while flexibly balancing performance indices like minimum-velocity two-norm, infinity-norm, and repetitive motion. Compared to task-space pose control, SCNN offers lower dimensionality and computational cost, making it suitable for tasks like collaborative painting or formation-based choreography.

(2) Unlike prior works that impose restrictions on multi-arm topology Refs. [6,14,15], the proposed SCNN has no special requirements. It employs a master-slave strategy where leaders receive desired trajectories and followers obtain information only from adjacent robots via coupling connections, allowing arbitrary inter-robot connectivity. Coupling connections. The connections between robots can be configured arbitrarily.

(3) Based on the system s communication topology, the SCNN solves the multi-objective optimization problem. It achieves joint speed synchronization through internal state coupling without adding explicit constraints. Comparative experiments with Zeroing Neural Networks demonstrate SCNN s superior performance in balancing multiple indicators.

2 Problem Formulation

In this section, the collaboration problem for multiple distributed robotic arms is formulated.

2.1 Distributed Multiple Robotic Arm Collaboration System

Consider a system of n identically configured arms, whose kinematics are described by:

$$g_i(\Theta_i(t)) = c_i(t), i = 1, 2, \cdots, n, \tag{1}$$

where $g_i(\cdot) : \mathbb{R}^k \to \mathbb{R}^m$ denotes the nonlinear mapping of the ith robotic arm from the joint state to the Cartesian space, $\Theta_i(t) = [\theta_{i1}(t), \theta_{i2}(t), \cdots, \theta_{ik}(t)]^{\mathrm{T}} \in \mathbb{R}^k$ denotes the joint state vector of the ith robotic arm, and $c_i(t) \in \mathbb{R}^m$ denotes the end-effector trajectory of the ith robotic arm in the world coordinate system.

In a distributed multiple robotic arm system, the leader must track a reference trajectory, while the followers collaborate with it to complete the task. Thus, the trajectory tracking constraint for the leader can be expressed as follows:

$$\kappa_i \left(g_i(\Theta_i(t)) - c_{\mathrm{R}}(t) \right) = 0, \tag{2}$$

where $c_{\mathrm{R}}(t)$ denotes reference trajectory. When the ith robotic arm is the leader, $\kappa_i = 1$, otherwise, $\kappa_i = 0$.

According to the graph theory, Eqn. (2) can be rewritten as:

$$\mathcal{K} \otimes I_m \left(G(\bar{\Theta}) - \mathrm{I}_n \otimes c_{\mathrm{R}}(t)\right) = 0, \tag{3}$$

where $\mathcal{K} = \mathrm{diag}\{\kappa_1, \kappa_2, \cdots, \kappa_n\}$, $\bar{\Theta} = \left[\Theta_1^{\mathrm{T}}, \Theta_2^{\mathrm{T}}, \cdots, \Theta_n^{\mathrm{T}}\right]^{\mathrm{T}}$, I_m denotes m-dimensional identity matrix, $\mathrm{I}_n = [1, 1, \cdots, 1]^{\mathrm{T}} \in R^n$, and $G = \left[g_1^{\mathrm{T}}, g_2^{\mathrm{T}}, \cdots, g_n^{\mathrm{T}}\right]^{\mathrm{T}}$. In addition, the derivation of Eqn. (2) yields a velocity constraint of the leader as follows:

$$\kappa_i \left(j_i(\Theta_i(t))\dot{\Theta}_i(t) - \dot{c}_{\mathrm{R}}(t)\right) = 0, \tag{4}$$

where $j_i(\Theta_i(t)) = \partial g_i / \partial \Theta_i$ is the Jacobian matrix of the ith robotic arm. According to the graph theory, Eqn. (4) can be rewritten as:

$$\mathcal{K} \otimes I_m \left(J(\bar{\Theta})\dot{\bar{\Theta}} - \mathrm{I}_n \otimes \dot{c}_{\mathrm{R}}(t)\right) = 0, \tag{5}$$

where $J = \left[j_1^{\mathrm{T}}, j_2^{\mathrm{T}}, \cdots, j_n^{\mathrm{T}}\right]^{\mathrm{T}}$.

When the leader starts to work, the followers need to collaborate with the leader to accomplish the task. That is, the positional coupling constraints between the arms are satisfied as follows:

$$g_i(\Theta_i(t)) \sum_{j \in \Omega_i} a_{ij} = \sum_{j \in \Omega_i} a_{ij} \left(g_j(\Theta_j(t)) + l_{ij}\right), \tag{6}$$

where Ω_i denotes the set of robotic arms connected to the ith manipulator, and l_{ij} denotes the positional relationship between the ith and the jth manipulators. According to the graph theory, (6) can be rewritten as:

$$\mathcal{L} \otimes I_m G(\bar{\Theta}) = \bar{l}, \tag{7}$$

where $\bar{l} = [\sum_{j \in \Omega_i} l_{1j}^{\mathrm{T}}, \sum_{j \in \Omega_i} l_{2j}^{\mathrm{T}}, \cdots, \sum_{j \in \Omega_i} l_{nj}^{\mathrm{T}}]^{\mathrm{T}}$.

The velocity coupling constraints between the multiple robotic arms are satisfied as follows:

$$\dot{g}_i(\Theta_i(t)) \sum_{j \in \Omega_i} a_{ij} = \sum_{j \in \Omega_i} a_{ij} \dot{g}_j(\Theta_j(t)). \tag{8}$$

According to the graph theory, (8) can be rewritten as:

$$\mathcal{L} \otimes I_m J(\bar{\Theta})\dot{\bar{\Theta}} = 0. \tag{9}$$

The joint angle and velocity limits are unified into a velocity-level inequality:

$$L\dot{\bar{\Theta}} \leq M, \tag{10}$$

where L and M are matrices constructed from the joint limits.

2.2 Collaboration Motion Generation For Distributed Multiple Robotic Arm System

The minimum velocity norm is considered as an optimization objective while multiple robotic arms collaborate to accomplish the task. According to the collaboration constraints (3), (5), (7), (9), and joint physical constraints (10), the collaborative motion generation for distributed multiple robotic arm system can be simplified as:

$$\begin{aligned}
\min.\ & \omega_1 \frac{1}{2}\dot{\bar{\Theta}}^T\dot{\bar{\Theta}} + \omega_2 s + \omega_3 \frac{\mu}{2}(\bar{\Theta} - \bar{\Theta}(0))^T(\bar{\Theta} - \bar{\Theta}(0)),\\
\text{s.t. } & \eta\,(A+B)\,G(\bar{\Theta}) + (A+B)\,J(\bar{\Theta})\dot{\bar{\Theta}}(t)\\
& \quad - \eta AC - A\dot{C} - \eta\bar{l} = 0,\\
& L\dot{\bar{\Theta}}(t) \leq M,\\
& \begin{bmatrix} I_{nk} \\ -I_{nk} \end{bmatrix}\dot{\bar{\Theta}}(t) + \begin{bmatrix} -\mathbf{1} \\ -\mathbf{1} \end{bmatrix} s(t) \leq M.
\end{aligned} \tag{11}$$

Here, the matrices $A = \mathcal{K} \otimes I_m$, $B = \mathcal{L} \otimes I_m$ are defined by the communication topology, while $C = \mathrm{I}_n \otimes c_{\mathrm{R}}(t)$ and $\dot{C} = \mathrm{I}_n \otimes \dot{c}_{\mathrm{R}}(t)$ describe the desired trajectory and its derivative, respectively. $\eta > 0$ is the convergence parameter, $\omega_1, \omega_2, \omega_3 \geq 0$ are weighting coefficients balancing different performance indices, $\mu > 0$ is a penalty parameter for repetitive motion, $\bar{\Theta}(0)$ is the initial joint angle vector, $s(t)$ is an auxiliary scalar variable introduced to handle the infinity-norm, and $\mathbf{1} \in \mathbb{R}^{nk}$ is an all-ones vector. In this optimization problem, the equality constraint corresponds to the original end-effector collaboration requirement, the first inequality constraint describes the physical joint limits, and the newly introduced second inequality constraint ensures $s(t) \geq \|\dot{\bar{\Theta}}(t)\|_\infty$. This pivotal transformation makes minimizing the auxiliary variable $s(t)$ mathematically equivalent to minimizing the infinity-norm of the joint velocities $\|\dot{\bar{\Theta}}(t)\|_\infty$, thereby converting the non-smooth infinity-norm minimization objective into a standard quadratic programming problem.

3 State-Coupled Neural Network

In this section, an SCNN is designed to solve the optimization problem (11). A Lagrange form of problem (11) is given as follows:

$$\begin{aligned}
&\mathbb{L}(\dot{\bar{\Theta}}(t), s(t), \lambda(t), \nu(t)) = \omega_1 \frac{1}{2}\dot{\bar{\Theta}}^T(t)\dot{\bar{\Theta}}(t) + \omega_2 s(t)\\
&+ \omega_3 \frac{\mu}{2}(\bar{\Theta}(t) - \bar{\Theta}(0))^T(\bar{\Theta}(t) - \bar{\Theta}(0))\\
&+ \lambda^T(t)(\eta(A+B)G + (A+B)J(\bar{\Theta})\dot{\bar{\Theta}}(t) - \eta AC - A\dot{C} - \eta\bar{l}\\
&+ \nu^T(t)(L\dot{\bar{\Theta}}(t) - M) + \kappa^T(\begin{bmatrix} I_{nk} \\ -I_{nk} \end{bmatrix}\dot{\bar{\Theta}}(t) + \begin{bmatrix} -\mathbf{1}_{nk} \\ -\mathbf{1}_{nk} \end{bmatrix} s(t))
\end{aligned} \tag{12}$$

where $\lambda \in \mathbb{R}^{nm}$, $\nu \in \mathbb{R}^{2nk}$ and $\kappa \in \mathbb{R}^{2nk}$ are Lagrange multipliers.

Lemma 1. *The vector* $\dot{\hat{\Theta}}^* \in \mathbb{R}^{nk}$ *and the scalar* $s^* \in \mathbb{R}$ *are the optimal solution of (11) if and only if there exist* $\lambda^* \in \mathbb{R}^{nm}$, $\nu^* \in \mathbb{R}^{2nk}$ *and* $\kappa^* \in \mathbb{R}^{2nk}$ *satisfy the following Karush-Kuhn-Tucker (KKT) conditions [16], i.e.,*

$$\begin{cases} \omega_1\dot{\bar{\Theta}}^* + \omega_3\mu(\bar{\Theta} - \bar{\Theta}(0)) + [(A+B)J]^T \lambda^* + L^T\nu^* + \begin{bmatrix} I_{nk} - I_{nk} \end{bmatrix} \kappa^* = 0, \\ \omega_2 - \begin{bmatrix} \mathbf{1}_{nk}^T \ \mathbf{1}_{nk}^T \end{bmatrix} \kappa^* = 0, \\ \eta(A+B)G + (A+B)J\dot{\hat{\Theta}}^* - \eta AC - A\dot{C} - \eta\bar{l} = 0, \\ M - L\dot{\hat{\Theta}}^* \geq 0, \quad (\nu^*)^T(L\dot{\hat{\Theta}}^* - M) = 0, \\ -\begin{bmatrix} I_{nk} \\ -I_{nk} \end{bmatrix} \dot{\hat{\Theta}}^* - \begin{bmatrix} -\mathbf{1}_{nk} \\ -\mathbf{1}_{nk} \end{bmatrix} s^* \geq 0, \\ (\kappa^*)^T \left(\begin{bmatrix} I_{nk} \\ -I_{nk} \end{bmatrix} \dot{\hat{\Theta}}^* + \begin{bmatrix} -\mathbf{1}_{nk} \\ -\mathbf{1}_{nk} \end{bmatrix} s^* \right) = 0. \end{cases} \tag{13}$$

Combining the KKT condition (13) with the Fischer-Burmeister nonlinear complementarity function, the following system of equations is obtained:

$$\begin{cases} \omega_1\dot{\bar{\Theta}}^* + \omega_3\mu(\bar{\Theta} - \bar{\Theta}(0)) + [(A+B)J]^T\lambda^* + L^T\nu^* + [I_{nk} - I_{nk}]\,\kappa^* = 0, \\ \omega_2 - \begin{bmatrix} \mathbf{1}_{nk}^T - \mathbf{1}_{nk}^T \end{bmatrix} \kappa^* = 0, \\ \eta(A+B)G + (A+B)J\dot{\bar{\Theta}}^* - \eta AC - A\dot{C} - \eta\bar{l} = 0, \\ \zeta_{FB}^{\sigma}\left(M - L\dot{\bar{\Theta}}^*, \nu^*\right) = 0, \\ \zeta_{FB}^{\sigma}\left(-\begin{bmatrix} I_{nk} \\ -I_{nk} \end{bmatrix} \dot{\bar{\Theta}}^* - \begin{bmatrix} -\mathbf{1}_{nk} \\ -\mathbf{1}_{nk} \end{bmatrix} s^*, \kappa^* \right) = 0. \end{cases} \tag{14}$$

The error function be denoted by $E(t) = \left[\varepsilon_1^{\mathrm{T}}(t), \varepsilon_2^{\mathrm{T}}(t), \varepsilon_3^{\mathrm{T}}(t), \varepsilon_4^{\mathrm{T}}(t), \varepsilon_5(t), \varepsilon_6^{\mathrm{T}}(t)\right]^{\mathrm{T}}$, where

$$\begin{aligned} \varepsilon_1(t) &= \omega_1\dot{\bar{\Theta}} + \omega_3\alpha(\bar{\Theta} - \bar{\Theta}(0)) + [(A+B)J]^{\mathrm{T}}\lambda + L^{\mathrm{T}}\nu + [I_{nk}, -I_{nk}]\kappa \\ \varepsilon_2(t) &= \eta(A+B)G + (A+B)J\dot{\bar{\Theta}} - \eta AC - A\dot{C} - \eta\bar{l} \\ \varepsilon_3(t) &= \zeta_{\mathrm{FB}}^{\sigma}(M - L\dot{\bar{\Theta}}, \nu) \\ \varepsilon_4(t) &= \mathcal{L} \otimes I_k\dot{\bar{\Theta}} \\ \varepsilon_5(t) &= \omega_2 - [\mathbf{1}_{nk}^{\mathrm{T}}, \mathbf{1}_{nk}^{\mathrm{T}}]\kappa \\ \varepsilon_6(t) &= \zeta_{\mathrm{FB}}^{\sigma}(-\begin{bmatrix} I_{nk} \\ -I_{nk} \end{bmatrix} \dot{\bar{\Theta}} - \begin{bmatrix} -\mathbf{1}_{nk} \\ -\mathbf{1}_{nk} \end{bmatrix} s, \kappa) \end{aligned}$$

Define the augmented state vector $y = \left[\dot{\bar{\Theta}}^{\mathrm{T}}, s, \lambda^{\mathrm{T}}, \nu^{\mathrm{T}}, \kappa^{\mathrm{T}}\right]^{\mathrm{T}} \in \mathbb{R}^{3nk+nm+1}$. The matrix form of the error function is as follows:

$$E(t) = R_1 y + R_2, \tag{15}$$

By deriving the error function $E(t)$, it can ultimately be expressed as a linear form with respect to the state variable y and its derivatives:

$$\dot{E}(t) = Q_1\dot{y} + Q_2 y + Q_3, \tag{16}$$

where

$$Q_1 = \begin{bmatrix} I_{nk} & \mathbf{0} & [(A+B)J]^{\mathrm{T}} & L^{\mathrm{T}} & [I_{nk}, -I_{nk}] \\ \mathbf{0} & 0 & \mathbf{0} & \mathbf{0} & \mathbf{0} \\ -L + D_1 L & \mathbf{0} & \mathbf{0} & I_{2nk} - D_2 & \mathbf{0} \\ \mathcal{L} \otimes I_k & \mathbf{0} & \mathbf{0} & \mathbf{0} & \mathbf{0} \\ \mathbf{0}^{\mathrm{T}} & 0 & \mathbf{0}^{\mathrm{T}} & \mathbf{0}^{\mathrm{T}} & -[\mathbf{1}_{nk}^{\mathrm{T}}, \mathbf{1}_{nk}^{\mathrm{T}}] \\ \Lambda & \Upsilon & \mathbf{0} & \mathbf{0} & I_{2nk} - D_4 \end{bmatrix}$$

$$Q_2 = \begin{bmatrix} \mathbf{0} & \mathbf{0} & [(A+B)\dot{J}]^{\mathrm{T}} & \mathbf{0} & \mathbf{0} \\ (A+B)\dot{J} + \eta(A+B)J & \mathbf{0} & \mathbf{0} & \mathbf{0} & \mathbf{0} \\ \mathbf{0} & \mathbf{0} & \mathbf{0} & \mathbf{0} & \mathbf{0} \\ \mathbf{0} & \mathbf{0} & \mathbf{0} & \mathbf{0} & \mathbf{0} \\ \mathbf{0}^{\mathrm{T}} & 0 & \mathbf{0}^{\mathrm{T}} & \mathbf{0}^{\mathrm{T}} & \mathbf{0}^{\mathrm{T}} \\ \mathbf{0} & \mathbf{0} & \mathbf{0} & \mathbf{0} & \mathbf{0} \end{bmatrix}$$

$$Q_3 = \begin{bmatrix} \omega_3 \mu \dot{\Theta} \\ -A\ddot{C} - \eta A\dot{C} \\ \dot{M} - D_1 \dot{M} \\ \mathbf{0}_{N_1 \times 1} \\ 0 \\ \mathbf{0}_{N_3 \times 1} \end{bmatrix}, y = \begin{bmatrix} \dot{\Theta} \\ s \\ \lambda \\ \nu \\ \kappa \end{bmatrix}.$$

Among them, $\Lambda = (I - D_3)\begin{bmatrix} -I_{nk} \\ I_{nk} \end{bmatrix}$, $\Upsilon = (I - D_3)\begin{bmatrix} \mathbf{1}_{nk} \\ \mathbf{1}_{nk} \end{bmatrix}$.

According to the error system (16), the enhanced SCNN is designed as follows

$$\dot{E}(t) = -\gamma \Psi(E(t)) - \rho \Gamma E(t) - K E(t), \tag{17}$$

where $\gamma > 0$ is a given constant parameter used to scale the convergence rate of the network [17]. $\Psi(E(t)) = [\varphi(e_1(t)), \varphi(e_2(t)), \cdots, \varphi(e_{3nk+nm+1}(t))]^{\mathrm{T}} : \mathbb{R}^{3nk+nm+1} \to \mathbb{R}^{3nk+nm+1}$ denotes the activation function, which is a monotonically increasing odd function. In this paper, a linear activation function is used, i.e. $\Psi(E(t)) = E(t)$. $e_j(t), j = 1, 2, \cdots, 3nk + nm + 1$ denotes the jth element of the vector $E(t)$. $K \in \mathbb{R}^{(3nk+nm+1)\times(3nk+nm+1)}$ is a feedback gain matrix. ρ is the coupling strength of the network, and $\Gamma = [\mathbf{0}, \mathbf{0}; \mathbf{0}, I_{nk}] \in \mathbb{R}^{(3nk+nm+1)\times(3nk+nm+1)}$.

Since Γ is a singular matrix, the SCNN is strictly speaking partially state coupling. To ensure the stability of the SCNN, this feedback term $KE(t)$ is added to the SCNN, where the feedback gain matrix K is a nonsingular matrix.

Substituting (16) into (17), the dynamic equation for the SCNN is obtained as

$$Q_1\dot{y} = -\gamma \Psi(E(t)) - Q_2 y - Q_3 - \rho \Gamma E(t) - K E(t). \tag{18}$$

The differential equation (18) is the SCNN model proposed in this paper for solving the multiple robotic arms synchronized collaboration problem (11), and its algorithmic flowchart is shown in Fig. 1. Then, the stability theorem of SCNN (18) is given.

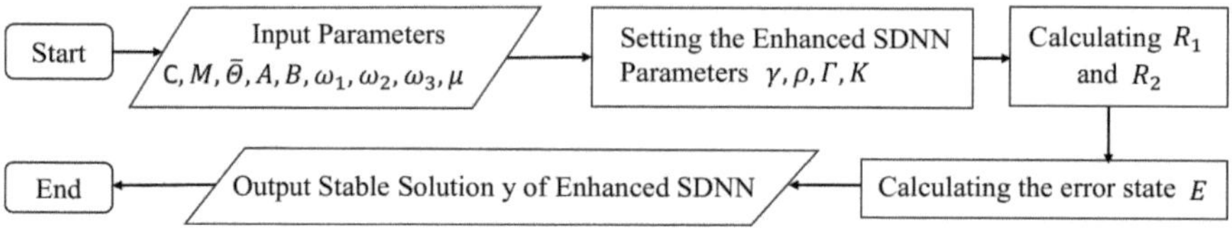

Fig. 1. The algorithmic flowchart of the SCNN.

Theorem 1. *The F-B function is used to map a time-varying multi-objective optimization problem (11) to the system of nonlinear equations (14), and* $\psi(\cdot)$ *is a monotonically-increasing odd activation function. Given scalar* $\rho > 0$ *and matrix* $\Gamma \in \mathbb{R}^{(4nk+nm)\times(3nk+nm+1)}$. *If there exist diagonal matrix* $P > 0 \in \mathbb{R}^{(3nk+nm+1)\times(3nk+nm+1)}$ *and* $K \in \mathbb{R}^{(4nk+nm)\times(3nk+nm+1)}$ *satisfying the following linear matrix inequality (LMI)*

$$-\rho P\Gamma - \rho\Gamma^{\mathrm{T}}P - PK - K^{\mathrm{T}}P < 0, \tag{19}$$

then the state variable y *of the SCNN (18), starting from any initial state* $y(0) \in \mathbb{R}^{3nk+nm+1}$, *asymptotically converges to the unique theoretical solution* y^*. *In addition, the solution* $\dot{\Theta}^*$ *of (11) is the first* $n \times k$ *elements of* y^*.

Proof. Construct the Lyapunov function candidate as

$$V(t) = E^{\mathrm{T}}(t)PE(t). \tag{20}$$

The derivation of $V(t)$ along the solution of system (17) is obtained as:

$$\begin{aligned}\dot{V}(t) &= 2E^{\mathrm{T}}(t)P\dot{E}(t)\\ &= -2\gamma E^{\mathrm{T}}(t)P\Psi(E(t)) - 2E^{\mathrm{T}}(t)\left(\rho P\Gamma + PK\right)E(t).\end{aligned} \tag{21}$$

According to the nature of the activation function $\Psi(\cdot)$, the following inequality holds:

$$-2\gamma E^{\mathrm{T}}(t)P\Psi(E(t)) \le 0. \tag{22}$$

The linear matrix inequality (19) guarantees that,

$$-2E^{\mathrm{T}}(t)\left(\rho P\Gamma + PK\right)E(t) \le 0. \tag{23}$$

From (22) and (23), we have $\dot{V}(t) < 0$, for $t \in [0, +\infty), E(t) \neq 0$. According to Lynapnov's theory, the SCNN (17) asymptotically converges to 0. Then, based on Theorem 1, $y^* - y$ converges to 0, i.e., the SCNN (17) can converge to the optimal solution of the optimization problem (11). The proof is completed. □

4 Simulation Results

The effectiveness of the proposed enhanced SCNN is validated through numerical simulations. A distributed five-robotic-arm system is used to accomplish synchronized drawing of star-shaped trajectories, demonstrating its multi-objective optimization and synchronization capabilities. All simulations are conducted on a MATLAB platform.

4.1 Synchronized Collaboration of Distributed Five Robotic Arm System

In this simulation, a distributed five robotic arm system is used to accomplish synchronized drawing of star-shaped trajectories. The physics limits of robotic arms are set as $\Theta_i^+ = 5\,\text{rad}, i = 1, 2, \cdots, 5$, $\Theta_i^- = -5\,\text{rad}$, $\dot{\Theta}_i^+ = 2\,\text{rad/s}$, $\dot{\Theta}_i^- = -2\,\text{rad/s}$. Task time is set to 5 s. The bases of the five robotic arms are mounted at positions $(0, 2, 0)$, $(0, 1, 0)$, $(0, 0, 0)$, $(0, -1, 0)$, and $(0, -2, 0)$ in the world coordinate system. It is worth noting that the SCNN (18) proposed in this paper is applicable to arbitrary topologies. The Laplace matrix of the system is set as $\mathcal{L} = [2, -1, -1, 0, 0; -1, 2, 0, -1, 0; -1, 0, 2, 0, -1; 0, -1, 0, 1, 0; 0, 0, -1, 0, 1]$.

SCNN (18) parameters: $\gamma = 1.5$, $\rho = 5$, and $K = 0.01I_{135}$. According to the LMI toolbox of Matlab, the condition (19) is feasible and $P = [2.7835I_{105}, \mathbf{0}; \mathbf{0}, 0.2967I_{30}]$.

The simulation results of the distributed five robotic arms synchronized to draw the star-shaped trajectories are shown in Figs. 2–5. The motion trajectories of the five robotic arms to accomplish the task are given in Fig. 2, where R_1 is the leader, R_2, R_3, and R_4 are the follower. Based on the communication topology between the robotic arms, the synchronization errors between the robotic arms are defined as $\text{E}_{\Theta 1} = 2\Theta_1 - \Theta_2 - \Theta_3$, $\text{E}_{\Theta 2} = 2\Theta_2 - \Theta_1 - \Theta_4$, $\text{E}_{\Theta 3} = 2\Theta_3 - \Theta_1 - \Theta_5$, $\text{E}_{\Theta 4} = \Theta_4 - \Theta_2$, and $\text{E}_{\Theta 5} = \Theta_5 - \Theta_3$, where $\text{E}_{\Theta i} = [\text{E}_{\theta i1}, \text{E}_{\theta i2}, \cdots, \text{E}_{\theta i6}]$ denotes the angular synchronization error of ith robotic arm. The velocity synchronization error is denoted as $\text{E}_{\dot{\Theta} i} = [\text{E}_{\dot{\theta} i1}, \text{E}_{\dot{\theta} i2}, \cdots, \text{E}_{\dot{\theta} i6}]$. The joint angle and velocity synchronization errors are shown in Figs. 3 (a)–(e) show that all angle synchronization errors remain below 4×10^{-9}. Similarly, Fig. 3 (f)–(j) indicate that all velocity synchronization errors are under 4×10^{-8}. These results demonstrate that the SCNN effectively maintains joint synchronization.

The leader R_1 tracking star-shaped trajectory errors are shown in Fig. 4, where Fig. 4 (a) and (b) show the tracking position error (TPE) and tracking velocity error (TVE) of the end-effector in the world coordinate system, respectively. From this, it can be concluded that the trajectory tracking of the leader is successful. Second, P_i is defined as the position coordinate of the end-effector of the ith arm. According to the communication topology between the robotic arms, the collaborative position errors of five end-effectors are denoted as $\text{E}_{\text{C}1} = 4\text{P}_1 - \text{P}_2 - \text{P}_3 - \text{P}_4 - \text{P}_5 - l_{12} - l_{13} - l_{14} - l_{15}$, $\text{E}_{\text{C}2} = 4\text{P}_2 - \text{P}_1 - \text{P}_3 - \text{P}_4 - \text{P}_5 - l_{21} - l_{24} - l_{23} - l_{25}$, $\text{E}_{\text{C}3} = 2\text{P}_3 - \text{P}_1 - \text{P}_2 - l_{31} - l_{32}$, $\text{E}_{\text{C}4} = 2\text{P}_4 - \text{P}_1 - \text{P}_2 - l_{41} - l_{42}$, $\text{E}_{\text{C}5} = 2\text{P}_5 - \text{P}_1 - \text{P}_2 - l_{51} - l_{52}$,$\text{E}_{\text{C}6} = 4\text{P}_6 - \text{P}_7 - \text{P}_8 - \text{P}_9 - \text{P}_10 - l_{67} - l_{68} - l_{69} - l_{610}$,

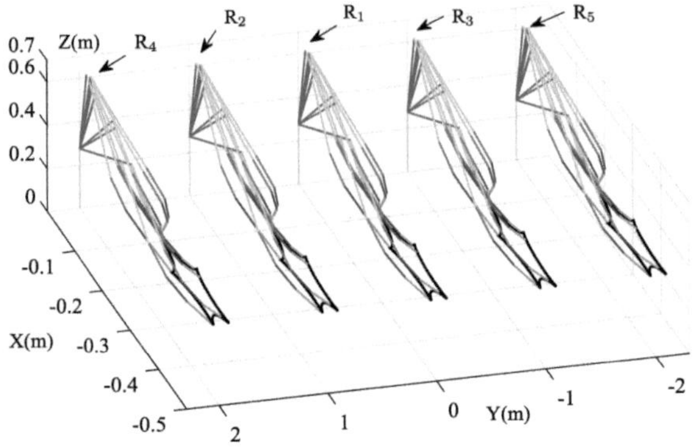

Fig. 2. Motion trajectories of the five robotic arms completing the task.

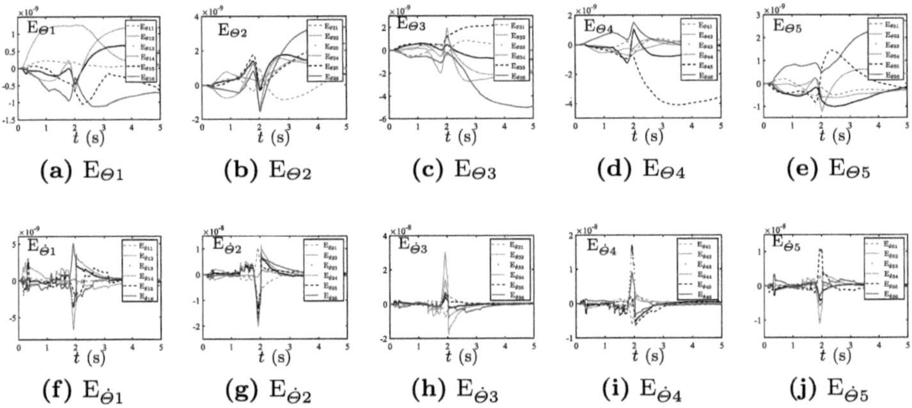

Fig. 3. Joint angle synchronization errors and joint velocity synchronization errors between robotic arms. (a) $E_{\Theta 1}$. (b) $E_{\Theta 2}$. (c) $E_{\Theta 3}$. (d) $E_{\Theta 4}$. (e) $E_{\Theta 5}$. (f) $E_{\dot{\Theta}1}$. (g) $E_{\dot{\Theta}2}$. (h) $E_{\dot{\Theta}3}$. (i) $E_{\dot{\Theta}4}$. (j) $E_{\dot{\Theta}5}$.

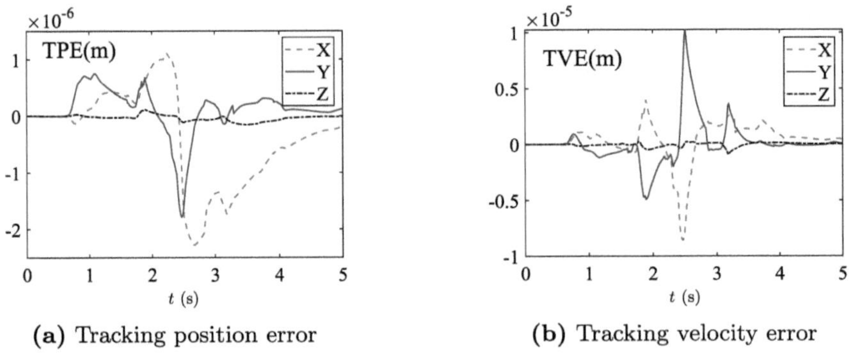

Fig. 4. Leader tracking star-shaped trajectory errors.

$E_{C6} = 4P_6 - P_7 - P_8 - P_9 - P_10 - l_{76} - l_{78} - l_{79} - l_{710}$, $E_{C8} = 2P_8 - P_6 - P_7 - l_{86} - l_{87}$, $E_{C9} = 2P_9 - P_6 - P_7 - l_{96} - l_{97}$, and $E_{C10} = 2P_10 - P_6 - P_7 - l_{106} - l_{107}$ where $E_{Ci}, i = 1, 2, \cdots, 10$ denotes the collaborative position error between the ith robotic arm and the coupled robotic arm. Figures 5 (a)–(e) and (f)-(j) present the collaborative position and velocity errors of the five end-effectors, respectively. Both error sets remain extremely small, with position errors below 1.5×10^{-9} and velocity errors below 5×10^{-9}, confirming successful collaboration at both the positional and velocity levels. In summary, the results from Fig. 2 to Fig. 5 demonstrate that the SCNN (18) ensures synchronized collaboration across joint angles, joint velocities, end-effector position, and end-effector velocity.

Table 1 summarizes the repetitive motion performance. After tracing the star-shaped trajectory, the maximum joint angle deviation is 2.68×10^{-2} rad, with an average of 9.51×10^{-3} rad across all 30 joints. These small deviations confirm that the enhanced SCNN effectively returns the joints near their initial configuration, achieving high-precision repetitive motion.

Table 1. Overall Performance of Repetitive Motion Planning

Performance Metric	Value (*rad*)
Maximum absolute joint error,max $\|\bar{\Theta}(T) - \bar{\Theta}(0)\|$	$\mathbf{2.68 \times 10^{-2}}$
Average absolute joint error, mean$\|\bar{\Theta}(T) - \bar{\Theta}(0)\|$	$\mathbf{9.51 \times 10^{-3}}$

The dynamic process of repetitive motion planning is visualized in Figs. 6 (a)–(e), which shows the evolution of the joint angle deviation norm $\|\bar{\theta}_i(t) - \bar{\theta}_i(0)\|_2$ for all five robotic arms. The deviation curves for all arms exhibit the same trend: an initial rise followed by a controlled decline, converging to a small value by the task's end. This visually confirms that the repetitive motion term $\omega_3\mu(\bar{\Theta} - \bar{\Theta}(0))$

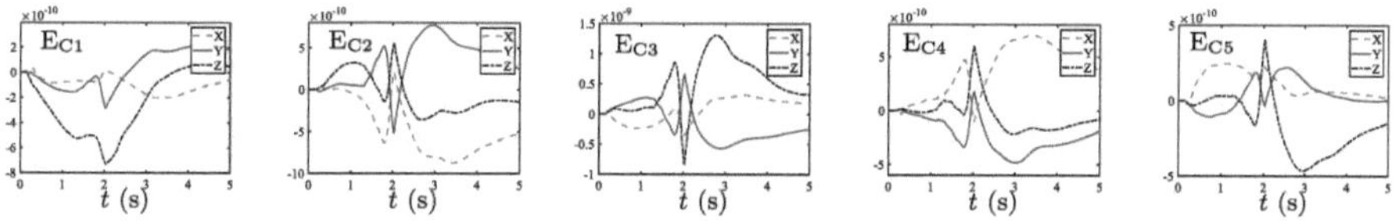

(a) E_{C1} **(b)** E_{C2} **(c)** E_{C3} **(d)** E_{C4} **(e)** E_{C5}

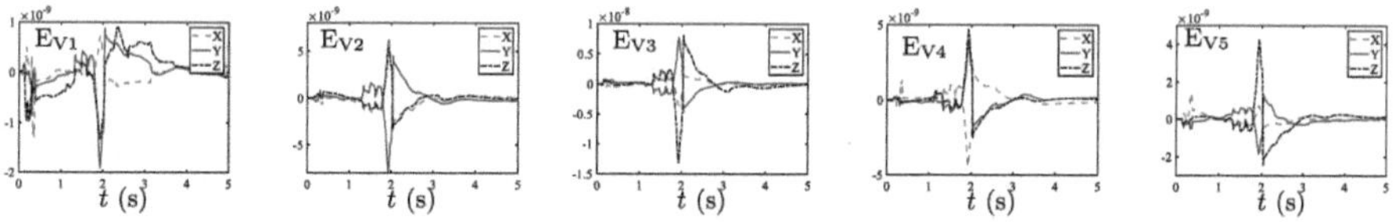

(f) E_{V1} **(g)** E_{V2} **(h)** E_{V3} **(i)** E_{V4} **(j)** E_{V5}

Fig. 5. Collaboration errors of distributed five robotic arms system.(a) E_{C1}. (b) E_{C2}. (c) E_{C3}. (d) E_{C4}. (e) E_{C5}. (f) E_{V1}. (g) E_{V2}. (h) E_{V3}. (i) E_{V4}. (j) E_{V5}.

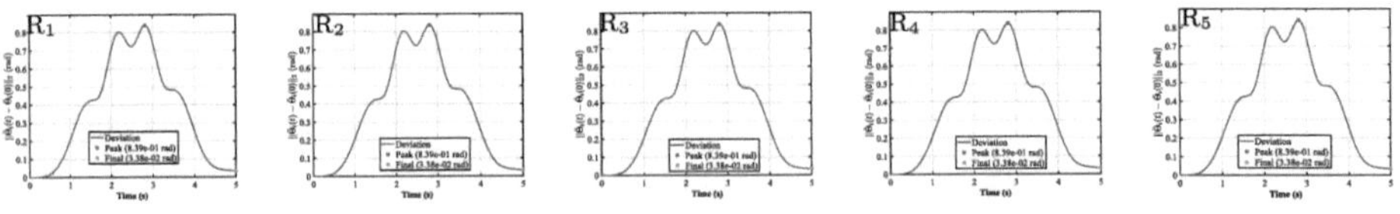

(a) R 1 **(b)** R 2 **(c)** R 3 **(d)** R 4 **(e)** R 5

Fig. 6. Evolution of joint angle deviation norm $\|\bar{\theta}_i(t) - \bar{\theta}_i(0)\|_2$ for repetitive motion validation.

continuously acts to suppress joint angle drift throughout the task. The near-overlap of all five curves further demonstrates the joint-space synchronization achieved by the SCNN.

5 Conclusion

This paper proposes a state-coupled neural network (SCNN) to solve the synchronization and collaboration problem of distributed multi-robot arms (DMRAs). This method constructs the problem as a multi-objective quadratic programming framework that integrates the minimum velocity two-norm, infinity norm, and repetitive motion planning. The SCNN, based on the communication topology design of DMRAs, synchronizes the joint movements of the robotic arms through its internal partial state coupling mechanism without adding additional optimization constraints. The simulation results of five robotic arms drawing a star-shaped trajectory show that the SCNN can effectively ensure the collaboration accuracy of the end effector and the synchronization of the joint space (including angles and velocities), and successfully achieve the high-precision repetitive motion planning of joint returning to the initial position.

References

1. Ning, B., Han, Q.-L., Lu, Q., Sanjayan, J.: Cooperative control of multirobot systems subject to control gain uncertainty. IEEE Trans. Industr. Inf. **19**(4), 5367–5376 (2022)
2. Chen, B., Zhang, Y., Xia, X., Martinez-Garcia, M., Jombo, G.: Knowledge sharing enabled multirobot collaboration for preventive maintenance in mixed model assembly. IEEE Trans. Industr. Inf. **18**(11), 8098–8107 (2022)
3. Xu, Z., Zhao, L.: Distributed adaptive gain-varying finite-time event-triggered control for multiple robot manipulators with disturbances. IEEE Trans. Ind. Inf. **19**(9), 9302–9313 (2023)
4. Li, S., Chen, S., Liu, B., Li, Y., Liang, Y.: Decentralized kinematic control of a class of collaborative redundant manipulators via recurrent neural networks. Neurocomputing **91**, 1–10 (2012)
5. Ren, Y., Sosnowski, S., Hirche, S.: Fully distributed cooperation for networked uncertain mobile manipulators. IEEE Trans. Rob. **36**(4), 984–1003 (2020)

6. Li, S., He, J., Li, Y., Rafique, M.U.: Distributed recurrent neural networks for cooperative control of manipulators: a game-theoretic perspective. IEEE Trans. Neural Netw. Learn. Syst. **28**(2), 415–426 (2016)
7. Jin, L., Li, S., Xiao, L., Lu, R., Liao, B.: Cooperative motion generation in a distributed network of redundant robot manipulators with noises. IEEE Trans. Syst. Man Cybern. Syst. **48**(10), 1715–1724 (2017)
8. Jin, L., Li, S., Luo, X., Li, Y., Qin, B.: Neural dynamics for cooperative control of redundant robot manipulators. IEEE Trans. Industr. Inf. **14**(9), 3812–3821 (2018)
9. Zhang, X., Lu, W., Su, M., Xu, W.: Research on synchronous control strategy of robot arm based on cross-coupling control. Int. J. Innov. Comput. Inf. Control **16**(6), 1987–2005 (2020)
10. Shang, W., Cong, S., Zhang, Y., Liang, Y.: Active joint synchronization control for a 2-DOF redundantly actuated parallel manipulator. IEEE Trans. Control Syst. Technol. **17**(2), 416–423 (2008)
11. Almusawi, A. R., Dülger, L. C., and Kapucu, S.: A new artificial neural network approach in solving inverse kinematics of robotic arm. Computational Intelligence and Neuroscience (2016). https://doi.org/10.1155/2016/5720163
12. He, W., Sun, Y., Yan, Z., Yang, C., Li, Z., Kaynak, O.: Disturbance observer-based neural network control of cooperative multiple manipulators with input saturation. IEEE Trans. Neural Netw. Learn. Syst. **31**(5), 1735–1746 (2019)
13. Zhang, Z., Zheng, L., Chen, Z., Kong, L., Karimi, H.R.: Mutual-collision-avoidance scheme synthesized by neural networks for dual redundant robot manipulators executing cooperative tasks. IEEE Trans. Neural Netw. Learn. Syst. **32**(3), 1052–1066 (2020)
14. Zhang, J., Jin, L., Cheng, L.: RNN for perturbed manipulability optimization of manipulators based on a distributed scheme: a game-theoretic perspective. IEEE Trans. Neural Netw. Learn. Syst. **31**(12), 5116–5126 (2020)
15. Li, S., Cui, H., Li, Y., Liu, B., Lou, Y.: Decentralized control of collaborative redundant manipulators with partial command coverage via locally connected recurrent neural networks. Neural Comput. Appl. **23**, 1051–1060 (2013)
16. Li, W., Ma, X., Luo, J., Jin, L.: A strictly predefined-time convergent neural solution to equality-and inequality-constrained time-variant quadratic programming. IEEE Trans. Syst. Man Cybern. Syst. **51**(7), 4028–4039 (2019)
17. Zhang, Z., Lu, Y., Zheng, L., Li, S., Yu, Z., Li, Y.: A new varying-parameter convergent-differential neural-network for solving time-varying convex QP problem constrained by linear-equality. IEEE Trans. Autom. Control **63**(12), 4110–4125 (2018)

Zeroing Neural Network Control for Underwater Manipulators with Unknown Kinematics

Haode Ye, Dan Xiang, Junliu Zhong, Jing Ling, Shuo Yang, and Yaqiong Ding(✉)

School of Artificial Intelligence, Guangzhou Maritime University, Guangzhou 510725, China
dingyaqiong@gzmtu.edu.cn

Abstract. This paper addresses the trajectory-tracking problem of underwater manipulators with unknown kinematics. To achieve accurate tracking without requiring exact prior kinematic information, a Jacobian-estimation-based zeroing neural network (ZNN) control scheme is developed. Specifically, a tracking error model is first established, based on which a ZNN design formula is introduced to derive the tracking control law. Then, an online Jacobian estimation mechanism is constructed by defining a Jacobian estimation residual and enforcing its exponential decay. With the estimated Jacobian matrix incorporated into the controller, the proposed method can generate the generalized velocity input $\dot{v}(t)$ in real time. Theoretical analysis shows that the Jacobian estimation residual converges exponentially and that the tracking error converges asymptotically to zero. Simulation results further verify the effectiveness and feasibility of the proposed method for underwater manipulator trajectory tracking.

Keywords: Underwater manipulator · Zeroing neural network (ZNN) · Trajectory tracking · Jacobian estimation · Unknown kinematics

1 Introduction

In recent years, trajectory tracking of underwater manipulator systems has attracted increasing attention because of its importance in underwater intervention, exploration, and maintenance tasks [1–8]. However, owing to the complexity and uncertainty of the marine environment, as well as the difficulty in obtaining an accurate mathematical model, achieving high-precision tracking for underwater manipulator systems remains a challenging problem [9–14]. In particular, the kinematic relationship of such systems is often difficult to characterize accurately in practical applications, which significantly degrades the tracking performance of conventional model-based control approaches.

Recent advances have further improved underwater manipulator control from different perspectives. Fixed-time and finite-time sliding mode methods have

Z. Zhang (Ed.): ISNN 2026, LNCS 16848, pp. 246–256, 2027.
https://doi.org/10.1007/978-981-92-5870-3_20

been developed to enhance convergence speed and robustness under model uncertainties and external disturbances. For example, Huang et al. proposed an adaptive nonsingular fixed-time sliding mode controller for moving-base underwater flexible manipulators [21], Zaare and Soltanpour designed a predefined-time nonsingular terminal sliding mode control scheme with uncertainty and velocity observers for underwater robot manipulators [22], and Wang et al. developed a global fast terminal sliding mode controller based on a finite-time extended state observer for underwater manipulators [23]. Meanwhile, reinforcement-learning-based methods have also been investigated for underwater manipulation, including reinforcement-learning-based visual servoing for underwater vehicle dual-manipulator systems [13] and deep reinforcement learning control for underwater manipulator systems under environmental disturbances and actuator faults [24]. In addition, digital-twin-assisted frameworks have recently been introduced for autonomous underwater grasping and intelligent operation, such as actor–critic learning via digital twins [25] and digital twin-based stress prediction with reinforcement learning [26].

Although these methods provide valuable solutions, several limitations still remain. First, fixed-time and finite-time sliding mode control methods usually require explicit kinematic or dynamic models, and their performance often depends on carefully designed observers, switching terms, and robust compensation structures. This may increase implementation complexity and may also lead to chattering or parameter-tuning difficulties in practical underwater operations. Second, reinforcement-learning-based controllers can improve adaptability, but they usually require sufficient training data, simulation environments, and carefully designed reward functions. Therefore, their closed-loop convergence and generalization properties are not always easy to guarantee theoretically. Third, digital-twin-based frameworks can improve simulation fidelity and safety evaluation, but they generally rely on the construction and calibration of high-fidelity virtual models, which may be costly when the manipulator parameters, hydrodynamic coefficients, and environmental conditions are uncertain.

In parallel, zeroing neural network (ZNN) has emerged as an efficient framework for solving time-varying problems and real-time robot control [15]. Meanwhile, recent ZNN-based studies have emphasized robustness, noise suppression, inverse-free design, and model-free implementation, and have demonstrated strong performance in manipulator trajectory tracking and motion control [16–18]. These developments indicate that ZNN is a promising candidate for online control problems involving unknown Jacobian-related mappings [16–20].

Nevertheless, current ZNN studies are still mainly developed for terrestrial or generic robot manipulators, whereas dedicated ZNN-based trajectory-tracking schemes for underwater manipulators with unknown kinematics remain limited. On the other hand, recent underwater manipulation research has paid more attention to visual servoing, interaction control, or system integration than to Jacobian-estimation-based ZNN tracking under unknown kinematics. Moreover, compared with recent sliding-mode, reinforcement-learning, and digital-twin-based underwater manipulation methods, the use of ZNN for online Jacobian estimation and theoretically analyzable tracking control under unknown underwater kinematics has not been sufficiently investigated. Therefore, there remains a need for a control framework that can simultaneously estimate the Jacobian online and generate generalized velocity inputs in real time without requiring exact prior kinematic information.

To address this issue, this paper develops a ZNN-based control framework for underwater manipulator systems with unknown kinematics. By utilizing time-derivative information together with tracking-error feedback, the Jacobian matrix is estimated online, such that accurate trajectory tracking can be achieved without requiring an exact kinematic model. On this basis, a corresponding tracking controller is constructed within the proposed framework. Furthermore, theoretical analysis is provided to show that the tracking error asymptotically converges to zero. The main contributions of this study are summarized as follows.

1. A Jacobian-estimation-based ZNN tracking framework is proposed for underwater manipulators with unknown kinematics.
2. The convergence and effectiveness of the proposed method are theoretically and numerically validated.

The rest of this paper is arranged as follows. Section 2 presents the mathematical modeling of the trajectory-tracking problem for underwater manipulators. Section 3 introduces the ZNN tracking controller based on Jacobian estimation. Section 4 provides theoretical analysis. Section 5 presents the simulation results, and Sect. 6 concludes the paper.

2 Problem Formulation

For the square case of an underwater manipulator system with n generalized position variables and an n-dimensional desired trajectory, the corresponding kinematic equation is given by

$$\dot{\boldsymbol{x}}_a(t) = \boldsymbol{J}(\boldsymbol{v}, t)\dot{\boldsymbol{v}}(t), \tag{1}$$

where $\boldsymbol{x}_a(t) \in \mathbb{R}^n$ denotes the actual trajectory, $\dot{\boldsymbol{x}}_a(t) \in \mathbb{R}^n$ denotes the actual velocity, $\dot{\boldsymbol{v}}(t) \in \mathbb{R}^n$ denotes the generalized velocity, and $\boldsymbol{J}(\boldsymbol{v}, t) \in \mathbb{R}^{n\times n}$ is the unknown Jacobian matrix.

The objective is to design $\dot{\boldsymbol{v}}(t)$ such that the actual trajectory $\boldsymbol{x}_a(t)$ tracks the desired trajectory $\boldsymbol{x}_d(t)$.

Define the trajectory-tracking error as

$$\boldsymbol{e}(t) = \boldsymbol{x}_a(t) - \boldsymbol{x}_d(t). \tag{2}$$

Differentiating (2), we obtain

$$\dot{\boldsymbol{e}}(t) = \boldsymbol{J}(\boldsymbol{v}, t)\dot{\boldsymbol{v}}(t) - \dot{\boldsymbol{x}}_d(t). \tag{3}$$

Then, we adopt the ZNN design formula

$$\dot{\boldsymbol{e}}(t) = -\lambda \boldsymbol{e}(t), \tag{4}$$

where $\lambda > 0$ is a positive design parameter.

Substituting (3) into (4), we obtain

$$\boldsymbol{J}(\boldsymbol{v}, t)\dot{\boldsymbol{v}}(t) = \dot{\boldsymbol{x}}_d(t) - \lambda \boldsymbol{e}(t). \tag{5}$$

3 Jacobian-Estimation-Based Tracking Control

Since the true Jacobian matrix $\boldsymbol{J}(\boldsymbol{v}, t)$ is unknown, an estimated Jacobian matrix is defined as

$$\hat{\boldsymbol{J}}(t) \in \mathbb{R}^{n\times n}. \tag{6}$$

Define the Jacobian estimation residual by

$$\boldsymbol{e}_J(t) = \hat{\boldsymbol{J}}(t)\dot{\boldsymbol{v}}(t) - \dot{\boldsymbol{x}}_a(t). \tag{7}$$

To drive the estimation residual to zero, the ZNN design formula is used as

$$\dot{\boldsymbol{e}}_J(t) = -\eta \boldsymbol{e}_J(t), \tag{8}$$

where $\eta > 0$ is a positive design parameter.

Differentiating (7) gives

$$\dot{\boldsymbol{e}}_J(t) = \dot{\hat{\boldsymbol{J}}}(t)\dot{\boldsymbol{v}}(t) + \hat{\boldsymbol{J}}(t)\ddot{\boldsymbol{v}}(t) - \ddot{\boldsymbol{x}}_a(t). \tag{9}$$

Substituting (7) and (9) into (8), we obtain

$$\dot{\hat{\boldsymbol{J}}}(t)\dot{\boldsymbol{v}}(t) = -\eta\big(\hat{\boldsymbol{J}}(t)\dot{\boldsymbol{v}}(t) - \dot{\boldsymbol{x}}_a(t)\big) - \hat{\boldsymbol{J}}(t)\ddot{\boldsymbol{v}}(t) + \ddot{\boldsymbol{x}}_a(t). \tag{10}$$

When $\|\dot{\boldsymbol{v}}(t)\|_2 \neq 0$, a constructive Jacobian update law is chosen as

$$\dot{\hat{\boldsymbol{J}}}(t) = \frac{\Big(-\eta\big(\hat{\boldsymbol{J}}(t)\dot{\boldsymbol{v}}(t) - \dot{\boldsymbol{x}}_a(t)\big) - \hat{\boldsymbol{J}}(t)\ddot{\boldsymbol{v}}(t) + \ddot{\boldsymbol{x}}_a(t)\Big)\dot{\boldsymbol{v}}^T(t)}{\|\dot{\boldsymbol{v}}(t)\|_2^2}. \tag{11}$$

When $\|\dot{\boldsymbol{v}}(t)\|_2 = 0$, we set

$$\dot{\hat{\boldsymbol{J}}}(t) = \boldsymbol{0}. \tag{12}$$

Substituting the estimated Jacobian matrix into (5), we obtain

$$\hat{\boldsymbol{J}}(t)\dot{\boldsymbol{v}}(t) = \dot{\boldsymbol{x}}_d(t) - \lambda \boldsymbol{e}(t). \tag{13}$$

If $\hat{\boldsymbol{J}}(t)$ is nonsingular, then

$$\dot{\boldsymbol{v}}(t) = \hat{\boldsymbol{J}}^{-1}(t)\big(\dot{\boldsymbol{x}}_d(t) - \lambda \boldsymbol{e}(t)\big). \tag{14}$$

4 Theoretical Analysis

In this section, the convergence properties of the proposed Jacobian-estimation-based ZNN controller are analyzed. Since the control law in (14) uses the inverse of $\hat{\boldsymbol{J}}(t)$, we assume that $\hat{\boldsymbol{J}}(t)$ remains nonsingular during the control process. In addition, $\boldsymbol{x}_d(t)$, $\boldsymbol{x}_a(t)$, $\dot{\boldsymbol{x}}_a(t)$, and $\ddot{\boldsymbol{x}}_a(t)$ are assumed to be bounded and continuously differentiable. Moreover, when $\|\dot{\boldsymbol{v}}(t)\|_2 = 0$, it is additionally assumed that $\ddot{\boldsymbol{v}}(t) = \mathbf{0}$ and $\ddot{\boldsymbol{x}}_a(t) = \mathbf{0}$. This assumption guarantees the consistency of the piecewise Jacobian update law in (11) and (12).

Theorem 1. *For the Jacobian estimation residual defined in* (7)*, the update laws* (11) *and* (12) *guarantee exponential convergence of* $\boldsymbol{e}_J(t)$ *to zero, i.e.,*

$$\boldsymbol{e}_J(t) = e^{-\eta t}\boldsymbol{e}_J(0). \tag{15}$$

Thus, we have

$$\|\boldsymbol{e}_J(t)\|_2 \le \|\boldsymbol{e}_J(0)\|_2 e^{-\eta t}. \tag{16}$$

Proof. Differentiating (7) yields

$$\dot{\boldsymbol{e}}_J(t) = \dot{\hat{\boldsymbol{J}}}(t)\dot{\boldsymbol{v}}(t) + \hat{\boldsymbol{J}}(t)\ddot{\boldsymbol{v}}(t) - \ddot{\boldsymbol{x}}_a(t). \tag{17}$$

When $\|\dot{\boldsymbol{v}}(t)\|_2 \neq 0$, substituting (11) into (17) gives

$$\dot{\boldsymbol{e}}_J(t) = -\eta\boldsymbol{e}_J(t). \tag{18}$$

When $\|\dot{\boldsymbol{v}}(t)\|_2 = 0$, it follows from (1) that $\dot{\boldsymbol{x}}_a(t) = \mathbf{0}$, and thus

$$\boldsymbol{e}_J(t) = \hat{\boldsymbol{J}}(t)\dot{\boldsymbol{v}}(t) - \dot{\boldsymbol{x}}_a(t) = \mathbf{0}. \tag{19}$$

Together with (12) and the additional assumption $\ddot{\boldsymbol{v}}(t) = \mathbf{0}$ and $\ddot{\boldsymbol{x}}_a(t) = \mathbf{0}$, one still has

$$\dot{\boldsymbol{e}}_J(t) = \mathbf{0} = -\eta\boldsymbol{e}_J(t). \tag{20}$$

Therefore, (18) holds for all $t \ge 0$. Solving this linear system directly leads to (15) and (16). ∎

Theorem 2. *If the Jacobian estimation is exact, i.e.,* $\hat{\boldsymbol{J}}(t) = \boldsymbol{J}(\boldsymbol{v}, t)$ *for all* $t \ge 0$*, then the tracking error* $\boldsymbol{e}(t)$ *defined in* (2) *converges exponentially to zero according to*

$$\boldsymbol{e}(t) = e^{-\lambda t}\boldsymbol{e}(0), \tag{21}$$

and thus

$$\|\boldsymbol{e}(t)\|_2 \le \|\boldsymbol{e}(0)\|_2 e^{-\lambda t}. \tag{22}$$

Proof. If $\hat{\boldsymbol{J}}(t) = \boldsymbol{J}(\boldsymbol{v}, t)$, then (13) is equivalent to

$$\boldsymbol{J}(\boldsymbol{v}, t)\dot{\boldsymbol{v}}(t) = \dot{\boldsymbol{x}}_d(t) - \lambda\boldsymbol{e}(t). \tag{23}$$

Substituting (23) into (3) yields

$$\dot{\boldsymbol{e}}(t) = \boldsymbol{J}(\boldsymbol{v}, t)\dot{\boldsymbol{v}}(t) - \dot{\boldsymbol{x}}_d(t) = -\lambda\boldsymbol{e}(t). \tag{24}$$

Solving (24) gives (21). The norm inequality (22) follows immediately. ∎

Theorem 3. *Under the assumptions stated at the beginning of this section, and under the Jacobian estimation laws* (11) *and* (12) *together with the control law* (14), *the tracking error* $\boldsymbol{e}(t)$ *defined in* (2) *satisfies*

$$\dot{\boldsymbol{e}}(t) = -\lambda \boldsymbol{e}(t) - \boldsymbol{e}_J(t). \tag{25}$$

Furthermore, by Theorem 1,

$$\boldsymbol{e}_J(t) = e^{-\eta t}\boldsymbol{e}_J(0). \tag{26}$$

Hence, the tracking error admits the explicit solution

$$\boldsymbol{e}(t) = \begin{cases} e^{-\lambda t}\boldsymbol{e}(0) - \dfrac{e^{-\eta t} - e^{-\lambda t}}{\lambda - \eta}\boldsymbol{e}_J(0), & \lambda \neq \eta, \\ e^{-\lambda t}\big(\boldsymbol{e}(0) - t\,\boldsymbol{e}_J(0)\big), & \lambda = \eta. \end{cases} \tag{27}$$

Consequently, $\boldsymbol{e}(t) \to \boldsymbol{0}$ *as* $t \to \infty$. *Moreover, for* $\lambda \neq \eta$,

$$\|\boldsymbol{e}(t)\|_2 \leq \|\boldsymbol{e}(0)\|_2 e^{-\lambda t} + \frac{\|\boldsymbol{e}_J(0)\|_2}{|\lambda - \eta|}\left|e^{-\eta t} - e^{-\lambda t}\right|, \tag{28}$$

and for $\lambda = \eta$,

$$\|\boldsymbol{e}(t)\|_2 \leq \big(\|\boldsymbol{e}(0)\|_2 + t\|\boldsymbol{e}_J(0)\|_2\big)e^{-\lambda t}. \tag{29}$$

Therefore, the tracking error converges asymptotically to zero.

Proof. From (3), the tracking-error dynamics can be written as

$$\dot{\boldsymbol{e}}(t) = \boldsymbol{J}(\boldsymbol{v}, t)\dot{\boldsymbol{v}}(t) - \dot{\boldsymbol{x}}_d(t). \tag{30}$$

Using (13), which is equivalent to (14) under the nonsingularity assumption of $\hat{\boldsymbol{J}}(t)$, one has

$$\hat{\boldsymbol{J}}(t)\dot{\boldsymbol{v}}(t) = \dot{\boldsymbol{x}}_d(t) - \lambda \boldsymbol{e}(t). \tag{31}$$

Substituting (31) into (30) gives

$$\dot{\boldsymbol{e}}(t) = \boldsymbol{J}(\boldsymbol{v}, t)\dot{\boldsymbol{v}}(t) - \hat{\boldsymbol{J}}(t)\dot{\boldsymbol{v}}(t) - \lambda \boldsymbol{e}(t). \tag{32}$$

On the other hand, from the definition of the Jacobian estimation residual (7) and the kinematic equation (1), one has

$$\boldsymbol{e}_J(t) = \hat{\boldsymbol{J}}(t)\dot{\boldsymbol{v}}(t) - \dot{\boldsymbol{x}}_a(t) = \hat{\boldsymbol{J}}(t)\dot{\boldsymbol{v}}(t) - \boldsymbol{J}(\boldsymbol{v}, t)\dot{\boldsymbol{v}}(t). \tag{33}$$

Therefore,

$$\boldsymbol{J}(\boldsymbol{v}, t)\dot{\boldsymbol{v}}(t) - \hat{\boldsymbol{J}}(t)\dot{\boldsymbol{v}}(t) = -\boldsymbol{e}_J(t). \tag{34}$$

Substituting (34) into (32) yields (25).

By Theorem 1, the Jacobian estimation residual satisfies (26). Thus, (25) becomes

$$\dot{\boldsymbol{e}}(t) + \lambda \boldsymbol{e}(t) = -e^{-\eta t}\boldsymbol{e}_J(0). \tag{35}$$

Multiplying both sides of (35) by $e^{\lambda t}$ yields

$$\frac{d}{dt}\left(e^{\lambda t}\boldsymbol{e}(t)\right) = -e^{(\lambda-\eta)t}\boldsymbol{e}_J(0). \tag{36}$$

Integrating (36) over $[0, t]$, with s denoting the dummy integration variable, we obtain

$$e^{\lambda t}\boldsymbol{e}(t) - \boldsymbol{e}(0) = -\int_0^t e^{(\lambda-\eta)s}\boldsymbol{e}_J(0)\,ds. \tag{37}$$

If $\lambda \neq \eta$, then

$$\int_0^t e^{(\lambda-\eta)s}\,ds = \frac{e^{(\lambda-\eta)t} - 1}{\lambda - \eta}, \tag{38}$$

and substituting (38) into (37) gives (27). Applying the triangle inequality then yields (28).

If $\lambda = \eta$, then

$$\int_0^t e^{(\lambda-\eta)s}\,ds = \int_0^t 1\,ds = t, \tag{39}$$

which together with (37) gives (27). Hence, (29) follows.

Since $e^{-\lambda t} \to 0$, $e^{-\eta t} \to 0$, and $te^{-\lambda t} \to 0$ as $t \to \infty$, it follows that $\boldsymbol{e}(t) \to \mathbf{0}$. Therefore, the tracking error converges asymptotically to zero. ■

5 Simulation

In this section, to verify the effectiveness of the proposed ZNN-based tracking scheme for underwater manipulators with unknown kinematics, a numerical simulation is carried out in MATLAB. The kinematic model of the manipulator is described by (1) as follows:

$$\dot{\boldsymbol{x}}_a(t) = \boldsymbol{J}(\boldsymbol{v}, t)\dot{\boldsymbol{v}}(t),$$

where $\boldsymbol{x}_a(t) \in \mathbb{R}^3$ denotes the actual end-effector trajectory, $\boldsymbol{v}(t) \in \mathbb{R}^3$ denotes the generalized position variable, $\dot{\boldsymbol{v}}(t) \in \mathbb{R}^3$ denotes the generalized velocity, and $\boldsymbol{J}(\boldsymbol{v}, t) \in \mathbb{R}^{3\times 3}$ is the unknown time-varying Jacobian matrix. The desired trajectory is formulated as

$$\boldsymbol{x}_d(t) = \begin{bmatrix} x_d(t) \\ y_d(t) \\ z_d(t) \end{bmatrix} = \begin{bmatrix} 0.68\left(1 - e^{-\frac{4.2}{T}t}\right) + 0.035e^{-0.45t}\sin(1.1t) \\ 0.10e^{-\frac{3.2}{T}t}\sin\left(\frac{8}{T}t\right) \\ -0.22\left(1 - e^{-\frac{3.6}{T}t}\right) + 0.055e^{-\frac{3.2}{T}t}\cos\left(\frac{8}{T}t\right) \end{bmatrix}. \tag{40}$$

In the simulation, the total duration is set to $T = 10$ s and the sampling interval is chosen as $\Delta t = 0.01$ s. The initial conditions are set as $\boldsymbol{x}_a(0) = [0,\ 0,\ 0]^\top$ and $\boldsymbol{v}(0) = [0,\ 0,\ 0]^\top$. The initial estimate of the Jacobian matrix is chosen as

$$\hat{\boldsymbol{J}}(0) = \boldsymbol{J}(\boldsymbol{v}(0), 0) + \mathrm{randn}(3, 3),$$

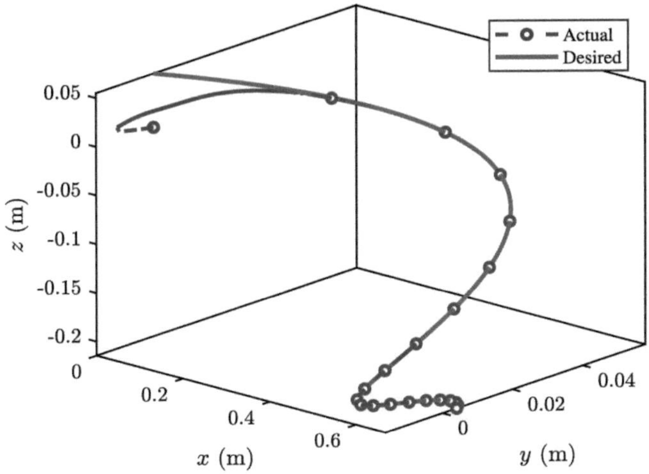

Fig. 1. Three-dimensional tracking trajectories $\boldsymbol{x}_a(t)$ and $\boldsymbol{x}_d(t)$ for $\boldsymbol{x}_a(0) = [0, 0, 0]^\top$, $\boldsymbol{x}_d(0) = [0, 0, 0.055]^\top$, and $\boldsymbol{v}(0) = [0, 0, 0]^\top$ under $\lambda = 10$ and $\eta = 30$.

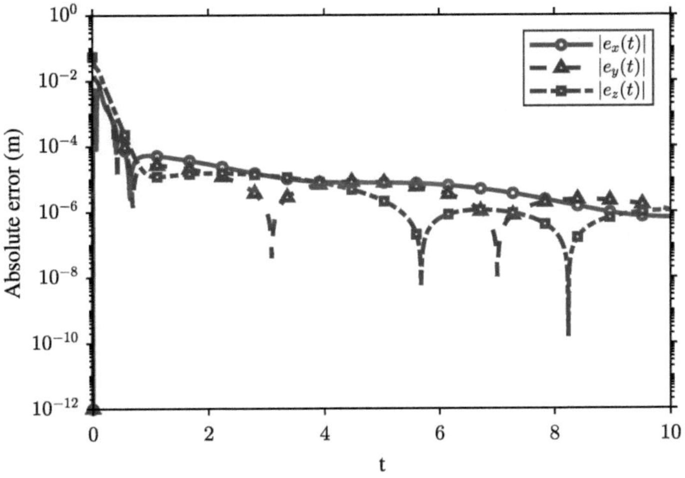

Fig. 2. Absolute tracking errors on a logarithmic scale for $\boldsymbol{x}_a(0) = [0, 0, 0]^\top$, $\boldsymbol{x}_d(0) = [0, 0, 0.055]^\top$, and $\boldsymbol{v}(0) = [0, 0, 0]^\top$ under $\lambda = 10$ and $\eta = 30$.

where randn(3, 3) denotes a 3×3 random matrix.

From Fig. 1, it can be observed that the actual trajectory closely tracks the desired trajectory within a short time. Figure 2 shows that the absolute tracking errors $|e_x(t)|$, $|e_y(t)|$, and $|e_z(t)|$ decrease rapidly to small magnitudes. Figure 3 depicts the evolution of the generalized position states $v_1(t)$, $v_2(t)$, and $v_3(t)$. The state trajectories remain bounded and smooth during the entire simulation, confirming that the generated control action is physically reasonable.

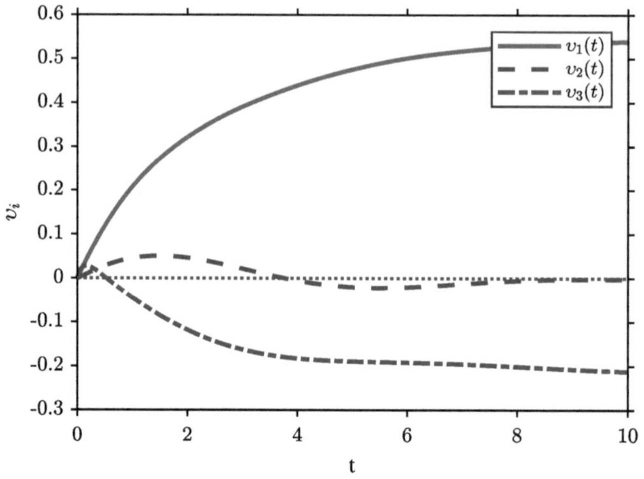

Fig. 3. Time responses of $v_1(t)$, $v_2(t)$, and $v_3(t)$ for $\boldsymbol{x}_a(0) = [0, 0, 0]^\top$ and $\boldsymbol{v}(0) = [0, 0, 0]^\top$ under $\lambda = 10$ and $\eta = 30$.

Overall, the simulation results verify that the proposed ZNN-based controller can realize accurate trajectory tracking for underwater manipulators even when the Jacobian matrix is unknown and time-varying. Meanwhile, the online Jacobian estimation mechanism enables the controller to maintain good transient and steady-state performance without requiring prior exact kinematic information.

6 Conclusion

A ZNN-based trajectory-tracking framework for underwater manipulators with unknown kinematics has been presented. By introducing a Jacobian estimation residual and enforcing a ZNN evolution equation on it, an online Jacobian estimator is obtained. The estimated Jacobian is then directly embedded into the control law to realize trajectory tracking. The resulting scheme preserves the exponential tracking property in the ideal case and yields a perturbed exponentially stable error system in the practical case.

Acknowledgments. This work is supported in part by the Special Projects in universities' key fields of Guangdong Province (2023ZDZX3017), in part by the Guangdong Key Discipline Research Enhancement Project (20240ZDJS055), in part by the 2024 Tertiary Education Scientific Research Project of Guangzhou Municipal Education Bureau (2024312026), and in part by the Guangdong Key Discipline Research Capacity Building Project (2024ZDJS055, 2025ZDJS129).

References

1. Morgan, E., Carlucho, I., Ard, W., Barbalata, C.: Autonomous underwater manipulation: current trends in dynamics, control, planning, perception, and future directions. Curr. Robot. Rep. **3**, 187–198 (2022)
2. Mao, J., Song, G., Hao, S., Zhang, M., Song, A.: Development of a lightweight underwater manipulator for delicate structural repair operations. IEEE Robot. Autom. Lett. **8**(10), 6563–6570 (2023)
3. Huang, H., et al.: A review on visual servoing for underwater vehicle manipulation systems automatic control and case study. Ocean Eng. **260**, 112065 (2022)
4. Santos, P.C., et al.: Fully automatic visual servoing control for underwater vehicle manipulator systems based on a heuristic inverse kinematics. J. Intell. Robot. Syst. **107**, 42 (2023)
5. Huang, H., et al.: Learning strategies for underwater robot autonomous manipulation control. J. Franklin Inst. **361**(7), 106773 (2024)
6. Gao, T., Yan, J., Yang, X., Chen, C., Guan, X.: Integrated target recognition and grasp control for the underwater vehicle-manipulator system. IEEE Trans. Ind. Electron. **71**(10), 12732–12743 (2024)
7. Liu, J., Chen, G., Zhang, F., Gao, J.: Whole-body vision/force control for an underwater vehicle-manipulator system with smooth task transitions. J. Mar. Sci. Eng. **13**(8), 1447 (2025)
8. Tu, Z., Zhu, Y., Wu, X., Yang, C.: A unified shared control architecture for underwater vehicle-manipulator systems using task priority. Front. Inf. Technol. Electron. Eng. **26**(8), 1411–1427 (2025)
9. Sun, L., Wang, Y., Hui, X., Ma, X., Bai, X., Tan, M.: Underwater robots and key technologies for operation control. Cyborg Bionic Syst. **5**, 0089 (2024)
10. Liu, W., Xu, J., Li, L., Zhang, K., Zhang, H.: Adaptive model predictive control for underwater manipulators using Gaussian process regression. J. Mar. Sci. Eng. **11**(9), 1641 (2023)
11. Xue, H., Huang, J.: Dynamics and control of underwater manipulators considering interaction between soft-body link and fluid flow. J. Sound Vib. **570**, 118022 (2024)
12. Yang, X., Chen, C., Yan, J., Hua, C., Guan, X.: Model-free control of underwater vehicle-manipulator system interacting with unknown environments. Ocean Eng. **295**, 116924 (2024)
13. Wang, Y., Gao, J.: Reinforcement-learning-based visual servoing of underwater vehicle dual-manipulator system. J. Mar. Sci. Eng. **12**(6), 940 (2024)
14. Hu, X., Shen, C., Huang, F., Mei, D., Chen, Z.: Precise tracking of underwater electric manipulator via model-based adaptive robust control with joint-friction compensation. ISA Trans. **166**, 266–279 (2025)
15. Wang, T., Zhang, Z., Huang, Y., Liao, B., Li, S.: Applications of zeroing neural networks: a survey. IEEE Access **12**, 51346–51363 (2024)
16. Chen, J., Pan, Y., Zhang, Y., Li, S., Yang, M.: Inverse-free Jacobian-estimated zeroing neural network model for mechanical arm path tracking with minimal joint motion. Expert Syst. Appl. **296**, 129152 (2026)
17. Jin, J., Zhao, L., Chen, L., Chen, W.: A robust zeroing neural network and its applications to dynamic complex matrix equation solving and robotic manipulator trajectory tracking. Front. Neurorobot. **16**, 1065256 (2022)
18. Yang, M., Yu, P., Tan, N.: Discrete integral-type zeroing neurodynamics for robust inverse-free and model-free motion control of redundant manipulators. Comput. Electr. Eng. **118**, 109344 (2024)

19. Yu, P., Tan, N., Zhong, Z., Liao, S.: Model-free kinematic control of redundant manipulators with simultaneous joint-physical-limit and joint-angular-drift handling. ISA Trans. **139**, 635–649 (2023)
20. Xie, Z., Jin, L.: A fuzzy neural controller for model-free control of redundant manipulators with unknown kinematic parameters. IEEE Trans. Fuzzy Syst. **32**(3), 1589–1601 (2024)
21. Huang, H., Tang, G., Chen, H., Wang, J., Han, L., Xie, D.: Adaptive non-singular fixed-time sliding mode control of moving-base underwater flexible manipulators. Nonlinear Dyn. **112**, 4409–4433 (2024)
22. Zaare, S., Soltanpour, M.R.: Uncertainty and velocity observer-based predefined-time nonsingular terminal sliding mode control of the underwater robot manipulators. Eur. J. Control. **75**, 100939 (2024)
23. Wang, R., Huang, W., Wu, J., Wang, H., Li, J.: Global fast terminal sliding mode control of underwater manipulator based on finite-time extended state observer. J. Mar. Sci. Eng. **13**(6), 1038 (2025)
24. Pan, S., Cao, F.: Deep reinforcement learning-based control strategy for underwater manipulator systems. IFAC-PapersOnLine **59**(35), 430–435 (2025)
25. Yang, X., Gao, J., Wang, P., Long, W., Fu, C.: Social learning with actor-critic for dynamic grasping of underwater robots via digital twins. Ocean Eng. **306**, 118070 (2024)
26. Yang, X., Gao, J., Wang, P., Li, Y., Wang, S., Li, J.: Digital twin-based stress prediction for autonomous grasping of underwater robots with reinforcement learning. Expert Syst. Appl. **267**, 126164 (2025)

An Improved Anti-interference Motion Control Scheme for Differential-Wheeled Mobile Manipulators

Xiyuan Zhang, Qiuyan Li, Jifan Yang, Kai Tang, and Dongsheng Guo(✉)

School of Information and Communication Engineering,
Hainan University, Haikou, China
gdongsh2022@hainanu.edu.cn

Abstract. Wheeled mobile manipulators (WMM) are adopted to perform tasks such as handling, picking, and spraying through the coordination between the mobile platform and the manipulator. In complex scenarios, WMM are subjected to not only the noise of their own systems but also the uncertainty of the external environment. Under such conditions, great challenges are posed to the motion control of WMM in meeting the required accuracy. The differential-wheeled mobile manipulator (DWMM), a typical type of WMM, is considered in this paper. An improved anti-interference motion control scheme is proposed to tackle complex noise disturbances. Based on the internal model principle, the integral term of the error and the adaptive noise compensation model are introduced into the motion control scheme. The effectiveness and practicability of the proposed scheme are verified through theoretical analysis and experimental validation.

Keywords: Wheeled mobile manipulators · Motion control · Anti-interference · Internal model principle

1 Introduction

Wheeled mobile manipulators (WMM) can execute various operational tasks in scenarios such as home services, logistics transportation, and spraying manufacturing through the coupled coordination between the mobile platform and the manipulator [1–3]. With the development of embodied intelligence technology, the application scenarios of WMM have been continuously expanding, and motion control, as a core component of WMM, is also facing increasingly stringent technical requirements. Kinematic modeling serves as the foundation for the motion control of WMM and exerts an important influence on the design

This paper is supported by the National Natural Science Foundation of China (62463004), International Science and Technology Cooperation Program of Hainan Province (GHYF2025032), and Scientific Research Fund of Hainan University (KYQD(ZR)23025).

Z. Zhang (Ed.): ISNN 2026, LNCS 16848, pp. 257–266, 2027.
https://doi.org/10.1007/978-981-92-5870-3_21

of motion control schemes [4–7]. At present, the corresponding modeling methods are mainly classified into two categories. In the first method, the WMM is decomposed into the mobile platform and the manipulator, which are modeled separately and then integrated through constraint conditions. The other method takes the dynamic coupling between the mobile platform and the manipulator into consideration and establishes the kinematic model by regarding the WMM as a unified whole.

In the design of motion control schemes, the separate kinematic modeling method tends to ignore the inherent coupling characteristics of the system, requiring independent control of the mobile platform and the manipulator, which partially restricts the control performance of WMM. In contrast, the integrated kinematic modeling method is more consistent with the requirements for the cooperative operation capability of WMM. Therefore, many scholars have carried out extensive design work and related research on motion control schemes based on this modeling method. Cang et al. [8] proposed velocity-level and acceleration-level kinematic modeling and trajectory tracking schemes for two-wheeled differential mobile manipulators, with effectiveness validated via simulations and physical experiments. Nie et al. [9] proposed a prescribed performance control scheme for redundant WMM, realizing high-accuracy smooth trajectory tracking under multiple constraints. Shafei et al. [10] proposed a recursive Gibbs–Appell predictive control method for mobile manipulators, enabling synchronous base-end-effector trajectory tracking under model uncertainties.

During the operation of WMM, noise interference is inevitably encountered. The noise sources of WMM not only include internal noise generated by their own structural components such as mechanical transmission and motor operation, but also encompass external noise introduced by external environmental disturbances and frictional contact with the ground. In the field of motion control, noise interference has always been a crucial challenge that urgently needs to be addressed, and researchers have proposed numerous motion control methods to solve this problem. Korayem et al. [11] developed an optimal sliding mode control method for WMM under uncertainties and wheel slippage, effectively suppressing slippage-induced chattering and tracking errors. Xing et al. [12] presented a virtual decomposition control-based trajectory tracking controller for WMM to address noise-related issues, with its effectiveness validated via experiments.

Although numerous anti-noise motion control schemes have been proposed [13–15], the uncertainties induced by the inherent strong coupling characteristics of the system, coupled with external disturbances caused by complex environmental changes, further increase the difficulty of solving this problem. Therefore, the design of an efficient and practical motion control scheme has important research value but also poses practical challenges. To address the aforementioned issues, this paper proposes an improved anti-disturbance motion control scheme suitable for differential WMM (DWMM). By virtue of the internal model principle, both the integral error term and the adaptive noise compensation model are incorporated into the proposed motion control scheme. The main contributions of this paper are as follows:

1) In this paper, an improved anti-interference motion control scheme is proposed for DWMM.
2) In this paper, the proposed scheme is applied to the Lebai DWMM in various noise environments to perform simulation experiments, thereby validating its effectiveness.

2 DWMM Kinematic Model

This section presents a systematic overview of the kinematic modeling for the DWMM. For the DWMM, the end-effector position in Cartesian space is inherently related to the joint configuration via the following differentiable nonlinear mapping:

$$\boldsymbol{p} = g(\boldsymbol{\theta}) \tag{1}$$

where $\boldsymbol{p}$ denotes the end-effector position vector, $\boldsymbol{\theta} = (\varphi_1, \varphi_2, q_1, ..., q_i)^T$ represents the joint and wheel position vector, $\varphi = (\varphi_1, \varphi_2)$ denotes the angles of the left and right wheels, and $q = (q_1, ..., q_i)$ represents the joint angles of the manipulator. $g(\cdot)$ is a differentiable nonlinear function derived from the inherent structural parameters of the DWMM.

By taking the time derivative of Eq. (1), the velocity-level mapping between the end-effector and joint space is obtained as:

$$\dot{\boldsymbol{p}} = J\dot{\boldsymbol{\theta}} \tag{2}$$

where $\dot{\boldsymbol{p}}$ is the end-effector velocity vector, J represents the Jacobian matrix, and $\dot{\boldsymbol{\theta}} = (\dot{\varphi}_1, \dot{\varphi}_2, \dot{q}_1, ..., \dot{q}_i)^T$ denotes the joint and wheel velocity vector. $\dot{\varphi} = (\dot{\varphi}_1, \dot{\varphi}_2)$ denotes the angular velocities of the left and right wheels, and $\dot{q} = (\dot{q}_1, ..., \dot{q}_i)$ represents the angular velocity of the manipulator joints.

The Jacobian matrix J can be further decomposed into the product of two independent matrices, as follows:

$$J = \mathcal{F}\mathcal{M} \tag{3}$$

where $\mathcal{F}$ denotes the Jacobian matrix of the manipulator mounted on the mobile platform, and $\mathcal{M}$ represents the kinematic transformation matrix. For the differential wheeled mobile platform, there exists a linear mapping between the Cartesian velocity of the mobile base $\dot{\boldsymbol{p}}_c$ and the wheel angular velocities $\dot{\varphi}$, which is determined by the geometric structure of the chassis: $\dot{\boldsymbol{p}}_c = D\dot{\varphi}$, where D is the platform kinematic matrix. Based on the above relationship, the transformation matrix $\mathcal{M}$ is constructed as a block diagonal matrix:

$$\mathcal{M} = \begin{bmatrix} D & \mathbf{0} \\ \mathbf{0} & I \end{bmatrix} \tag{4}$$

where I is the n-dimensional identity matrix.

From the above derivation and construction, the Jacobian matrix J integrates the mobile base and the manipulator into a unified system, which explicitly characterizes the instantaneous response of the end-effector velocity in Cartesian space to the motions of the mobile base and manipulator joints.

3 Improved Anti-interference Motion Control Scheme

In this section, an improved anti-interference motion control scheme is proposed, which can effectively cope with the influence of different noises and enhance the motion accuracy and robustness of DWMM.

In DWMM motion control, the pseudo-inverse method based on inverse kinematics has garnered extensive attention owing to its fast solution speed. At the joint velocity level, this method is typically expressed as:

$$\dot{\boldsymbol{\theta}} = J^{\dagger}\dot{\boldsymbol{p}} \tag{5}$$

where $J^{\dagger} = J^{T}\left(JJ^{T}\right)^{-1}$ denotes the pseudoinverse of the Jacobian matrix. DWMM inevitably suffers from noise interference in Cartesian coordinates, while Eq. (5) neglects this effect. Considering noise interference, the method can be expressed as:

$$\dot{\boldsymbol{\theta}} = J^{\dagger}(\dot{\boldsymbol{p}} + \boldsymbol{\xi}(t)), \tag{6}$$

where $\boldsymbol{\xi}(t)$ denotes additive noise.

To eliminate the effect of $\boldsymbol{\xi}(t)$, a noise rejection motion control scheme for DWMM is proposed, which integrates error correction and its integral part. The corresponding expression of this scheme is as follows:

$$\dot{\boldsymbol{\theta}} = J^{\dagger}\left(\dot{\boldsymbol{p}} - k_p\boldsymbol{\varepsilon}(t) - k_i\int_0^t \boldsymbol{\varepsilon}(\tau)\mathrm{d}\tau + \boldsymbol{\xi}(t)\right), \tag{7}$$

where $\boldsymbol{\varepsilon}(t) = g(\boldsymbol{\theta}) - \boldsymbol{p}$ represents the error, and $k_p, k_i > 0$ are design parameters. In fact, the noise interference suffered by DWMM in practical operation scenarios is highly complex, characterized by the superposition of multiple types of noise such as measurement noise and environmental disturbances. Based on the aforementioned noise rejection motion control scheme, an improved anti-noise motion control algorithm is further designed to cope with such complex composite noise interference and improve the motion control accuracy and robustness of DWMM. The improved scheme equation is shown as follows:

$$\begin{cases} \dot{\boldsymbol{\theta}}(t) = \boldsymbol{J}^{\dagger}\left(\dot{\boldsymbol{p}} - k_p\boldsymbol{\varepsilon}(t) - k_i\int_0^t \boldsymbol{\varepsilon}(\tau)\mathrm{d}\tau - \left(\mathbf{1}_n^{\top}\otimes\boldsymbol{I}_3\right)\boldsymbol{\Phi}(t) + \boldsymbol{\xi}(t)\right) \\ \dot{\boldsymbol{\Phi}}(t) = \boldsymbol{\Psi}(t) + k_{\lambda}\boldsymbol{\Omega}\left(\mathbf{1}_n\otimes\boldsymbol{\varepsilon}(t)\right) \\ \dot{\boldsymbol{\Psi}}(t) = -\boldsymbol{\Omega}\boldsymbol{\Phi}(t) \end{cases} \tag{8}$$

where $\boldsymbol{\Phi}(t)$ and $\boldsymbol{\Psi}(t)$ are the adaptive compensation vector and auxiliary state vector, respectively, which are vertically constructed by concatenating n groups of 3-dimensional compensation components $\phi_i(t)$ for each frequency and the corresponding auxiliary states $\psi_i(t)$ $(i = 1, 2, \ldots, n)$. $\dot{\boldsymbol{\Phi}}(t)$ and $\dot{\boldsymbol{\Psi}}(t)$ are their time derivatives. The diagonal matrix is $\Omega = \mathrm{diag}(4\pi^2 f_1^2 I_3, 4\pi^2 f_2^2 I_3, \ldots, 4\pi^2 f_n^2 I_3)$, where diag(·) is a function for constructing diagonal matrices, I_3 is the 3rd-order identity matrix, and f_i is the frequency of the i-th noise. $k_\lambda > 0$ is the adaptive learning gain, $\otimes$ denotes the Kronecker product, and $\mathbf{1}_n$ is the n-dimensional all-ones column vector.

4 Simulation Experiment

In this section, the trajectory tracking performance of the mobile manipulator under three different noise environments is presented. The Lebai DWMM is adopted as the experimental object to verify the effectiveness of the proposed motion control scheme. In the simulation environment, the initial angles of the Lebai DWMM are set to $[0; 0; 0; \pi/6; \pi/4; 0; \pi/3; 0]$. The desired trajectory is represented as a cardioid curve, with the mathematical expression given as follows:

$$\begin{bmatrix} r_x \\ r_y \\ r_z \end{bmatrix} = \begin{bmatrix} R\cos(\phi)(1+\cos(\phi)) - 2R + i_{x0} \\ R\cos(\alpha)(1+\cos(\phi))\sin(\phi) + i_{y0} \\ R\sin(\alpha)(1+\cos(\phi))\sin(\phi) + i_{z0} \end{bmatrix},$$

with $\phi = 2\pi\sin^2(\pi t/2T)$, the radius and deflection angle are set to $R = 0.05$ and $\alpha = \pi/6$. The noise encountered by mobile manipulators in practical operation scenarios can be divided into three categories, as detailed below.

In the first case, the noise is modeled as harmonic noise: $\xi_1 = 0.1\sin(4\pi t + 0.1)$. Figure 1 presents the simulation results of the Lebai DWMM using scheme (8) for cardioid trajectory tracking under noise ξ_1. Figure 1(a) shows the time-domain waveform of ξ_1, where the equal-amplitude and periodic oscillatory characteristics can be clearly observed. Figure 1(b) illustrates the motion process of the mobile manipulator performing the cardioid trajectory tracking task. It can be seen from Fig. 1(c) that the end-effector of the mobile manipulator can closely follow the desired trajectory. Figure 1(d) shows that the Cartesian error between the end-effector and the desired trajectory is maintained within the range of $[-4, 3.5] \times 10^{-7}$ m. These results sufficiently verify the suppression capability and robustness of scheme (8) against harmonic noise.

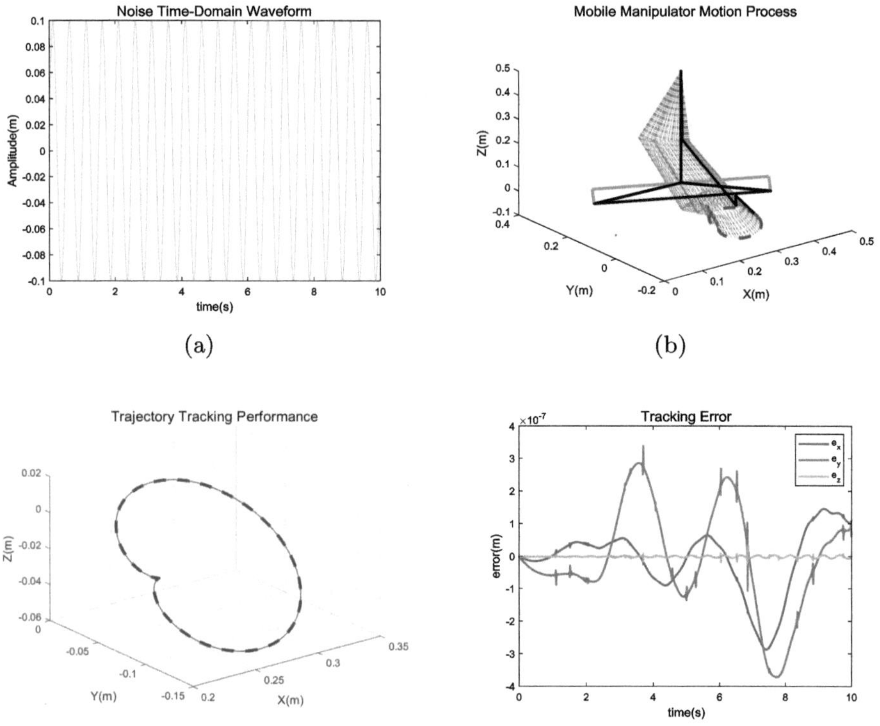

(a) (b)

(c) (d)

Fig. 1. Simulation results of Lebai DWMM tracking the cardioid trajectory under noise ξ_1

In the second case, the noise model is extended to harmonic noise with time-varying components: $\xi_2 = 0.1\sin(4\pi t + 0.1) + 0.01t + 0.05$. Figure 2 shows the simulation results under noise ξ_2. Figure 2(a) depicts the time-domain waveform of ξ_2, which adds a steady-state bias of 0.05 and a time-varying term $0.01t$ based on ξ_1. Figure 2(b) displays the motion process of the mobile manipulator in the trajectory tracking task. As shown in Fig. 2(c), the end-effector can still track the desired trajectory closely. The Cartesian curve tracking error remains within $[0, 5] \times 10^{-6}$ m. These results validate the effectiveness and robustness of scheme (8) in suppressing time-varying harmonic noise.

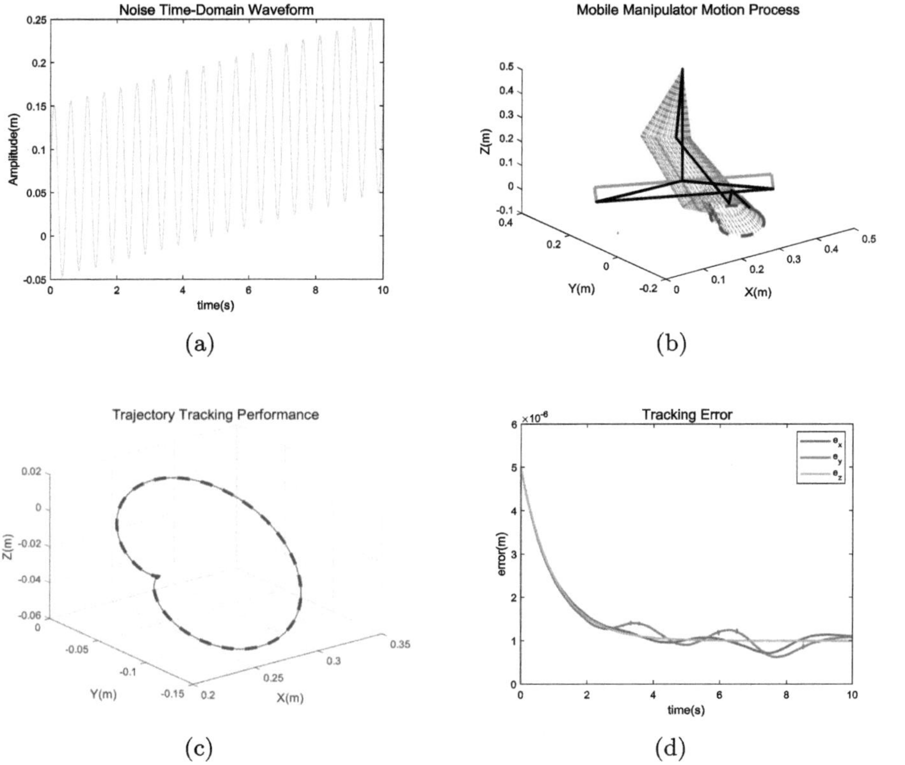

Fig. 2. Simulation results of Lebai DWMM tracking the cardioid trajectory under noise ξ_2

In the third case, a random noise component is further introduced into the noise model: $\xi_3 = 0.1\sin(4\pi t + 0.1) + 0.01t + 0.05 + 0.03rand$, where rand is a random number in the range [0, 1]. Based on ξ_2, this noise is superimposed with a random noise of amplitude 0.03, which fully covers three typical noise characteristics: periodic oscillation, time-varying noise, and random burst interference. Figure 3 presents the simulation results of Lebai DWMM tracking the cardioid trajectory under noise ξ_3. The results show that even in complex noise environments, the tracking error of the proposed motion control scheme remains at a small magnitude, which demonstrates its stable and excellent anti-noise performance.

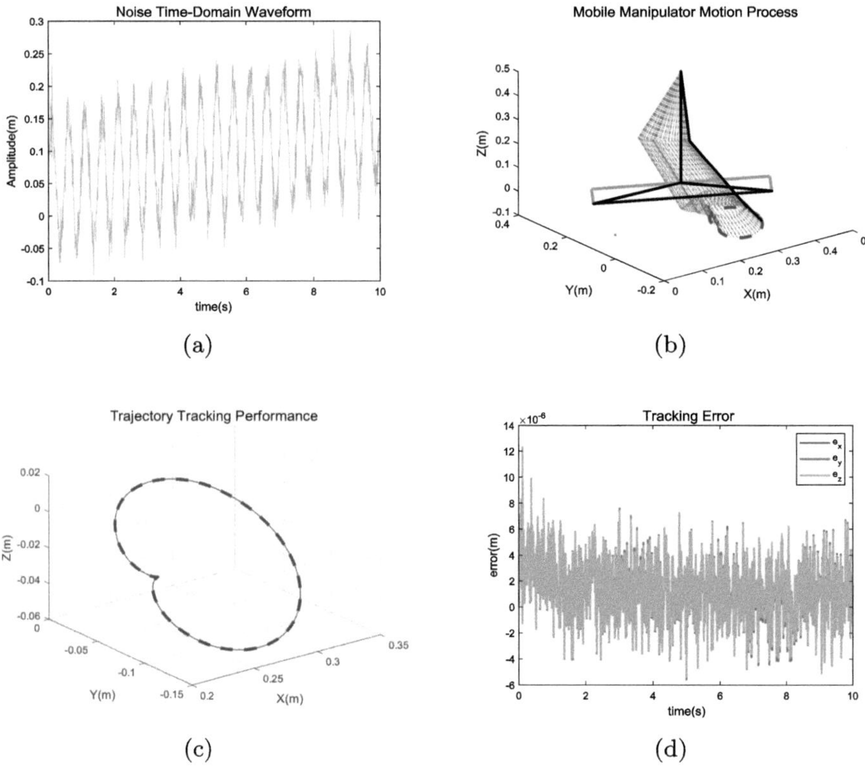

Fig. 3. Simulation results of Lebai DWMM tracking the cardioid trajectory under noise ξ_3

To verify the superiority of our proposed scheme, quantitative comparison experiments with control scheme (7) are conducted under the same parameter settings k_p, k_i and noise conditions. Figure 4 presents the simulation results of scheme (7) tracking the cardioid trajectory under noise ξ_3. Compared with scheme (7), the scheme proposed in this paper reduces the tracking error by half, demonstrating significantly better anti-interference capability.

Overall, the simulation results under the above three different noise environments comprehensively validate the effectiveness and accuracy of the proposed motion control scheme, which shows good potential for practical applications in real-world scenarios.

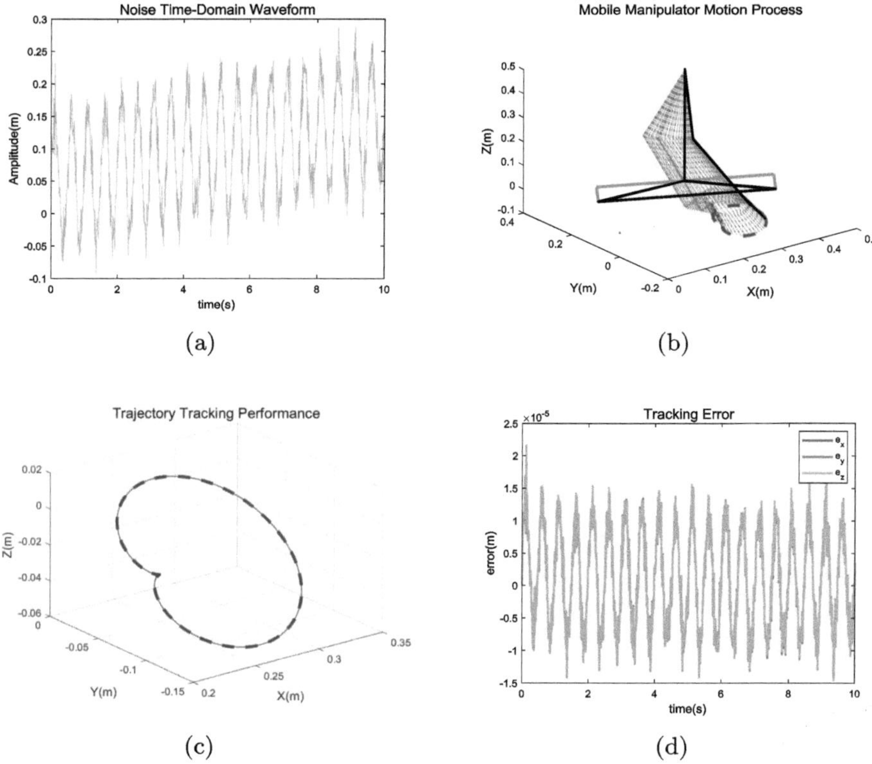

Fig. 4. Simulation results of scheme (7) tracking the cardioid trajectory under noise ξ_3

5 Conclusion

This paper proposes an improved anti-disturbance motion control scheme for DWMM to mitigate the challenges posed by system noise and external environmental uncertainties in complex operational scenarios. Focusing on a typical DWMM as the research platform, the proposed control law introduces the tracking error integral term and adaptive noise compensation model, so as to suppress the negative impacts of noise and external disturbances effectively. The effectiveness and practicality of the control scheme are validated through rigorous theoretical analysis and comprehensive simulation studies, which demonstrate its superior disturbance rejection and tracking performance. Meanwhile, subsequent research will involve real-robot experiments on different types of mobile manipulators to comprehensively demonstrate the practical utility of the proposed scheme.

References

1. Xing, H., et al.: An admittance-controlled wheeled mobile manipulator for mobility assistance: human-robot interaction estimation and redundancy resolution for enhanced force exertion ability. Mechatronics **74**, 102497 (2021)
2. Cebollada, S., Payá, L., Flores, M., Peidró, A., Reinoso, O.: A state-of-the-art review on mobile robotics tasks using artificial intelligence and visual data. Expert Syst. Appl. **167**, 114195 (2021)
3. Fragapane, G., de Koster, R., Sgarbossa, F., Strandhagen, J.O.: Planning and control of autonomous mobile robots for intralogistics: literature review and research agenda. Eur. J. Oper. Res. **294**, 405–426 (2021)
4. Korayem, M.H., Dehkordi, S.F., Ghobadi, N.: Wheel slippage compensation in mobile manipulators through combined kinematic, dynamic, and sliding mode control. Arabian J. Sci. Eng. **49**, 11565–11585 (2024)
5. Picard, G., Lenain, R., Mezouar, Y., Thuilot, B., Laneurit, J.: Multi-trajectory approach for a generic coordination paradigm of wheeled mobile manipulators. IEEE Robot. Autom. Lett. **7**, 2329–2336 (2022)
6. Khan, A.H., Li, S., Chen, D., Liao, L.: Tracking control of redundant mobile manipulator: an RNN based metaheuristic approach. Neurocomputing **400**, 272–284 (2020)
7. Sun, Z., Tang, S., Zhou, Y., Yu, J., Li, C.: A GNN for repetitive motion generation of four-wheel omnidirectional mobile manipulator with nonconvex bound constraints. Inf. Sci. **607**, 537–552 (2022)
8. Cang, N., et al.: Kinematic modeling and trajectory tracking of two-wheeled mobile robot manipulators at the acceleration level. IEEE Access **12**, 159138–159151 (2024)
9. Nie, J., et al.: An HQP-based obstacle avoidance control scheme for redundant mobile manipulators under multiple constraints. IEEE Trans. Ind. Electron. **70**, 6004–6016 (2023)
10. Shafei, A., Mirzaeinejad, H.: A novel recursive formulation for dynamic modeling and trajectory tracking control of multi-rigid-link robotic manipulators mounted on a mobile platform. Proc. Inst. Mech. Eng. I J. Syst. Control Eng. **235**, 1204–1217 (2021)
11. Korayem, M.H., Ghobadi, N., Dehkordi, S.F.: Designing an optimal control strategy for a mobile manipulator and its application by considering the effect of uncertainties and wheel slipping. Optim. Control Appl. Methods **42**, 1487–1511 (2021)
12. Xing, H., Xu, Y., Ding, L., Chen, J., Gao, H., Tavakoli, M.: Trajectory tracking control of wheeled mobile manipulators with joint flexibility via virtual decomposition approach. IEEE Trans. Autom. Sci. Eng. **22**, 11808–11825 (2025)
13. Shafei, H.R., Bahrami, M., Talebi, H.A.: Trajectory tracking of an uncertain wheeled mobile robotic manipulator with a hybrid control approach. J. Brazilian Soc. Mech. Sci. Eng. **42**, 156–166 (2020)
14. Peng, J., Yang, Z., Wang, Y., Zhang, F., Liu, Y.: Robust adaptive motion/force control scheme for crawler-type mobile manipulator with nonholonomic constraint based on sliding mode control approach. ISA Trans. **92**, 166–179 (2019)
15. Baek, J., Jin, M., Han, S.: A new adaptive sliding-mode control scheme for application to robot manipulators. IEEE Trans. Ind. Electron. **63**, 3628–3637 (2016)

Control Systems

Distributed Nash Equilibrium Seeking for Constrained N-Coalition Game

Yiyao Xu[1], Chengshuo Huang[1], Na Liu[2], and Sitian Qin[1](✉)

[1] School of Mathematical Sciences, Harbin Institute of Technology, Weihai, China
qinsitian@163.com

[2] School of Mathematical Sciences, and the Institute of Mathematics and Interdisciplinary Sciences, Tianjin Normal University, Tianjin, China

Abstract. In this paper, an N-coalition game is investigated, where each player is subjected to a local inequality constraint. A two-time scale structure algorithm has been designed to seek the generalized Nash equilibrium (GNE). In the decision-making layer, an integral term that changes automatically according to the degree of deviation from the feasible region is used to prompt the states to develop towards the constraint set. Moreover, it can be derived that the state reaches the feasible set in a finite time. By using LaSalle's Invariance Principle, it can be deduced that the algorithm completes the tracking of the GNE. Finally, an example is presented respectively to confirm the reliability of the theoretical results.

1 Introduction

In multi-player interactive systems, individual strategies are often not isolated but deeply influenced by group relationships. The N-coalition game model formulated by [1] serves as a crucial framework for precisely such scenarios. The core of the N-coalition game lies in the relationship of "intra-coalition cooperation and inter-coalition competition" among multiple players. In this game-theoretic model, each coalition is treated as a virtual player whose objective is to minimize its own cost function, which defined as the sum of the cost functions of the players within the coalition. A growing body of real-world applications demonstrates that this model provides vital theoretical support for addressing multi-player scenarios, and playing an irreplaceable role in fields such as healthcare networks [2], smart grids [3], transportation networks [4], and formation problem [5]. Therefore, in recent years, the problem of seeking Nash equilibrium in N-coalition games has received increasing attention [6–8].

However, most existing research addressing constrained N-coalition game problems based on primal-dual methods [9–12], which requires auxiliary variables. This significantly increases computational dimensionality when dealing with large-scale or high-dimensional problems. While some penalty methods offer an alternative, they still suffer from the disadvantage of requiring predefined penalty parameters that depend on centralized information. To overcome these

Z. Zhang (Ed.): ISNN 2026, LNCS 16848, pp. 269–280, 2027.
https://doi.org/10.1007/978-981-92-5870-3_22

disadvantage, and inspired by [13], this paper introduces an integral term that varies with the degree of constraint violation to handle the inequality constraints.

Based on the above discussion, this paper studies the Nash equilibrium seeking problem for constrained multi-cluster games. Compared to related results, the main contributions of this paper are: The N-coalition game considered in this paper involves local inequality constraints for each player, making it more aligned with the requirements of practical applications. Compared to [9–12], this paper utilizes a penalty method to transform the problem into an unconstrained one. Compared with existing primal-dual methods that require auxiliary variables and penalty methods relying on prefixed parameters, the proposed two-time-scale algorithm uses an adaptive time-varying integral penalty to handle local inequality constraints in a fully distributed manner, without centralized information. We prove that states enter the feasible set in finite time and converge exponentially to the GNE by singular perturbation analysis and LaSalle's invariance principle.

The remainder of this paper is arranged as follows. In Sect. 2, some preliminary knowledge is given, while the NE seeking problem is formulated. The distributed algorithm is designed, whose convergence is proven in Sect. 3. In Sect. 4, the resource allocation problem is given to verify the effectiveness of the algorithm. And a strictly monotone game mapping problem is presented to find whether the proposed algorithm can handle the N-coalition game problems. Finally, Sect. 5 concludes the paper.

Notation: $\mathbb{R}, \mathbb{R}^n$ and $\mathbb{R}^{n\times m}$ are the set of real numbers, the n-dimensional Euclidean space and $n \times m$ real matrices, respectively. Let $\mathrm{col}\,(x_1, x_2, \cdots, x_n) = \left[x_1^T, x_2^T, \cdots, x_n^T\right]^T$ represent a column vector. Define $1_n, 0_n \in \mathbb{R}^n$ as a n-dimensional column vector with all elements being 1,0. $\otimes$ denotes the Kronecker product. sign$(\cdot)$ represents the the sign function. $\lambda_{min}(P)$ denotes the smallest eigenvalue of matrix P. $\lambda_{\min}(P_{ij})$ ($\lambda_{\min}(P_i)$) denotes the smallest eigenvalue among all matrices P_{ij} (P_i).

2 Preliminaries and Problem Formulation

A. Graph Theory

An undirected graph defined as $\mathcal{G} = \{\mathcal{V}, \mathcal{E}\}$, where $\mathcal{V} = \{1, 2, \cdots, N\}$ is the set of edges, and $\mathcal{E} \subseteq \mathcal{V} \times \mathcal{V}$ is the node set. $\mathcal{A} = [a_{ij}] \in \mathbb{R}^{N\times N}$ represents the adjacency matrix of $\mathcal{G}$, where $a_{ij} = a_{ji} = 1$ if $(i, j) \in \mathcal{E}$, and $a_{ij} = 0$ otherwise. $L = \mathcal{D} - \mathcal{A}$ is the Laplacian matrix of $\mathcal{G}$, where $\mathcal{D} = \mathrm{diag}\left\{d_1, \cdots, d_N\right\}$ with $d_i = \sum_{j=1}^N a_{ij}$. For an undirected connected graph, the eigenvalues of L can be ordered by $0 = \lambda_1 < \lambda_2 \leqslant \cdots \leqslant \lambda_N$.

B. Multi-coalition Game Problem

Consider a multi-coalition game with N coalitions over an undirected connected graph $\mathcal{G}_0$, and each coalition contains n_i players over an undirected connected graph $\mathcal{G}_i$. The action of player j in coalition i is denoted as $x_{ij} \in \mathbb{R}$. Each coalition is regarded as a virtual player to compete with other coalitions, and $\mathbf{x}_i = (x_{i1}, x_{i2}, \cdots, x_{in_i}) \in \mathbb{R}^{n_i}$ stands for the action set of the players in coalition i. The entire action profile is $\mathbf{x} = \text{col}(\mathbf{x}_1, \mathbf{x}_2, \cdots, \mathbf{x}_N) \in \mathbb{R}^n$, where $n = \sum_{i=1}^{N} n_i$. Denote $\mathbf{x}_{-i} = \text{col}(\mathbf{x}_1, \cdots, \mathbf{x}_{i-1}, \cdots, \mathbf{x}_{i+1}, \cdots, \mathbf{x}_N)$ as the entire action profile except the actions of the coalition i. The objective function of coalition i is $f_i(\mathbf{x}_i, \mathbf{x}_{-i}) = \sum_{j=1}^{n_i} f_{ij}(\mathbf{x}_i, \mathbf{x}_{-i})$, with $f_{ij}(\mathbf{x}_i, \mathbf{x}_{-i})$ being the cost function of player j in coalition i. The players in coalition i can be regarded as a distributed optimization problem, which cooperate with each other to minimize the objective function of the coalition.

With limited resource, restricted production capacity, rigorous regulations in practical application, players are often constrained by certain constraints when making decisions. We consider a class of local inequality constraint for the decision variables of the players, which can be formulated as $S_{ij} = \{x_{ij} \in \mathbb{R} : g_{ij}(x_{ij}) = (g_{ij}^1(x_{ij}), g_{ij}^2(x_{ij}), \cdots, g_{ij}^{p^2}(x_{ij})) \leq 0_{p^2}\}$, where $g_{ij}^m(x_{ij})$ is convex but not necessary smooth. Specifically, the target of coalition i is to minimize the following constrained N-coalition game problem:

$$\begin{aligned} \min_{\mathbf{x}_i \in \mathbb{R}^{n_i}} \quad & f_i(\mathbf{x}_i, \mathbf{x}_{-i}) \\ \text{s.t.} \quad & x_{ij} \in S_{ij}, \end{aligned} \tag{1}$$

Define $G_{ik}(x_{ik}) = \sum_{m=1}^{p^2} \max\{0, g_{ik}^m(x_{ik})\}$, $Q_{G_{ij}}^{+} = \{m \in \{1 \cdots p^2\} : g_{ij}^m(x_{ij}) > 0\}, Q_{G_{ij}}^0 = \{m \in \{1 \cdots p^2\} : g_{ij}^m(x_{ij}) = 0\}$.

The Nash equilibrium of the N-coalition game (1) is defined as follows.

Definition 1. *For a given N-coalition game like (1), a strategy profile* $\mathbf{x}^* = \text{col}(\mathbf{x}_1^*, \mathbf{x}_2^*, ...\mathbf{x}_N^*) \in \mathbb{R}^n$ *is said to be a NE if*

$$f_i(\mathbf{x}_i^*, \mathbf{x}_{-i}^*) \leq f_i(\mathbf{x}_i, \mathbf{x}_{-i}^*) \quad \forall \mathbf{x}_i \in S_i \tag{2}$$

where $S_i = S_{i1} \times S_{i2} \times \cdots \times S_{in_i}$ *is the feasible set of the players in coalition* i.

Define the pseudo-gradient mapping $F(\mathbf{x}) : \mathbb{R}^n \to \mathbb{R}^n$ as

$$F(\mathbf{x}) \triangleq \prod_{i=1}^{N} \nabla_{x_{ij}} f_i(\mathbf{x}_i, \mathbf{x}_{-i}).$$

The following standard assumptions are needed for the later discussion.

Assumption 1. *The cost function* $f_{ij}(\mathbf{x}) : \mathbb{R}^{np^1} \to \mathbb{R}$ *is twice continuously differentiable and convex in* x_{ij} *for every* $\mathbf{x}_{-ij}$.

Assumption 2. $F(\mathbf{x})$ *is* μ*-strongly monotone, i.e.,* $(F(\mathbf{x}) - F(\mathbf{y}))^T(\mathbf{x} - \mathbf{y}) \geqslant \mu\|\mathbf{x} - \mathbf{y}\|^2$, $\forall \mathbf{x}, \mathbf{y} \in \mathbb{R}^{np^1}$.

Assumption 3. *There exists* $\hat{\mathbf{x}}$ *which is an interior point of the feasible region, that is* $g_{ij}(x_{ij}) < 0_{p^2}$, *and there exists* $R_{ij} > 0$ *satisfying* $S_{ij} = \{x_{ij} : g_{ij}(x_{ij}) \leq 0_{p^2}\} \subseteq B(\hat{q}_{ij}, R_{ij})$.

The above assumptions are essential for the existence and uniqueness of the NE.

3 Main Results

3.1 Distributed Algorithm Design

- To find the NE of problem (1) under imperfect information condition, the following two-time scale distributed algorithm is designed.

$$\dot{x}_{ij} \in - E(G_{ij}(x_{ij}))(n_i z_{ijj}) - \int_0^t \mathrm{sign}(G_{ij}(x_{ij}(s)))ds\partial G_{ij}(x_{ij}) \tag{3a}$$

$$\varepsilon \dot{z}_{ijk} = - \left[z_{ijk} + \sum_{l=1}^{n_i} a^i_{ji}(z_{ijk} - z_{ilk}) + \sum_{l=1}^{n_i} a^i_{ji}(r_{ijk} - r_{ilk}) - \frac{\partial f_{ij}(y^{ij})}{\partial x_{ik}}\right] \tag{3b}$$

$$\varepsilon \dot{r}_{ijk} = \sum_{l=1}^{n_i} a^i_{jl}(z_{ijk} - z_{ilk}) \tag{3c}$$

$$\dot{y}^{ij}_{mr} = - \sum_{u=1}^{N}\sum_{v=1}^{n_u} a^{ij}_{uv}(y^{ij}_{mr} - y^{uv}_{mr}),\ ij \neq mr \tag{3d}$$

where y^{ij}_{mr} is player ij's estimation of player mr, $y^{ij}_{ij} = x_{ij}$, ε is a small positive parameter. $E(s) : [0, +\infty) \to [0, 1]$ is a function defined as follows,

$$E(s) = \begin{cases} 0 & s > 1 \\ -s & 0 \leq s \leq 1 \end{cases}.$$

To facilitate subsequent proofs, the following notation is defined.
$y_{ij} = (y^{11}_{ij}, y^{12}_{ij}, \cdots, y^{i,j-1}_{ij}, y^{i,j+1}_{ij}, y^{N,n_N}_{ij})$ is the profile of estimation vectors for player ij from all other players.
$y^{ij} = (y^{ij}_{11}, y^{ij}_{12}, \cdots, y^{ij}_{i,j-1}, x_{ij}, y^{ij}_{i,j+1}, \cdots, y^{ij}_{N,n_N})$ represents the estimation for other players by player ij.
$\bar{y}_{ij} = y_{ij} - x_{ij}1_{n-1}$ is the estimator error of player ij, $\bar{y} = \mathrm{col}(\bar{y}_{11}, \cdots, \bar{y}_{Nn_N})$, $z_{ij} = (z_{ij1}, z_{ij2}, \cdots, z_{ijn_i})$, $r_{ij} = (r_{ij1}, r_{ij2}, \cdots, r_{ijn_i})$.

3.2 Convergence Analysis

For the third subsystem, let $\tau = \frac{t}{\varepsilon}$, and denote $z_i = (z^T_{i1}, z^T_{i2}, \cdots, z^T_{in_i})^T$, $r_i = (r^T_{i1}, r^T_{i2}, \cdots, r^T_{in_i})^T$, $y_i = \mathrm{col}(y^{i1}, y^{i2}, \cdots, y^{in_i})$, $y = \mathrm{col}(y_1, y_2, \cdots, y_N)$. We can obtain the boundary system of (3b) (3c) (3d) in the τ-time scale,

$$\begin{aligned} \frac{dz_i}{d\tau} &= - \Big((I_{n_i^2} + L_i \otimes I_{n_i})z_i + ((L_i \otimes I_{n_i}))r_i - h_i(y_i)\Big) \\ \frac{dr_i}{d\tau} &= (L_i \otimes I_{n_i})z_i \\ \frac{d\bar{y}_{ij}}{d\tau} &= - (L_{ij} + B_{ij})\bar{y}_{ij} \end{aligned} \tag{4}$$

where L_{ij} is Laplacian matrix of communication graph $\mathcal{G}_{ij} = \{\mathcal{V}_{ij}, \mathcal{E}_{ij}\}$, where $\mathcal{V}_{ij} = \mathcal{V} \setminus \{ij\}$, $B_{ij} = \mathrm{diag}\{a_{ij}^{uv}\}_{uv \neq ij}$, $h_i(y_i) = \left[\frac{\partial f_{i1}(y^{i1})}{\partial x_{i1}}, \cdots, \frac{\partial f_{i1}(y^{i1})}{\partial x_{in_i}}, \frac{\partial f_{i2}(y^{i2})}{\partial x_{i1}}, \cdots, \frac{\partial f_{in_i}(y^{in_i})}{\partial x_{in_i}}\right]$. According to Theorem 4 in [14], it can be obtained that $L_{ij} + A_{ij}$ is symmetric and positive definite.

Let $\Omega_i = (\Omega_{i1}, \Omega_{i2})$, where $\Omega_{i1} \in R^{n_i^2 \times (n_i^2 - n_i)}, \Omega_{i2} \in R^{n_i^2 \times n_i}$, where $\Omega_{i1}^T(L_i \otimes I_{n_i})$ is row full rank, $\Omega_{i2}^T(L_i \otimes I_{n_i}) = 0_{n_i \times n_i^2}$; For $[s_i^T, \tilde{s}_i^T]^T = \Omega_i^T r_i$, we can obtain

$$\begin{aligned}\frac{dz_i}{d\tau} &= -(I_{n_i}^2 + L_i \otimes I_{n_i})z_i - (L_i \otimes I_{n_i})\Omega_{i1}s_i + h_i(y_i) \\ \frac{ds_i}{d\tau} &= \Omega_{i1}^T(L_i \otimes I_{n_i})z_i\end{aligned} \tag{5}$$

Define $z_i^q(y_i), s_i^q(y_i)$ as the quasi-steady state, which satisfy the following equations.

$$W_i \begin{pmatrix} z_i^q(y_i) \\ s_i^q(y_i) \end{pmatrix} + \begin{pmatrix} h_i(y_i) \\ 0 \end{pmatrix} = 0 \tag{6}$$

where

$$W_i = -\begin{pmatrix} I_{n_i}^2 + L_i \otimes I_{n_i} & (L_i \otimes I_{n_i})\Omega_{i1} \\ -\Omega_{i1}^T(L_i \otimes I_{n_i}) & 0 \end{pmatrix}$$

According to [1], W_i is Hurwitz. Hence, there exist positive-definite matrices P_{i1}, P_{i2} satisfying $W_i^T P_{i1} + P_{i1} W_i = -P_{i2}$. Let $\lambda_i = [z_i^T, s_i^T]^T$, $\lambda_i^q = [z_i^T, s_i^T]^T$ $\psi_i = \lambda_i - \lambda_i^q$, $\psi = \{\psi_1, \psi_2, \cdots, \psi_N\}$. According to (6) and Assumption 1, there exists a positive constant η_1 such that $\left\|\frac{\partial \lambda_i^q(y_i)}{\partial x}\right\| \leq \eta_1$. Moreover, combing (5) and (6), we can get

$$\frac{d\psi_i}{d\tau} = W_i \psi_i - \frac{d\lambda_i^q(y_i)}{d\tau} \tag{7}$$

Then the (4) and (7) can be rewritten as,

$$\frac{d\bar{y}_{ij}}{d\tau} = -\,Q_{ij}\bar{y}_{ij} \tag{8a}$$

$$\frac{d\psi_i}{d\tau} = W_i \psi_i - \frac{d\lambda_i^q(y_i)}{d\tau} \tag{8b}$$

Denote $Q_{ij} = L_{ij} + B_{ij}$.

Theorem 3.1. *Under Assumption 1–3, $\Phi(t) = (\psi^T(t), \bar{y}_{11}^T(t), \cdots, \bar{y}_{Nn_N}(t)^T)^T$ of (8) converge to $(0_{\bar{n}}^T, 0_{n(n-1)}^T))^T$ exponentially.*

Proof. Consider the following Lyapunov function

$$V = k_1 \sum_{i=1}^{N} \sum_{j=1}^{n_i} \bar{y}_{ij}^T W_{ij} \bar{y}_{ij} + \sum_{i=1}^{N} \psi_i^T P_{i1} \psi_i \tag{9}$$

where k_1 is positive constants to be defined later. Taking the time derivative of V along (8), it yields that,

$$\begin{aligned}\frac{dV}{d\tau} &= -2k_1\sum_{i=1}^{N}\sum_{j=1}^{n_i}\bar{y}_{ij}^T Q_{ij}Q_{ij}\bar{y}_{ij} - \sum_{i=1}^{N}\psi_i^T P_{i2}\psi_i - 2\sum_{i=1}^{N}\psi_i^T P_{i1}\frac{d\lambda_i^q(y_i)}{d\tau}\\ &\leqslant -2k_1\sum_{i=1}^{N}\sum_{j=1}^{n_i}\lambda_{\min}(Q_{ij}Q_{ij})\|\bar{y}_{ij}\|^2 - (\lambda_{\min}(P_{i2})-\eta_1\varepsilon)\|P_1\|\|\psi\|^2 + \frac{\eta_1}{\varepsilon}\|P_1\|\sum_{i=1}^{N}\|\frac{dy_i}{d\tau}\|^2\end{aligned} \tag{10}$$

where $0 < \varepsilon < \frac{\lambda_{\min}(P_{i2})}{\eta_1}$, $\|P_1\| = \max\{\|P_{11}\|, \cdots, \|P_{N1}\|\}$. It is obvious that $\sum_{i=1}^{N}\|\frac{dy_i}{d\tau}\|^2 = \|\frac{dy}{d\tau}\|^2$. Meanwhile, it can be obtained that $\|\frac{dy}{d\tau}\|^2 = \sum_{i=1}^{N}\sum_{j=1}^{n_i}\|\frac{d\bar{y}_{ij}}{d\tau}\|^2$. Choose $k_1 > \frac{\eta_1\|P_1\|}{2\varepsilon}\max_{ij}(\frac{\lambda_{\min}(Q_{ij}Q_{ij})}{\lambda_{\max}(Q_{ij}Q_{ij})})$. Then, it gets,

$$\begin{aligned}\frac{dV}{d\tau} &\leqslant -2k_1\sum_{i=1}^{N}\sum_{j=1}^{n_i}\lambda_{\min}(Q_{ij}Q_{ij})\|\bar{y}_{ij}\|^2 + \frac{\eta_1\|P_1\|}{\varepsilon}\sum_{i=1}^{N}\sum_{j=1}^{n_i}\lambda_{\max}(Q_{ij}Q_{ij})\|\bar{y}_{ij}\|^2\\ &\quad - (\lambda_{\min}(P_{i2})-\eta_1\varepsilon)\|P_1\|\|\psi\|^2\end{aligned} \tag{11}$$

It follows $\Phi(t)$ converge to $(0_{\bar{n}}^T, 0_{n(n-1)}^T))^T$ exponentially.

According to Theorem 3.1, the estimated actions y converge to actual actions $1_n \otimes x$, and z_{ijj} converges to the quasi-steady state $z_{ijj}^q = \frac{1}{n_i}\sum_{k=1}^{n_i}\frac{\partial f_{ij}(x)}{\partial x_{ik}}$. Then the reduced system of (3) can be written as follows,

$$\dot{x}_{ij} \in -E(G_{ij}(x_{ij}))\sum_{k=1}^{n_i}\frac{\partial f_{ik}(x)}{\partial x_{ij}} - \int_0^t \text{sign}(G_{ij}(x_{ij}(s)))ds\partial G_{ij}(x_{ij}) \tag{12}$$

Lemma 3.1. *Under Assumption 3, for any initial point $x(0) \in \mathbb{R}^n$, the state solution of the reduced system (12) exists globally and is bounded.*

Proof. From [15, Th. 2.3], there is a local state $x_{ij}(t)$ of (12) for $t \in [0, T)$. And, for almost everywhere on $[0, T)$, the state $x_{ij}(t)$ satisfying

$$\dot{x}_{ij} = -E(G_{ij}(x_{ij}))(\sum_{k=1}^{n_i}\frac{\partial f_{ik}(x)}{\partial x_{ij}}) - \int_0^t \text{sign}(G_{ij}(x_{ij}(s)))ds\gamma_{ij}(t) \tag{13}$$

where $\gamma_{ij}(t) \in \partial G_{ij}(x_{ij}(t))$. According to Lemma 1 in [16], $G_{ij}(x_{ij})$ is coercive. Then, there exits $r_{ij} > 0$ such that $S_{ij} = \{x_{ij} : G_{ij}(x_{ij}) = 0\} \subset \{x_{ij} : 0 \leq G_{ij}(x_{ij}) \leq 1\} \subset B(\hat{q}_{ij}, r_{ij})$. And it is easy to find a $\bar{r}_{ij} > 0$ satisfying $\{x_{ij} : g_{ij}(x_{ij}) \leq 0_{p^2}\}\bigcup\{x_{ij}(0)\} \subset B(\hat{q}_{ij}, \bar{r}_{ij})$. Then, we prove $x_{ij}(t) \in B(\hat{q}_{ij}, \bar{r}_{ij})$ for $t \in [0, T)$. Otherwise, there exits $T_1, T_2 \in [0, T)$ such that $x_{ij}(t)$ leaves $B(\hat{q}_{ij}, \bar{r}_{ij})$ at the moment of T_1, and $x_{ij}(t) \notin B(\hat{q}_{ij}, r_{ij})$ for $t \in (T_1, T_2]$, which yields $E(G_{ij}(x_{ij})) = 0$ for $t \in (T_1, T_2)$.

Meanwhile, since $x_{ij}(t) \notin S_{ij}$ then $Q^+_{G_{ij}} \neq \varnothing$, it obtains

$$\begin{aligned} G_{ij}(x_{ij}(t)) &= \sum_{p \in Q^+_{G_{ij}}} g^p_{ij}(x_{ij}(t)) \\ \gamma_{ij}(t) &= \sum_{p \in Q^+_{G_{ij}}} \gamma^p_{ij}(t) + \sum_{p \in Q^0_{G_{ij}}} \theta^p_{ij}\gamma^p_{ij}(t) \end{aligned} \tag{14}$$

where $\gamma^p_{ij}(t) \in \partial\big(\max\{0, g^p_{ij}(x_{ij}(t))\}\big), \theta^p_{ij} \in [0,1]$.

Considering the time derivative of $\frac{1}{2}\|x_{ij}(t) - \hat{q}_{ij}\|^2$, it holds almost everywhere for $t \in (T_1, T_2]$

$$\begin{aligned} &\frac{1}{2}\frac{d}{dt}\|x_{ij}(t) - \hat{q}_{ij}\|^2 \\ &= -\int_0^t \mathrm{sign}(G_{ij}(x_{ij}(s)))\mathrm{d}s(x_{ij}(t) - \hat{q}_{ij})^T\gamma_{ij}(t) \\ &\le -\int_0^t \mathrm{sign}(G_{ij}(x_{ij}(s)))\mathrm{d}s(\sum_{p \in Q^+_{G_{ij}}} (g^p_{ij}(x_{ij}(t)) - g^p_{ij}(\hat{q}_{ij}))) \le 0 \end{aligned}$$

which implies that $\|x_{ij}(T_2) - \hat{q}_{ij}\| \le \|x_{ij}(T_1) - \hat{q}_{ij}\| = \bar{r}_{ij}$, which contradicts with $x_{ij}(T_2) \notin B(\hat{q}_{ij}, \bar{r}_{ij})$. Hence for $\forall t \in [0, T)$, $x_{ij}(t) \in B(\hat{q}_{ij}, \bar{r}_{ij})$. Through extension theorem [15, Th. 2.3], the state $x_{ij}(t)$ exists globally and is bounded.

From Lemma 3.1, without loss of generality, we assume that for $t \ge 0$ $x_{ij}(t) \in B(\hat{q}_{ij}, \bar{r}_{ij})$. Similar to the discussion of Lemma 3.1, there exists $\gamma_{ij}(t) \in \partial G_{ij}(x_{ij}(t))$ satisfying (13) for a.e. $t \ge 0$.

Lemma 3.2. *For any $i = 1, \ldots, N$ and $j = 1, \ldots, n_i$, the state $x_{ij}(t)$ of (12) enters the constraint set S_{ij} in finite time and remains within the constraint set.*

Proof. By the global boundedness of $x_{ij}(t)$ and Assumption 1, it is easy to get $\frac{\partial f_i(x)}{\partial x_{ij}}$ is bounded with a positive constant l_{ij}. Moreover, since $\partial G_{ij}(x_{ij})$ is u.s.c, there is a constant $L_G > 0$ such that the inequality holds

$$\begin{aligned} \|\partial G_{ij}(x_{ij}(t))\| &= \sup\{\|\gamma_{ij}(t) : \gamma_{ij}(t) \in \partial G_{ij}(x_{ij}(t))\|\} \\ &\le L_G. \end{aligned}$$

For any $x_{ij}(t) \in B(\hat{q}_{ij}, \bar{r}_{ij}) \setminus S_{ij}$, we have

$$\gamma_{ij}(t)^T(x_{ij}(t) - \hat{q}_{ij}) \ge \sum_{p \in Q^+_{G_{ij}}} (g^p_{ij}(x_{ij}) - g^p_{ij}(\hat{q}_{ij})) \ge -\sum_{p \in Q^+_{G_{ij}}} g^p_{ij}(\hat{q}_{ij}) \triangleq \bar{G} > 0 \tag{15}$$

Combing with $\gamma_{ij}(t)^T(x_{ij}(t) - \hat{q}_{ij}) \le \|\gamma_{ij}(t)\|\bar{r}_{ij}$, one gets $\|\gamma_{ij}(t)\| > \frac{\bar{G}}{\bar{r}_{ij}}$.

Choosing a sufficiently large constant $\bar{T}$, we prove there is $T_1 > \bar{T}$ such that $x_{ij}(T_1) \in S_{ij}$. Otherwise, $x_{ij}(t) \notin S_{ij}$ for $t \geq \bar{T}$. Since $\bar{T}$ is sufficiently large, we know that

$$\frac{\bar{G}}{\bar{r}_{ij}}\Big[(\int_0^{\bar{T}} \text{sign}(G_{ij}(x_{ij}(s))ds)^{-1} l_{ij} - \frac{\bar{G}}{\bar{r}_{ij}}\Big] \triangleq b_0 < 0 \tag{16}$$

Consider $J_{ij}(t)$ with the following form

$$J_{ij}(t) = (\int_0^t \text{sign}(G_{ij}(x_{ij}(s))ds)^{-1} G_{ij}(x_{ij}(t) \tag{17}$$

According to Chain Rule, we obtain the time derivative of $J_{ij}(t)$ for a.e. $t \geq \bar{T}$,

$$\begin{aligned} \frac{dJ_{ij}(t)}{dt} =& -(\int_0^t \text{sign}(G_{ij}(x_{ij}(s))ds)^{-2} \times \text{sign}(G_{ij}(x_{ij}(t))G_{ij}(x_{ij}(t)) \\ &+ (\int_0^t \text{sign}(G_{ij}(x_{ij}(s))ds)^{-1} \gamma_{ij}(t)^\top \dot{x}_{ij}(t) \\ \leq& \frac{\bar{G}}{\bar{r}_{ij}}\Big[(\int_0^{\bar{T}} \text{sign}(G_{ij}(x_{ij}(s)))ds)^{-1} l_{ij} - \frac{\bar{G}}{\bar{r}_{ij}}\Big] < 0 \end{aligned} \tag{18}$$

Then integrating (18) from $\bar{T}$ to t, one gets $J_{ij}(t) \leq J_{ij}(\bar{T}) + b_0(t - \bar{T})$. When t is large enough, it is obvious $J_{ij}(t) < 0$, which contradicts to the nonnegativity of $J_{ij}(t)$. Thus, there is $T_1 > \bar{T}$ such that $x_{ij}(T_1) \in S_{ij}$.

Moreover, we can get $x_{ij}(t) \in S_{ij}$, for $t \geq T_1$. If not, there is a period $(t_1, t_2]$ in which $x_{ij}(t)$ leaves S_{ij} at t_1, and $x_{ij}(t) \notin S_{ij}$ for $t \in (t_1, t_2]$. Integrating (18) from t_1 to t_2, one have

$$J_{ij}(t_2) = -b_0(t_2 - t_1) < 0 \tag{19}$$

which contradicts with the positiveness of $J_{ij}(t)$, and further yields the state $x_{ij}(t)$ enters the feasible set at T_1 and remains there.

Since $x_{ij}(t) \in S_{ij}$ for $t \geq T_1$, there exists a constant $\epsilon_{ij}^* > 0$ such that $\int_0^t \text{sign}(G_{ij}(x_{ij}(s)))ds \equiv \epsilon_{ij}^* \quad \forall t \geq T_1$. And $E(G_{ij}(x_{ij}(t))) = 1, \forall t \geq T_1$. It follows that for almost every $t \geq T_1$, we have:

$$\dot{x}(t) \in -F(x) - \epsilon^* \partial G(x) \tag{20}$$

where $\epsilon^* = \text{diag}\{\epsilon_{ij}^*\}$, $\partial G(x) = \text{Col}(\partial G_{11}(x_{11}), \partial G_{12}(x_{12}), ..., \partial G_{Nn_N}(x_{Nn_N}))$.

Theorem 3.2. *Under the given assumptions, the state solution $x(t)$ of reduced system (20) converges to GNE x^* exponentially.*

Define $V(t, x) = \frac{1}{2}\|x(t) - x^*\|^2$. x^* is the GNE of problem (1) if and only if there exists $\alpha^* \in \mathbb{R}^{nP}$ such that

$$0 \in F(x^*) + \partial g(x^*)^T \alpha^* \quad \text{and} \quad {\alpha^*}^T g(x^*) = 0,$$

i.e., there exists $\rho^* \in \partial g(x^*)$ satisfying $0_{nP} = F(x^*) + \rho^{*T}\alpha^*$. Let $L_{(\#)}V$ denote the set-valued Lie derivative of V along (20). For any $\mathcal{D} \in L_{(\#)}V$, there exists $\tilde{r} \in \partial G(x)$. For $t \geq t^*$,

$$\begin{aligned}\mathcal{D} &= \langle x(t) - x^*, -(F(x) + \epsilon^* \tilde{r}(t) + F(x^*)) + \rho^{*T}\alpha^* \rangle \\ &\leq \epsilon^* \langle x^* - x(t), \tilde{r}(t) \rangle + \langle x(t) - x^*, \rho^{*T}\alpha^* \rangle - \langle x(t) - x^*, F(x) - F(x^*) \rangle \\ &\leq -\langle x - x^*, F(x) - F(x^*) \rangle \leq -\mu \|x - x^*\|^2.\end{aligned} \tag{21}$$

Therefore, the state solution $x(t)$ converges to x^* exponentially.

Theorem 3.3. *Under Assumption 1–3, the trajectory of $x(t)$ in the original system (3) asymptotically converges to the GNE of the game.*

Proof. According to Theorem 3.1 and Theorem 3.2, the asymptotical convergence of the trajectories of boundary-layer system and reduced system are obtained. Following directly from Proposition 2 in [17], the trajectory of $x(t)$ in the original system (3) asymptotically converges to the GNE of the game.

4 Numerical Experiment

We consider the energy resource allocation problem in electricity market among three generation systems, with each generation system consists of 5 subsidiaries [18,19]. Figure 1 depicts the communication condition between them. The cost function $f_i(q_i, q_{-i})$ of generation system i is the sum of its subsidiaries' cost functions, which can be expressed as

$$f_i(q_i, q_{-i}) = \sum_{j=1}^{5} \Big(C_{ij}(q_{ij}) - P_{ij}(q) q_{ij} \Big).$$

where q_{ij} is the number of the output power that produced by subsidiary j from company i. $C_{ij}(q_{ij}) = a_{ij}q_{ij}^2 + b_{ij}q_{ij} + c_{ij}$ is the cost of subsidiary ij, and $P_{ij}(q)q_{ij} = (P_1 - \sum_{u=1}^{3}\sum_{v=1}^{5} q_{uv})q_{ij}$ is the profit generated by subsidiary ij, with $P_1 = 980$. In addition, due to the limited production capacity and efficiency of the subsidiary, and in order to ensure the benefits of the generation system's normal operation, the output of each subsidiary is often subjected to the maximum and minimum production limits. In this example, we set the subsidiaries are subject to set constraints, that is $q_{ij} \in [\underline{q}_{ij}, \bar{q}_{ij}]$. It is easy to find the set constraint can be rewritten as the local inequality of q_{ij} as $g_{ij}(q_{ij}) = (q_{ij} - \bar{q}_{ij}, \underline{q}_{ij} - q_{ij}) \leq 0_2$. Further, the parameters are given in Table 1.

It is evident that the settings of the above constraint sets and the objective function all satisfy the assumptions proposed in this paper. Let $\varepsilon = 0.01$. Let $q(0)$ starts form (9, 10, 15, 40, 33, 20, 40, 55, 10, 20, 35, 35, 40, 20, 10). As can be seen from Fig. 2, the cost function of each generation system evolves toward minimization and eventually stabilizes. Figure 3, Fig. 4 and Fig. 5 demonstrate that all subsidiaries converge to the Generalized Nash Equilibrium (GNE) under the algorithm designed in this paper.

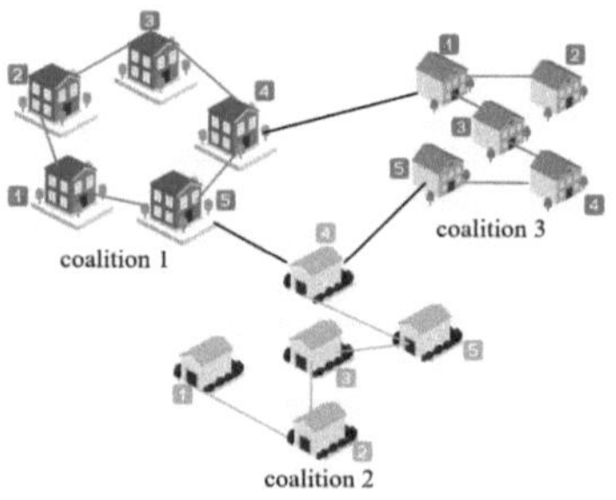

Fig. 1. Communication topology of the game.

Table 1. Parameters setting.

		a_m^n	b_m^n	c_m^n	$\bar{q}_{ij}$	$\underline{q}_{ij}$
Generation system 1	Subsidiary 1	1.3	2	3	55	40
	Subsidiary 2	2.1	0.1	4.2	65	45
	Subsidiary 3	1.7	1.2	1.4	50	33
	Subsidiary 4	2.4	3.1	2.1	40	25
	Subsidiary 5	3.5	2.2	4.3	35	20
Generation system 2	Subsidiary 1	2.7	2.7	1.8	40	15
	Subsidiary 2	2.4	1.3	2	54	45
	Subsidiary 3	2.6	2.1	2.2	65	50
	Subsidiary 4	3.6	2.5	0.1	30	20
	Subsidiary 5	4	1.9	1.9	40	20
Generation system 3	Subsidiary 1	1.9	1.2	2.7	35	25
	Subsidiary 2	3.1	1.4	3.3	45	30
	Subsidiary 3	3.8	1	3.7	55	15
	Subsidiary 4	3	2	1.6	65	35
	Subsidiary 5	2.1	1.9	4.5	50	25

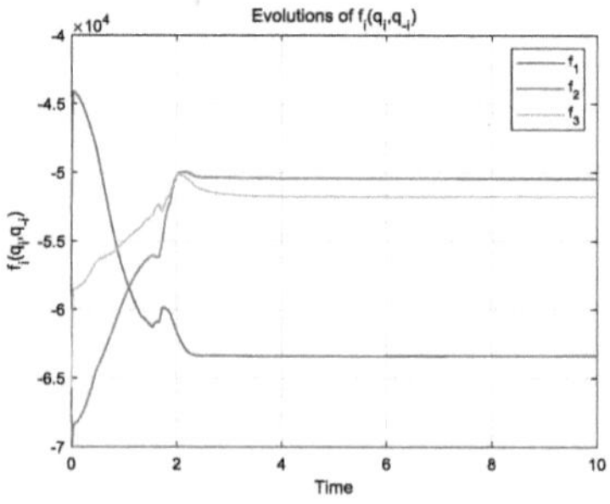

Fig. 2. The variation graph of the cost functions of three generation system.

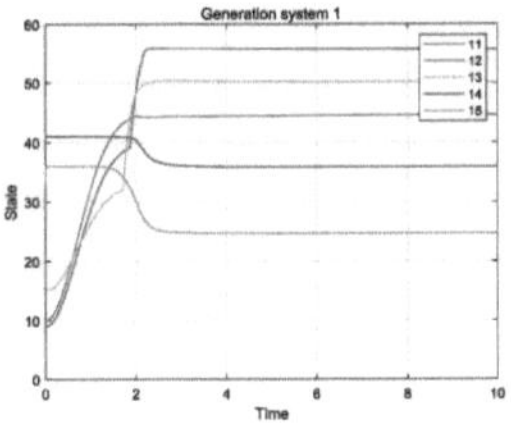

Fig. 3. Variation graph of power output values of each subsidiary in generation system 1.

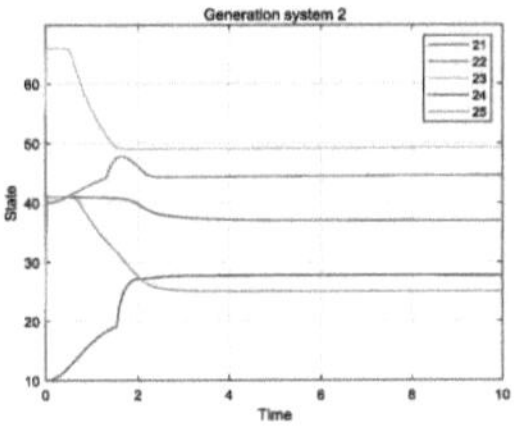

Fig. 4. Variation graph of power output values of each subsidiary in generation system 2.

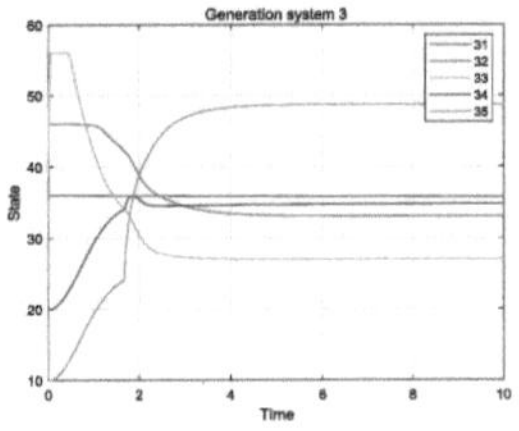

Fig. 5. Variation graph of power output values of each subsidiary in generation system 3.

5 Conclusions

This paper proposes a distributed two-time scale algorithm to address the constrained N-coalition game problem. In the case of imperfect information, the average consensus protocol is adopted to enable players to estimate the behaviors of other players. Meanwhile, the penalty method is utilized to transform the problem into an unconstrained one, and at the same time, an adaptive integral term is introduced to promote the progress towards the constraint set. Future work can focus on the game problems of players with coupled constraints.

References

1. Ye, M., Hu, G., Lewis, F.L.: Nash equilibrium seeking for N-coalition noncooperative games. Automatica **95**, 266–272 (2018). https://doi.org/10.1016/j.automatica.2018.05.020
2. Peng, T.-J.A., Bourne, M.: The coexistence of competition and cooperation between networks: implications from two Taiwanese healthcare networks. Br. J. Manag. **20**(3), 377–400 (2009)
3. Deng, Z.: Distributed algorithm design for aggregative games of Euler-Lagrange systems and its application to smart grids. IEEE Trans. Cybern. **52**(8), 8315–8325 (2022). https://doi.org/10.1109/TCYB.2021.3049462
4. Shehory, O., Kraus, S.: Methods for task allocation via agent coalition formation. Artif. Intell. **101**(1), 165–200 (1998). https://doi.org/10.1016/S0004-3702(98)00045-9
5. Sun, N., Trockel, W., Yang, Z.: Competitive outcomes and endogenous coalition formation in an n-person game. J. Math. Econ. **44**(7–8), 853–860 (2008)
6. Nian, X., Liu, D., Li, F.: Nash equilibrium seeking for multi-cluster games of second-order systems over weight-unbalanced digraphs. IEEE Trans. Circuits Syst. II Express Briefs **71**(4), 2209–2213 (2023)
7. Saad, W., Han, Z., Debbah, M., Hjorungnes, A., Basar, T.: Coalitional game theory for communication networks. IEEE Signal Process. Mag. **26**(5), 77–97 (2009)
8. Liu, F., Yu, J., Hua, Y., Dong, X., Li, Q., Ren, Z.: Dynamic generalized Nash equilibrium seeking for N-coalition noncooperative games. Automatica **147**, 110746 (2023)

9. Huang, B., Liu, Y., Lu, J., Jiang, Y., Gui, W.: Two-timescale distributed approach to Nash equilibrium seeking of Euler-Lagrange systems in multi-cluster nonconvex games. IEEE Trans. Autom. Control (2025)
10. Wei, Y., Zeng, X., Fang, H., Ding, Y., Ding, S.: Generalized Nash equilibrium seeking for directed nonsmooth multicluster games via a distributed Lipschitz algorithm. IEEE Trans. Control Netw. Syst. **11**(4), 2033–2042 (2024). https://doi.org/10.1109/tcns.2024.3372140
11. Sun, C., Hu, G.: Distributed generalized Nash equilibrium seeking of N-coalition games with inequality constraints. In: 2021 60th IEEE Conference on Decision and Control (CDC), pp. 215–220. IEEE (2021)
12. Liu, F., Wang, Q., Hua, Y., Dong, X., Ren, Z.: Distributed Nash equilibrium seeking for non-cooperative convex games with local constraints. In: 2021 40th Chinese Control Conference (CCC), pp. 7480–7485. IEEE (2021)
13. Zhou, H., Zeng, X., Hong, Y.: Adaptive exact penalty design for constrained distributed optimization. IEEE Trans. Autom. Control **64**(11), 4661–4667 (2019). https://doi.org/10.1109/TAC.2019.2902612
14. Khoo, S., Xie, L., Man, Z.: Robust finite-time consensus tracking algorithm for multirobot systems. IEEE/ASME Trans. Mechatron. **14**(2), 219–228 (2009). https://doi.org/10.1109/TMECH.2009.2014057
15. Aubin, J.-P., Cellina, A.: Differential inclusions with maximal monotone maps. In: Differential Inclusions: Set-Valued Maps and Viability Theory, pp. 139–171. Springer, Heidelberg (1984)
16. Ma, L., Bian, W.: A novel multiagent neurodynamic approach to constrained distributed convex optimization. IEEE Trans. Cybern. **51**(3), 1322–1333 (2019)
17. Teel, A.R., Moreau, L., Nesic, D.: A unified framework for input-to-state stability in systems with two time scales. IEEE Trans. Autom. Control **48**(9), 1526–1544 (2003). https://doi.org/10.1109/TAC.2003.816966
18. Nian, X., Niu, F., Yang, Z.: Distributed Nash equilibrium seeking for multicluster game under switching communication topologies. IEEE Trans. Syst. Man Cybern. Syst. **52**(7), 4105–4116 (2021)
19. Deng, Z., Liang, S.: Distributed algorithms for aggregative games of multiple heterogeneous Euler-Lagrange systems. Automatica **99**, 246–252 (2019). https://doi.org/10.1016/j.automatica.2018.10.041

Adaptive Event-Triggered Tracking Control of Nonlinear Networked Systems Under Dynamic Quantization

Xiujiao Li and Hailing Wang(✉)

School of Mathematics and Statistics, The Center for Applied Mathematics of Guangxi, Guangxi Normal University, Guilin 541006, China
wanghl@gxnu.edu.cn

Abstract. This paper studies the tracking control problem with network security for nonlinear networked systems subject to exogenous disturbances, bandwidth constraints and deception attacks. Firstly, the Takagi–Sugeno (T-S) fuzzy model is employed to approximate the nonlinear dynamics, and an integrated strategy combining dynamic quantization and adaptive event-triggered mechanism (AETM) is proposed. This strategy can not only compress data size, but also dynamically regulate the communication frequency, which effectively reduces the network burden. In this work, an improved two-sided looped-functional (TSLF) is constructed to fully acquire the state information during sampling intervals. On this basis, less conservative stability criteria expressed by linear matrix inequalities (LMIs) are derived. Finally, a permanent magnet synchronous motor (PMSM) is utilized for numerical simulation and experimental verification. The results illustrate that the proposed approach can markedly cut down network overhead while guaranteeing robust tracking performance.

Keywords: Tracking control · two-sided looped-functional · adaptive event-triggered mechanism · dynamic quantization

1 Introduction

The rapid development of networked control systems (NCSs) has highlighted the need for effective non-linear tracking control. The T-S fuzzy model provides a systematic approach to approximating nonlinear dynamics [7–9]. Compared to stabilization, tracking control is practically more significant but inherently more challenging due to reference-dependent dynamics and stricter performance requirements constraints.

A critical bottleneck in NCSs is the limited communication bandwidth. To alleviate network burdens, event-triggered mechanisms (ETM) have been extensively studied [3,6]. Specifically, AETM dynamically adjust triggering thresholds according to system state fluctuations, offering superior resource allocation over static ETM [9]. However, AETM only reduce transmission frequency. To further

Z. Zhang (Ed.): ISNN 2026, LNCS 16848, pp. 281–291, 2027.
https://doi.org/10.1007/978-981-92-5870-3_23

compress the data size of each transmission packet, dynamic quantization is an essential tool [1,2]. While combining AETM and dynamic quantization seems an intuitive solution for bandwidth-constrained NCSs, simultaneously guaranteeing tracking precision and stability under such a dual-restriction mechanism remains a non-trivial challenge.

Furthermore, data transmitted over open NCSs are highly vulnerable to cyber threats, such as deception attacks. In the presence of network imperfections and attacks, stability analysis typically relies on Lyapunov-Krasovskii functionals (LKFs). Recently, the TSLF has emerged as a powerful technique to reduce mathematical conservatism by incorporating state information from both sides of the sampling instants [5,10]. Yet, existing studies rarely integrate TSLF with the aforementioned AETM and quantization schemes for T-S fuzzy tracking control. Motivated by these gaps, this paper proposes an attack-resilient tracking control framework for T-S fuzzy systems. Distinct from existing works, the main contributions are summarized as follows:

1) Construct a novel quantized event-triggered communication scheme (QETCS). Communication resources are optimized in terms of both transmission frequency and data packet load, and the overall communication efficiency is consequently improved.
2) An improved TSLF is constructed to fully explore the state evolution information within sampling intervals, which effectively reduces the conservatism of LMIs-based stability criteria.

2 Problem Formulation

The i-th rule of the T-S fuzzy model representing the nonlinear networked system is formulated as:

Plant rule i: **IF** $\phi_1(t)$ is Ψ_{i1}, $\phi_2(t)$ is Ψ_{i2}, and $\cdots$ and $\phi_s(t)$ is Ψ_{is}, **THEN**

$$\begin{cases} \dot{x}(t) = A_i x(t) + B_i u(t) + B_{\kappa i}\kappa(t), \\ y(t) = C_i x(t), \end{cases} \tag{1}$$

where A_i, B_i, D_i, and C_i are the given constant matrices. $\Psi_{iv}(i = 1, \cdots, q; v = 1, \cdots, s)$ stands for the fuzzy set. Subsequently, through fuzzy aggregation, the system (1) is reformulated as:

$$\begin{cases} \dot{x}(t) = \sum_{i=1}^{r} \mu_i(\phi(t))[A_i x(t) + B_i u(t) + D_i \kappa(t)], \\ y(t) = \sum_{i=1}^{r} \mu_i(\phi(t)) C_i x(t), \end{cases} \tag{2}$$

where $\mu_i(\phi(t))$ are normalized membership functions satisfying $\sum_{i=1}^{r} \mu_i(\phi(t)) = 1$.

The reference signal $y_r(t)$, which constitutes the prescribed path that the system output $y(t)$ must adhere to, is given by the following expression:

$$\begin{cases} \dot{x}_r(t) = A_r x_r(t) + B_r r(t), \\ y_r(t) = C_r x_r(t), \end{cases} \tag{3}$$

where A_r, B_r and C_r are constant matrices.

In order to reduce transmission frequency and communication load, we introduce a dynamic quantizer. Let $\eta^T(t) = [x^T(t) \quad x_r^T(t)]$ be defined, and assume there exist $\mathcal{M} > 0$ and $\theta > 0$ for which:

$$|q_\theta(\eta(t), \theta(t)) - \eta(t)| \leq \Delta\theta(t), \quad \text{if } \Delta\theta(t) \leq |\eta(t)| \leq \mathcal{M}\theta(t),$$
$$|q_\theta(\eta(t), \theta(t)) - \eta(t)| > (\mathcal{M} - \Delta)\theta(t), \quad \text{if } |\eta(t)| > \mathcal{M}\theta(t),$$

where $\mathcal{M}$, $\theta(t)$, and Δ denote the quantizer range, dynamic parameter,and error bound, respectively. Following [12], the dynamically quantized output is:

$$q_\theta(\eta(t), \theta(t)) = \eta(t) + \Delta\bar{\vartheta}(t)\frac{\wp}{\mathcal{M}}\eta(t). \tag{4}$$

Aiming at communication resource conservation, this study leverages quantized states to formulate a QETCS.

$$\begin{aligned} t_{k+1} = t_k + \min_{s \in N} \Bigg\{ & \sum_{s=1}^{N-1} |\tau(t_k)\theta^T(t_k)\Phi\theta(t_k) \\ & \leq q_\theta^T(\rho(d_s), \theta(d_s))\Phi q_\theta(\rho(d_s), \theta(d_s)) \Bigg\}, \end{aligned} \tag{5}$$

where $q_\theta(\rho(d_s), \theta(d_s)) = q_\theta(\eta(d_s), \theta(d_s)) - q_\theta(\eta(t_k), \theta(t_k))$, $d_s = t_k + \sum_{s=1}^{N-1} l_s$, and $\tau(t_k)$ is the adaptive threshold parameter. Its adjustment rule is as follows:

$$\tau(t_k) = (1 - \text{sgn}[\mathcal{X}(t_k)]\mathcal{G}(t_k))\,\tau(t_{k-1}), \tag{6}$$

where

$$\mathcal{X}(t_k) = \|q_\theta(\eta(t_k), \theta(t_k))\| - \|q_\theta(\eta(t_{k-1}), \theta(t_{k-1}))\|,$$
$$\mathcal{K}(t_k) = \|q_\theta(\eta(t_k), \theta(t_k)) - q_\theta(\eta(t_{k-1}), \theta(t_{k-1}))\|,$$
$$\mathcal{G}(t_k) = \begin{cases} \frac{2}{\pi}\beta \arctan\frac{\mathcal{K}(t_k)}{\mathcal{X}(t_k)}, & \text{if } \mathcal{X}(t_k) \geq 0, \\ \frac{2}{\pi}\beta \arctan\frac{\mathcal{X}(t_k)}{\mathcal{X}(t_k)}, & \text{if } \mathcal{X}(t_k) < 0. \end{cases}$$

Moreover, signal transmission is often compromised by potential deception attacks in open network settings. When the system is subjected to such attacks, the attack signals may tamper with or replace the normal transmission data. To characterize the random nature of the attacks, a Bernoulli random variable $\xi \in \{0, 1\}$ is introduced to describe the occurrence of deception attacks, and its statistical characteristics are determined by the following expressions: $\text{Prob}\{\xi(t) = 1\} = \bar{\xi}, \quad \text{Prob}\{\xi(t) = 0\} = 1 - \bar{\xi}.$

Assumption 1. The deception attack $\mathcal{F}(q_\theta(\eta(t), \theta(t)))$ is assumed to satisfy the following norm-bounded constraint for a given constant matrix $\mathcal{H}$:

$$\|\mathcal{F}(q_\theta(\eta(t), \theta(t)))\|_2 \leq \|\mathcal{H} q_\theta(\eta(t), \theta(t))\|_2. \tag{7}$$

The fuzzy control input is given by:

$$u(t) = \sum_{j=1}^{r} \mu_j(\phi(t_k)) K_j \bar{u}(t), \tag{8}$$

where
$K_j = \begin{bmatrix} K_{j1} \; K_{j2} \end{bmatrix}$, $\bar{u}(t) = \Big[(1 - \xi(t)) q_\theta(\eta(t_k), \theta(t_k)) + \xi(t) \mathcal{F}\big(q_\theta(\eta(t_k), \theta(t_k))\big)\Big]$.

For further analysis, the interval $[t_k, t_{k+1})$ is divided into several subsets J_a, such that $[t_k, t_{k+1}) = \bigcup_{l=0}^{M-1} J_a$, where $J_a = [d_c, d_{c+1})$, $d_0 = t_k$, $d_M = t_{k+1}$, and the sampling interval $l_c = d_{c+1} - d_c \in [l_m, l_M]$. By defining $e(t) = y(t) - y_r(t)$, the resulting augmented closed-loop model is expressed as:

$$\begin{cases} \dot{\eta}(t) = \sum_{i=1}^{r} \sum_{j=1}^{r} \mu_i \mu_j \Big[\bar{A}_i \eta(t) + \bar{B}_i K_j \bar{u}(t) + \bar{D}_i \omega(t)\Big], \\ e(t) = \sum_{i=1}^{r} \mu_i \bar{C}_i \eta(t), \end{cases} \tag{9}$$

where

$$\bar{A}_i = \begin{bmatrix} A_i & 0 \\ 0 & A_r \end{bmatrix}, \bar{D}_i = \begin{bmatrix} D_i & 0 \\ 0 & B_r \end{bmatrix}, \bar{B}_i = \begin{bmatrix} B_i \\ 0 \end{bmatrix}, \bar{C}_i^T = \begin{bmatrix} C_i^T \\ -C_r^T \end{bmatrix}, \omega(t) = \begin{bmatrix} \kappa(t) \\ r(t) \end{bmatrix}.$$

3 Main Results

This paper introduces the following notations, to streamline the derivation:
$l_1(t) = t - d_c, \quad l_2(t) = d_{c+1} - t, \quad \nu^T = [\eta^T(d_c), \; \eta^T(d_{c+1})]$
$\sigma_1(t) = t - o_1 l_1(t), \quad \sigma_2(t) = t + o_2 l_2(t), \quad o_1, o_2 \in (0, 1)$

$$\zeta^T(t) = \Big[\eta^T(t), \; \eta^T(d_c), \; \eta^T(d_{c+1}), \; \eta^T(\sigma_1(t)), \; \eta^T(\sigma_2(t)), \dot{\eta}^T(t), \; \int_{\sigma_1(t)}^{t} \eta^T(s)\,\mathrm{d}s, \; \int_{d_c}^{\sigma_1(t)} \eta^T(s)\,\mathrm{d}s, \; \int_{t}^{\sigma_2(t)} \eta^T(s)\,\mathrm{d}s, \; \int_{\sigma_2(t)}^{d_{c+1}} \eta^T(s)\,\mathrm{d}s, \; \rho^T(d_c), \; \mathcal{F}^T\big(q_\theta(\eta(t_k), \theta(t_k))\big), \; \omega^T(t)\Big],$$

$l_m = [0_{n \times (m-1)n} \; I_n \; 0_{n \times (13-m)n}], \quad m = 1, 2, \ldots, 13.$

Theorem 1. For specified positive constants $\beta, \tau_0, \Delta, \mathcal{M}, \varepsilon$, $l_m \leq l_c \leq l_M$, and $o_1, o_2 \in (0,1)$, given the gain matrices K_j, the augmented closed-loop system is asymptotically stable if there exist symmetric positive definite matrices $\mathcal{Z}, \Phi$, a non-singular matrix $\mathfrak{W}$, and matrices $\mathcal{N}$, $\mathscr{J}_h, \mathscr{V}_h, \mathscr{L}_h, \mathcal{R}_h, \mathcal{S}_h, \mathcal{Y}_v$ of suitable dimensions, such that the ensuing LMIs are satisfied for $l_c \in \{l_m, l_M\}$ $(h = 1,2; v = 1,\ldots,8; i,j = 1,\ldots,r)$:

$$\begin{bmatrix} \Lambda_{ij} + l_c \Upsilon_1 & \Xi_1 & \Pi_{ij} \\ * & \Omega_1 & 0 \\ * & * & -I \end{bmatrix} < 0, \tag{10}$$

$$\begin{bmatrix} \Lambda_{ij} + l_c \Upsilon_2 & \Xi_2 & \Pi_{ij} \\ * & \Omega_2 & 0 \\ * & * & -I \end{bmatrix} < 0. \tag{11}$$

where

$$\begin{aligned}
\chi =& 1 + \Delta\bar{\vartheta}(t)\frac{\wp}{\mathcal{M}}, \\
\Lambda_{ij} =& Sym\Big[\iota_1^T \mathcal{Z}\iota_6 + \mathcal{Y}_1^T \iota_7 + \mathcal{Y}_2^T \iota_8 + \mathcal{Y}_3^T \iota_9 + \mathcal{Y}_4^T \iota_{10} + \mathcal{Y}_5^T(\iota_4 - \iota_2) + \mathcal{Y}_6^T(\iota_1 - \iota_4) \\
&+ \mathcal{Y}_7^T(\iota_5 - \iota_1) + \mathcal{Y}_8^T(\iota_3 - \iota_5) + (\iota_1^T + \epsilon\iota_6^T)\mathfrak{W}\Big(\bar{A}_i\iota_1 + \bar{D}_i\iota_{13} - \iota_6 \\
&+ \bar{B}_i K_j\Big(\chi(1-\bar{\xi})(\iota_2 - \iota_{11}) + \bar{\xi}\iota_{12}\Big)\Big)\Big] - \iota_7^T\mathcal{R}_1\iota_7 - \iota_8^T\mathcal{R}_2\iota_8 - \iota_9^T\mathcal{S}_1\iota_9 \\
&- \iota_{10}^T\mathcal{S}_2\iota_{10} - \chi^2\iota_{11}^T\Phi\iota_{11} + \tau_0(\iota_2 - \iota_{11})^T\frac{\wp}{\mathcal{M}}\Phi\frac{\wp}{\mathcal{M}}(\iota_2 - \iota_{11}) - \iota_{12}^T\mathcal{L}\iota_{12} \\
&+ \chi^2(\iota_2 - \iota_{11})^T\mathcal{H}^T\mathcal{L}\mathcal{H}(\iota_2 - \iota_{11}), \\
\Upsilon_1 =& -[\iota_2^T\ \iota_3^T]\mathcal{N}[\iota_2^T\ \iota_3^T]^T - [(1-o_2)\iota_5^T(\mathscr{V}_1 - \mathscr{V}_2)\iota_5 - \iota_1^T\mathscr{V}_1\iota_1] \\
&+ Sym[\iota_9^T\mathcal{S}_1((1-o_2)\iota_5 - \iota_1) - \iota_{10}^T\mathcal{S}_2(1-o_2)\iota_5] + \iota_6^T\mathscr{L}_2\iota_6, \\
\Upsilon_2 =& [\iota_2^T\ \iota_3^T]\mathcal{N}[\iota_2^T\ \iota_3^T]^T - [(1-o_1)\iota_4^T(\mathscr{J}_1 - \mathscr{J}_2)\iota_4 - \iota_1^T\mathscr{J}_1\iota_1] \\
&- Sym[\iota_7^T\mathcal{R}_1(\iota_1 - (1-o_1)\iota_4) - \iota_8^T\mathcal{R}_2(1-o_1)\iota_4] + \iota_6^T\mathscr{L}_1\iota_6, \\
\Xi_1 =& [\sqrt{l_c o_1}\mathcal{Y}_1^T\ \sqrt{l_c(1-o_1)}\mathcal{Y}_2^T\ \sqrt{l_c(1-o_1)}\mathcal{Y}_5^T\ \sqrt{l_c o_1}\mathcal{Y}_6^T], \\
\Xi_2 =& [\sqrt{l_c o_2}\mathcal{Y}_3^T\ \sqrt{l_c(1-o_2)}\mathcal{Y}_2^T\ \sqrt{l_c o_2}\mathcal{Y}_7^T\ \sqrt{l_c(1-o_2)}\mathcal{Y}_8^T], \\
\Omega_1 =& -\mathrm{diag}\{\mathscr{J}_1, \mathscr{J}_2, \mathscr{L}_1, \mathscr{L}_1\}, \Omega_2 = -\mathrm{diag}\{\mathscr{V}_1, \mathscr{V}_2, \mathscr{L}_2, \mathscr{L}_2\}, \Pi_{ij} = \bar{C}_i\iota_1.
\end{aligned}$$

Proof. The following LKF is established:

$$V(t) = V_0(t) + \sum_{g=1}^{5} V_g(t), \tag{12}$$

where

$$
\begin{aligned}
V_0(t) &= \eta^T(t)\mathcal{Z}\eta(t),\\
V_1(t) &= l_1(t)l_2(t)\nu^T\mathcal{N}\nu,\\
V_2(t) &= l_2(t)\left(\int_{\sigma_1(t)}^{t}\eta^T(s)\mathscr{J}_1\eta(s)\,ds + \int_{d_c}^{\sigma_1(t)}\eta^T(s)\mathscr{J}_2\eta(s)\,ds\right),\\
V_3(t) &= -l_1(t)\left(\int_{t}^{\sigma_2(t)}\eta^T(s)\mathscr{V}_1\eta(s)\,ds + \int_{\sigma_2(t)}^{d_{c+1}}\eta^T(s)\mathscr{V}_2\eta(s)\,ds\right),\\
V_4(t) &= l_2(t)\int_{d_c}^{t}\dot{\eta}^T(s)\mathscr{L}_1\dot{\eta}(s)\,ds - l_1(t)\int_{t}^{d_{c+1}}\dot{\eta}^T(s)\mathscr{L}_2\dot{\eta}(s)\,ds,\\
V_5(t) &= l_2(t)\Bigg[\Big(\int_{\sigma_1(t)}^{t}\eta(s)\,ds\Big)^T\mathcal{R}_1\Big(\int_{\sigma_1(t)}^{t}\eta(s)\,ds\Big)\\
&\quad + \Big(\int_{d_c}^{\sigma_1(t)}\eta(s)\,ds\Big)^T\mathcal{R}_2\Big(\int_{d_c}^{\sigma_1(t)}\eta(s)\,ds\Big)\Bigg]\\
&\quad - l_1(t)\Bigg[\Big(\int_{t}^{\sigma_2(t)}\eta(s)\,ds\Big)^T\mathcal{S}_1\Big(\int_{t}^{\sigma_2(t)}\eta(s)\,ds\Big)\\
&\quad + \Big(\int_{\sigma_2(t)}^{d_{c+1}}\eta(s)\,ds\Big)^T\mathcal{S}_2\Big(\int_{\sigma_2(t)}^{d_{c+1}}\eta(s)\,ds\Big)\Bigg].
\end{aligned}
$$

Based on (7) and [4], one can introduce a symmetric matrix $\mathcal{L}$ to obtain:

$$
\begin{aligned}
&\mathcal{F}^T\left(q_\theta(\eta(t_k),\theta(t_k))\right)\mathcal{L}\mathcal{F}\left(q_\theta(\eta(t_k),\theta(t_k))\right)\\
&- q_\theta^T(\eta(t_k),\theta(t_k))\mathcal{H}^T\mathcal{L}\mathcal{H}\,q_\theta(\eta(t_k),\theta(t_k)) \le 0.
\end{aligned}
\tag{13}
$$

For the system (9), a zero-value equality involving a non-singular matrix $\mathfrak{W}$ and a scalar ε is incorporated:

$$
2\sum_{i=1}^{q}\sum_{j=1}^{q}[\eta^T(t)+\varepsilon\dot{\eta}^T(t)]\mathfrak{W}\Big[\bar{A}_i\eta(t)+\bar{B}_iK_j\bar{u}(t)+\bar{D}_i\omega(t)-\dot{\eta}(t)\Big]=0. \tag{14}
$$

During the interval $t\in[d_c,d_{c+1})$, the criterion for event triggering is equivalently reformulated as:

$$
\tau_0\big(\eta(d_l)-\rho(d_l)\big)^T\frac{\wp}{M}\Phi\frac{\wp}{M}\big(\eta(d_l)-\rho(d_l)\big)-\chi^2\rho^T(d_l)\Phi\rho(d_l)>0. \tag{15}
$$

Differentiating the LKF (12) and taking its expectation, and then employing Lemma 1 from [5] to bound the integral terms of the obtained derivative, we obtain:

$$
E\{\dot{V}(t)+e^T(t)e(t)-\gamma^2\omega^T(t)\omega(t)\}\le\sum_{i=1}^{r}\sum_{j=1}^{r}\mu_i\mu_j\zeta^T(t)\Theta(t)\zeta(t), \tag{16}
$$

where

$$
\begin{aligned}
\Theta(t) &= \frac{l_1(t)}{l_c}\Theta_1^{ij} + \frac{l_2(t)}{l_c}\Theta_2^{ij},\\
\Theta_1^{ij} &= \Lambda_{ij} + \Pi_{ij} + l_c\left(\Upsilon_1 + \Gamma_1\right), \quad \Theta_2^{ij} = \Lambda_{ij} + \Pi_{ij} + l_c\left(\Upsilon_2 + \Gamma_2\right),\\
\Gamma_1 &= o_1\mathcal{Y}_1^T\mathscr{J}_1^{-1}\mathcal{Y}_1 + (1-o_1)\mathcal{Y}_2^T\mathscr{J}_2^{-1}\mathcal{Y}_2 + (1-o_1)\mathcal{Y}_5^T\mathcal{Q}_1^{-1}\mathcal{Y}_5 + o_1\mathcal{Y}_6^T\mathcal{Q}_1^{-1}\mathcal{Y}_6,\\
\Gamma_2 &= o_2\mathcal{Y}_3^T\mathscr{V}_1^{-1}\mathcal{Y}_3 + (1-o_2)\mathcal{Y}_4^T\mathscr{V}_2^{-1}\mathcal{Y}_4 + o_2\mathcal{Y}_7^T\mathcal{Q}_2^{-1}\mathcal{Y}_7 + (1-o_2)\mathcal{Y}_8^T\mathcal{Q}_2^{-1}\mathcal{Y}_8.
\end{aligned}
$$

By virtue of the convex combination approach and Schur complement, the conditions (10)–(11) ensure $E\{\dot{V}(t) + e^T(t)e(t) - \gamma^2\omega^T(t)\omega(t)\} < 0$, thus establishing the theorem. □

Theorem 2. Given the positive scalars $\beta, \tau_0, \Delta, \mathcal{M}, \varepsilon$, $o_1, o_2 \in (0,1)$, and $l_c \in [l_m, l_M]$, the system is stabilized if there exist symmetric positive definite matrices $\tilde{\mathcal{Z}}, \tilde{\Phi}$, a non-singular matrix $\mathfrak{W}$, and matrices $\tilde{\mathcal{N}}$, $\tilde{\mathscr{J}}_h, \tilde{\mathscr{V}}_h, \tilde{\mathscr{L}}_h, \tilde{\mathcal{R}}_h, \tilde{\mathcal{S}}_h, \tilde{\mathcal{Y}}_v, \tilde{K}_j$ with compatible dimensions, such that for $l_c \in \{l_m, l_M\}$, $h = 1, 2$, $v = 1, \ldots, 8$, and $i, j = 1, \ldots, r$, the subsequent LMIs are feasible:

$$
\begin{bmatrix} \tilde{\Lambda}_{ij} + l_c\tilde{\Upsilon}_1 & \tilde{\Xi}_1 & \tilde{\Pi}_{ij} \\ * & \tilde{\Omega}_1 & 0 \\ * & * & -I \end{bmatrix} < 0, \tag{17}
$$

$$
\begin{bmatrix} \tilde{\Lambda}_{ij} + l_c\tilde{\Upsilon}_2 & \tilde{\Xi}_2 & \tilde{\Pi}_{ij} \\ * & \tilde{\Omega}_2 & 0 \\ * & * & -I \end{bmatrix} < 0. \tag{18}
$$

where

$$
\begin{aligned}
\tilde{\Lambda}_{ij} =& Sym\Big[\iota_1^T\tilde{\mathcal{Z}}\iota_6 + \tilde{\mathcal{Y}}_1^T\iota_7 + \tilde{\mathcal{Y}}_2^T\iota_8 + \tilde{\mathcal{Y}}_3^T\iota_9 + \tilde{\mathcal{Y}}_4^T\iota_{10} + \tilde{\mathcal{Y}}_5^T(\iota_4 - \iota_2) + \tilde{\mathcal{Y}}_6^T(\iota_1 - \iota_4)\\
&+ \tilde{\mathcal{Y}}_7^T(\iota_5 - \iota_1) + \tilde{\mathcal{Y}}_8^T(\iota_3 - \iota_5) + (\iota_1^T + \varepsilon\iota_6^T)\big(\tilde{A}_i\tilde{U}^T\iota_1 + \tilde{B}_i\tilde{K}_j(\chi(1-\bar{\xi})(\iota_2 - \iota_{11})\\
&+ \bar{\xi}\iota_{12}) + \tilde{D}_i\tilde{U}^T\iota_{13} - \tilde{U}^T\iota_6\big)\Big] - \iota_7^T\tilde{R}_1\iota_7 - \iota_8^T\tilde{R}_2\iota_8 - \iota_9^T\tilde{S}_1\iota_9 - \iota_{10}^T\tilde{S}_2\iota_{10} - \iota_{12}^T\mathcal{L}\iota_{12}\\
&+ \tau_0(\iota_2 - \iota_{11})^T\frac{\wp}{\mathcal{M}}\tilde{\Phi}\frac{\wp}{\mathcal{M}}(\iota_2 - \iota_{11}) - \chi^2\iota_{11}^T\tilde{\Phi}\iota_{11} + \chi^2(\iota_2 - \iota_{11})^T\mathcal{H}^T\mathcal{L}\mathcal{H}(\iota_2 - \iota_{11}),\\
\tilde{\Upsilon}_1 =& -[\iota_2^T\ \iota_3^T]\tilde{\mathcal{N}}[\iota_2^T\ \iota_3^T]^T - [(1-o_2)\iota_5^T(\tilde{\mathscr{V}}_1 - \tilde{\mathscr{V}}_2)\iota_5 - \iota_1^T\tilde{\mathscr{V}}_1\iota_1]\\
&+ Sym[\iota_9^T\tilde{\mathcal{S}}_1((1-o_2)\iota_5 - \iota_1) - \iota_{10}^T\tilde{\mathcal{S}}_2(1-o_2)\iota_5] + \iota_6^T\tilde{\mathscr{L}}_2\iota_6,\\
\tilde{\Upsilon}_2 =& [\iota_2^T\ \iota_3^T]\tilde{\mathcal{N}}[\iota_2^T\ \iota_3^T]^T - [(1-o_1)\iota_4^T(\tilde{\mathscr{J}}_1 - \tilde{\mathscr{J}}_2)\iota_4 - \iota_1^T\tilde{\mathscr{J}}_1\iota_1]\\
&- Sym[\iota_7^T\tilde{\mathcal{R}}_1(\iota_1 - (1-o_1)\iota_4) - \iota_8^T\tilde{\mathcal{R}}_2(1-o_1)\iota_4] + \iota_6^T\tilde{\mathscr{L}}_1\iota_6,\\
\tilde{\Xi}_1 =& [\sqrt{l_c o_1}\tilde{\mathcal{Y}}_1^T\ \sqrt{l_c(1-o_1)}\tilde{\mathcal{Y}}_2^T\ \sqrt{l_c(1-o_1)}\tilde{\mathcal{Y}}_5^T\ \sqrt{l_c o_1}\tilde{\mathcal{Y}}_6^T],\\
\tilde{\Xi}_2 =& [\sqrt{l_c o_2}\tilde{\mathcal{Y}}_3^T\ \sqrt{l_c(1-o_2)}\tilde{\mathcal{Y}}_2^T\ \sqrt{l_c o_2}\tilde{\mathcal{Y}}_7^T\ \sqrt{l_c(1-o_2)}\tilde{\mathcal{Y}}_8^T],\\
\tilde{\Omega}_1 =& -diag\{\tilde{\mathscr{J}}_1, \tilde{\mathscr{J}}_2, \tilde{\mathscr{L}}_1, \tilde{\mathscr{L}}_1\}, \quad \tilde{\Omega}_2 = -diag\{\tilde{\mathscr{V}}_1, \tilde{\mathscr{V}}_2, \tilde{\mathscr{L}}_2, \tilde{\mathscr{L}}_2\},\ \tilde{\Pi}_{ij} = \bar{C}_i\mathfrak{W}^T\iota_1.
\end{aligned}
$$

The asymptotic stability of the resulting closed-loop tracking framework is guaranteed by the developed fuzzy control methodology, where the requisite gain matrices are synthesized via:

$$K_j = \tilde{K}_j \tilde{\mathfrak{W}}^{-T}. \tag{19}$$

Proof. By applying Lemma 1 from [12] and introducing a positive scalar ς to handle the uncertain term $\overline{\vartheta}(t)$, we conclude that:

$$\Delta\overline{\vartheta}(t)\frac{\wp}{\mathcal{M}} \leq \varsigma^{-1}\Delta^2 + \varsigma\frac{\wp^2}{\mathcal{M}^2}, \tag{20}$$

furthermore, the following variables are defined $\tilde{\mathfrak{W}} = \mathfrak{W}^{-1}$, $\tilde{\mathfrak{W}}_2 = \text{diag}\{\tilde{\mathfrak{W}}, \tilde{\mathfrak{W}}\}$, $\tilde{\mathfrak{W}}_3 = \text{diag}\{\underbrace{\tilde{\mathfrak{W}}, \ldots, \tilde{\mathfrak{W}}}_{12\text{ times}}, I\}$, $\tilde{\mathcal{Z}} = \tilde{\mathfrak{W}}\mathcal{Z}\tilde{\mathfrak{W}}^T$, $\tilde{\mathscr{J}}_h = \tilde{\mathfrak{W}}\mathscr{J}_h\tilde{\mathfrak{W}}^T$, $\tilde{\mathscr{V}}_h = \tilde{\mathfrak{W}}\mathscr{V}_h\tilde{\mathfrak{W}}^T$, $\tilde{\mathscr{L}}_h = \tilde{\mathfrak{W}}\mathscr{L}_h\tilde{\mathfrak{W}}^T$, $\tilde{\Phi} = \tilde{\mathfrak{W}}\Phi\tilde{\mathfrak{W}}^T$, $\tilde{\mathcal{R}}_h = \tilde{\mathfrak{W}}\mathcal{R}_h\tilde{\mathfrak{W}}^T$, $\tilde{\mathcal{S}}_h = \tilde{\mathfrak{W}}\mathcal{S}_h\tilde{\mathfrak{W}}^T$, $\tilde{\mathcal{Y}}_v = \tilde{\mathfrak{W}}\mathcal{Y}_v\tilde{\mathfrak{W}}_3^T$, $\tilde{\mathcal{N}} = \tilde{\mathfrak{W}}_2\mathcal{N}\tilde{\mathfrak{W}}_2^T$, and $\tilde{K}_j = K_j\tilde{\mathfrak{W}}^{-T}$. It is evident that by applying the congruence transformation, the corresponding results can be directly derived. □

4 Simulation Results

To validate the proposed control scheme, a PMSM [11] is adopted as the simulation example. Using the sector nonlinearity approach, the PMSM dynamics can be approximated by the following T-S fuzzy model:

$$\begin{cases} \dot{x}(t) = \sum_{i=1}^{2} \mu_i(z(t))[A_i x(t) + B_i u(t) + D_i \kappa(t)], \\ y(t) = \sum_{i=1}^{2} \mu_i(z(t)) C_i x(t). \end{cases} \tag{21}$$

where $\mu_1(z(t)) = \frac{1+x_3(t)/5}{2}$, $\mu_2(z(t)) = 1 - \mu_1(z(t))$. The relevant system coefficient matrices are given by:

$$A_1 = \begin{bmatrix} -1 & 5 & 0 \\ -5 & -1 & 1.1 \\ 0 & 5.46 & -5.46 \end{bmatrix}, \quad A_2 = \begin{bmatrix} -1 & 5 & 0 \\ 5 & -1 & 1.1 \\ 0 & 5.46 & -5.46 \end{bmatrix}, D_1 = D_2 = \begin{bmatrix} 0 & 0 & 0 \\ 1 & 0 & 0 \\ 1 & 0 & 1 \end{bmatrix},$$

$C_1 = C_2 = \begin{bmatrix} 1 & 1 & 0 \end{bmatrix}$, $B_1^T = B_2^T = \begin{bmatrix} 1 & 1 & 0 \end{bmatrix}$,

The trajectory of the specified reference system is formulated as:

$$\begin{cases} \dot{x}_r(t) = -x_r(t) + r(t) \\ y_r(t) = 0.4x_r(t) \end{cases} \tag{22}$$

Select parameters $\gamma = 0.7, \varepsilon = 0.5, \tau_0 = 0.9, \varsigma = 0.5, \mathcal{M} = 3, \wp = 2.65, \Delta = 0.3, o_1 = o_2 = 0.5, \xi = 0.3, l_m = 0.0001, l_M = 0.05$. The controller gains are:

$$K_{11} = [-4.7611 \quad -1.7693 \quad -0.1287], \quad K_{12} = 0.0001;$$
$$K_{21} = [-4.7611 \quad -1.7693 \quad -0.1287], \quad K_{22} = 0.0001.$$

To evaluate the proposed control strategy, the system is subjected to external disturbances $\kappa(t) = 5\cos(t)e^{-2t}$ and reference input $r(t) = 2\sin(t)e^{-t}$. The initial states are set as $x_r(0) = -2$ for the reference model and $x(0) = [1\ -4\ 4]^T$ for the controlled system. The deception attacks are characterized by the nonlinear mapping: $\mathcal{F}(q_\theta(\eta(t_k), \theta(t_k))) = -\tanh[1.2(q_\theta(\eta(t_k), \theta(t_k)))]$. The tracking performance and control signals are illustrated in Figs. 1–2, where the tracking error converges rapidly and the control input remains smooth. The release intervals for the proposed QETCS and the AETM in [3] are provided in Fig. 3 and Fig. 4, respectively.

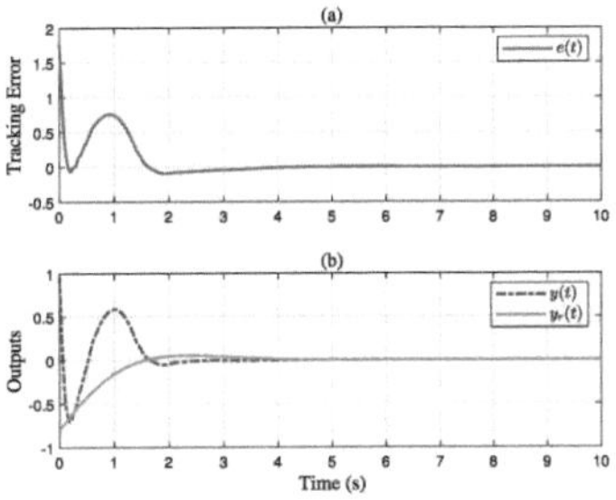

Fig. 1. Tracking performance.

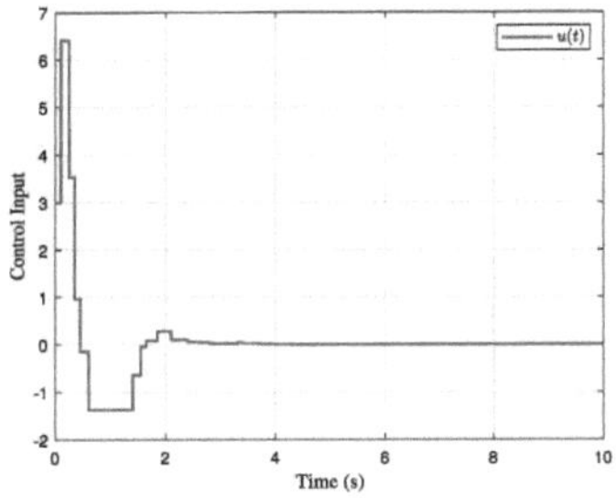

Fig. 2. Control input.

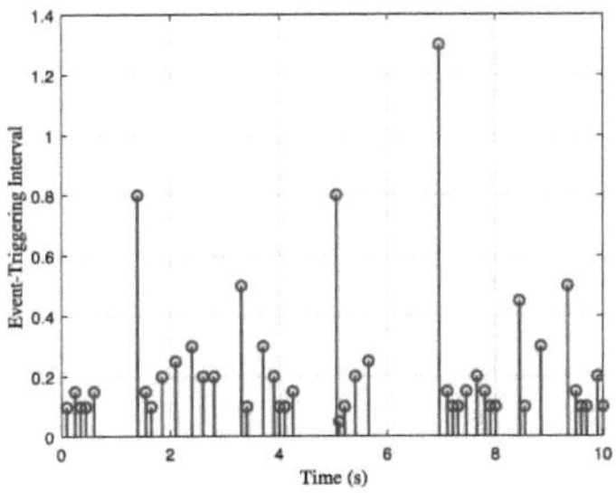

Fig. 3. QETCS.

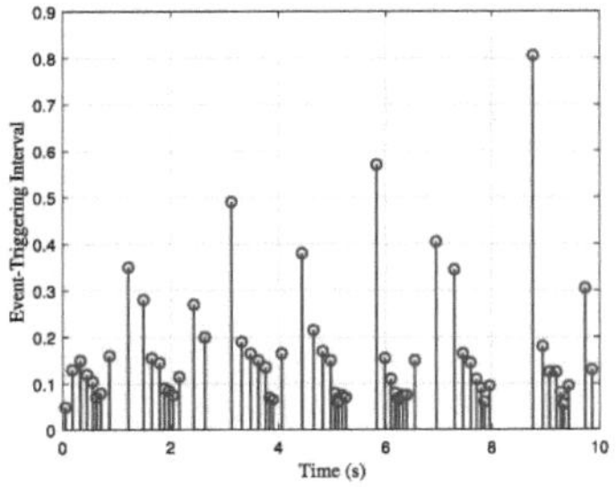

Fig. 4. AETM.

To further evaluate communication efficiency, different schemes are compared under fixed sampling interval $h = 0.05$ s, with results summarized in Table 1, including the number of transmitted packets(NTPs), the average transmission periods(ATPs) and the average data transmission rates (ADTRs). The proposed QETCS generates only 36 events, achieving an ADTR of 18.00%, which is significantly lower than that of fixed period sampling mode(FPSM), ETM, and AETM in [3]. This superiority stems from the adaptive triggering threshold, which effectively filters redundant transmissions while preserving high-fidelity performance.

Table 1. NTPs, ATPs and ADTRs for Different ETMs

Communication Scheme	NTP	ATP	ADTR
FPSM	200	0.0500	100%
ETM	95	0.1053	47.50%
AETM [3]	59	0.1695	29.50%
QETCS	36	0.2778	18.00%

5 Conclusion

This study develops a T-S fuzzy tracking framework for nonlinear systems under disturbances and network constraints. The proposed QETCS optimizes communication efficiency and resource allocation. A Bernoulli-modeled defense strategy counters random deception attacks, while a refined TSLF ensures augmented system stability. LMIs-based criteria facilitate controller synthesis, with simulation results confirming robust tracking and significantly reduced network overhead.

References

1. Chang, X.H., Yang, C., Xiong, J.: Quantized fuzzy output feedback $\mathcal{H}_\infty$ control for nonlinear systems with adjustment of dynamic parameters. IEEE Trans. Syst. Man Cybern. Syst. **49**(10), 2005–2015 (2019)
2. Li, Z.M., Chang, X.H., Xiong, J.: Event-based fuzzy tracking control for nonlinear networked systems subject to dynamic quantization. IEEE Trans. Fuzzy Syst. **31**(3), 941–954 (2023)
3. Liu, Y., Zhao, X., Park, J.H., Fang, F.: Fault-tolerant control for T-S fuzzy systems with an aperiodic adaptive event-triggered sampling. Fuzzy Sets Syst. **452**, 23–41 (2023)
4. Shan, Y., Xie, X., Sun, J., Park, J.H.: Observer-based adaptive event-triggered control for nonlinear networked systems under multiple cyber attacks. IEEE Trans. Fuzzy Syst. **32**(9), 5272–5284 (2024)
5. Shanmugam, L., Joo, Y.H.: Stability criteria for fuzzy-based sampled-data control systems via a fractional parameter-based refined looped Lyapunov functional. IEEE Trans. Fuzzy Syst. **30**(7), 2538–2549 (2022)
6. Tabuada, P.: Event-triggered real-time scheduling of stabilizing control tasks. IEEE Trans. Autom. Control **52**(9), 1680–1685 (2007)
7. Takagi, T., Sugeno, M.: Fuzzy identification of systems and its applications to modeling and control. IEEE Trans. Syst. Man Cybern. SMC **15**(1), 116–132 (1985)
8. Yang, Y., Li, J., Niu, Y.: Sliding mode quantized control of interval type-2 fuzzy systems under event-triggered scheme: stability analysis and performance optimization. IEEE Trans. Autom. Sci. Eng. **22**, 3295–3307 (2025)
9. You, Z., Yan, H., Zhang, H., Hu, Y., Zhu, S.: Aperiodic sampled-data-based control for T-S fuzzy systems: an improved fuzzy-dependent adaptive event-triggered mechanism. IEEE Trans. Fuzzy Syst. **31**(11), 4085–4096 (2023)
10. Zeng, H.B., Teo, K.L., He, Y.: A new looped-functional for stability analysis of sampled-data systems. Automatica **82**, 328–331 (2017)

11. Zeng, H.B., Zhu, Z.J., Peng, T.S., Wang, W., Zhang, X.M.: Robust tracking control design for a class of nonlinear networked control systems considering bounded package dropouts and external disturbance. IEEE Trans. Fuzzy Syst. **32**(6), 3608–3617 (2024)
12. Zhao, Q., Wu, X., Wang, H., Cheng, J.: Dynamic quantized feedback control for IT2 fuzzy systems with probabilistic event-based amidst dos attacks. Nonlinear Dyn. **113**(12), 14983–14995 (2025)

A Fault-Resilient Distributed Adaptive ZNN Estimator for Multisensor Systems Under Multiplicative Noises

Chenfu Yi(✉), Zhi Xie, Yuhuan Chen, and Yang Li

Guangdong Polytechnic Normal University, Guangzhou, China
yichenfu@163.com

Abstract. Existing Regularized Least-Squares (RLS) methods for multisensor state estimation under multiplicative noises suffer from numerical singularity due to ill-conditioned covariance matrices, and lack mechanisms to isolate sensor faults. To overcome these bottlenecks, we propose a Fault-Resilient Distributed Adaptive Zeroing Neural Network (FR-DAZNN) estimator. It reformulates static algebraic inversions into a dynamic integral-smoothing process. An adaptive-weighted energy function dynamically regularizes the gradient flow to prevent numerical divergence under severe noises. Furthermore, a temporal-consistency diagnostic indicator is integrated into the cyber layer to autonomously isolate malfunctioning sensor nodes. Lyapunov analysis proves the global exponential convergence of the network. Simulations demonstrate its superior structural robustness in suppressing transient noise and completely isolating faulty sensors, outperforming traditional analytical methods.

Keywords: Distributed estimation · zeroing neural network (ZNN) · multiplicative noises · adaptive regularization · fault resilience · physical isolation

1 Introduction

Since the seminal work on distributed Kalman filtering [1], distributed state estimation has become a cornerstone in modern multisensor networks. By enabling individual nodes to infer global states via local interactions, it offers enhanced scalability and resilience, finding extensive applications in autonomous tracking, smart grids, and complex multi-agent control [2–4]. Despite these advances, the practical deployment of these algorithms demands high precision under severe network uncertainties.

In practical environments, systems face severe challenges from complex stochastic uncertainties, particularly *multiplicative noises* originating from channel fading or sensor gain jitter [5,6]. These state-dependent noises are often coupled with other non-ideal factors like packet dropouts or malicious interference [7,8]. To address multiplicative noises, the Regularized Least-Squares (RLS)

Z. Zhang (Ed.): ISNN 2026, LNCS 16848, pp. 292–299, 2027.
https://doi.org/10.1007/978-981-92-5870-3_24

framework established by Rahmani *et al.* [9] is widely regarded as a benchmark. By transforming state-dependent variances into equivalent additive covariances, this RLS-based approach provides a theoretical optimum.

Despite its theoretical soundness, the conventional RLS method [9] suffers from two critical fragilities. First, under severe multiplicative noises, the equivalent covariance matrices easily become *severely ill-conditioned.* This pathological deformation of the data matrices leads to catastrophic numerical singularity and instantaneous divergence of the estimation process. Second, traditional decentralized RLS fusion blindly utilizes covariance-intersection weighting. Once a sensor suffers from non-Gaussian anomalies (e.g., hardware malfunction), its erroneous estimation unconditionally pollutes the entire network due to the lack of a targeted isolation mechanism.

To overcome numerical fragilities, Zeroing Neural Networks (ZNN) have been introduced as powerful continuous-time dynamic solvers [10–12]. Unlike static algebraic algorithms, ZNN approaches optimization through a continuous neurodynamic evolution, which inherently provides an integral-smoothing effect against stochastic disturbances. However, a fundamental limitation persists: if the underlying optimization landscape is severely distorted by ill-conditioned covariances under extreme multiplicative noises, the standard ZNN gradient flow will still suffer from instability and gradient explosion [13,14]. Consequently, there is an urgent need for an estimation architecture that structurally immunizes the system against both pathological noise environments and sensor faults.

To overcome these challenges, this article proposes a novel Fault-Resilient Distributed Adaptive ZNN (FR-DAZNN) estimator. By redesigning the conventional RLS framework [9] into a dynamically regularized neurodynamic process, the proposed method achieves inherent structural robustness. The main contributions are:

1. **Dynamic Regularization against Ill-Conditioning:** A novel adaptive-weighted energy function is designed to reformulate the local estimation. This dynamic regularization forces a positive lower bound on the eigenvalues of the gradient flow, strictly preventing numerical divergence under extreme multiplicative noises and overriding the pathological deformation of conventional RLS methods.
2. **Fault-Resilient Physical Isolation:** A temporal-consistency diagnostic indicator is coupled into the global energy landscape. It autonomously enforces a soft cut-off for faulty sensors, realizing physical isolation within the cyber topology to prevent error propagation.
3. **Rigorous Theoretical Guarantee:** Based on Lyapunov stability and Rayleigh quotient properties, the global exponential convergence of the FR-DAZNN is mathematically proven and subsequently validated by numerical simulations under severe non-Gaussian environments.

2 System Model and Problem Formulation

In practical multisensor networks, system dynamics and observations are often corrupted by state-dependent uncertainties. Motivated by the benchmark framework [9], we consider a discrete time-varying multisensor system with multiplicative noises:

$$\mathbf{x}_k = \mathbf{A}_k \mathbf{x}_{k-1} + \mathbf{w}_k \tag{1}$$

$$\mathbf{y}_{i,k} = \mathbf{C}_{i,k} \mathbf{x}_k + \mathbf{v}_{i,k}, \quad i = 1, 2, \ldots, N \tag{2}$$

where $\mathbf{x}_k \in \mathbb{R}^n$ and $\mathbf{y}_{i,k} \in \mathbb{R}^m$ are the target state and measurement vectors, respectively. The multiplicative noise phenomenon is captured by modeling the system matrices as $\mathbf{A}_k = \bar{\mathbf{A}}_k + \alpha_k \tilde{\mathbf{A}}_k$ and $\mathbf{C}_{i,k} = \bar{\mathbf{C}}_{i,k} + \gamma_{i,k} \tilde{\mathbf{C}}_{i,k}$. Here, $\bar{\mathbf{A}}_k$ and $\bar{\mathbf{C}}_{i,k}$ represent the deterministic nominal matrices, while $\tilde{\mathbf{A}}_k$ and $\tilde{\mathbf{C}}_{i,k}$ denote the known noise coefficient matrices dictating the structural intensity of the uncertainties.

The scalars α_k and $\gamma_{i,k}$ are zero-mean Gaussian multiplicative noises with variances σ_α^2 and $\sigma_{\gamma,i}^2$. The standard additive process and measurement noises are $\mathbf{w}_k \sim \mathcal{N}(\mathbf{0}, \mathbf{Q}_k)$ and $\mathbf{v}_{i,k} \sim \mathcal{N}(\mathbf{0}, \mathbf{R}_{i,k})$, with strictly positive definite covariances. All aforementioned noise sequences are assumed mutually independent.

Based on the localized observations $\{\mathbf{y}_{i,k}\}_{i=1}^N$, this paper aims to design robust filtering ($\hat{\mathbf{x}}_{k|k}$) and one-step prediction ($\hat{\mathbf{x}}_{k+1|k}$) estimators. Unlike conventional algebraic approaches that suffer from matrix ill-conditioning under severe noises, our objective is to achieve precise state estimation by developing a structurally resilient neural smoothing mechanism.

3 Robust State Estimation via Adaptive ZNN

Before addressing the network topology, we first establish a robust local estimator for a single sensor. Traditional algebraic filtering relies on matrix inversions, which suffer from numerical singularity under severe multiplicative noises. To overcome this, we reformulate the static estimation into a continuous-time neurodynamic search process.

3.1 Equivalent Noise Compensation and Energy Function

Following the RLS benchmark [9], the exact multiplicative noises are inaccessible, so their uncertainty effects are inherently compensated by an additional state-dependent variance term $\mathbf{\Sigma}_{\text{mult},k}$. The equivalent one-step prediction error covariance is defined as:

$$\mathbf{P}_{k|k-1} = \bar{\mathbf{A}}_k \mathbf{P}_{k-1} \bar{\mathbf{A}}_k^T + \mathbf{Q}_k + \mathbf{\Sigma}_{\text{mult},k} \tag{3}$$

Under extreme noises, the unweighted equivalent covariance matrices easily become severely ill-conditioned. To systematically avoid pathological inversions, we reformulate the traditional static objective into a dynamic search landscape by constructing an adaptive-weighted energy function $E_i(\mathbf{z})$ for the searching neural state $\mathbf{z}(t)$:

$$\begin{aligned} E_i(\mathbf{z}) = & \frac{\alpha_{obs,i}}{2} \|\mathbf{y}_{i,k} - \bar{\mathbf{C}}_{i,k} \mathbf{z}\|^2_{\mathbf{R}_{i,k}^{-1}} \\ & + \frac{\alpha_{pred,i}}{2} \|\mathbf{z} - \bar{\mathbf{A}}_k \hat{\mathbf{x}}_{k-1}\|^2_{\mathbf{P}_{k|k-1}^{-1}} \end{aligned} \tag{4}$$

where $\alpha_{obs,i}, \alpha_{pred,i} > 0$ provide real-time dynamic regularization, balancing the trust between current observations and kinetic predictions to prevent landscape distortion.

3.2 Proposed Local FR-DAZNN Model

To drive the neural state $\mathbf{z}_i(t)$ towards the global minimum, we take the exact gradient $\nabla E_i(\mathbf{z})$ and apply the standard ZNN evolution rule $\dot{\mathbf{z}}_i(t) = -\gamma_i \nabla E_i(\mathbf{z})$ (with convergence rate $\gamma_i > 0$). By grouping the coefficients of the unknown $\mathbf{z}(t)$, the continuous-time FR-DAZNN model is fully derived as:

$$\begin{aligned}\dot{\mathbf{z}}_i(t) = -\gamma_i\Big[&\Big(\alpha_{obs,i}\bar{\mathbf{C}}_{i,k}^T\mathbf{R}_{i,k}^{-1}\bar{\mathbf{C}}_{i,k} + \alpha_{pred,i}\mathbf{P}_{k|k-1}^{-1}\Big)\mathbf{z}_i(t) \\ &- \Big(\alpha_{obs,i}\bar{\mathbf{C}}_{i,k}^T\mathbf{R}_{i,k}^{-1}\mathbf{y}_{i,k} + \alpha_{pred,i}\mathbf{P}_{k|k-1}^{-1}\bar{\mathbf{A}}_k\hat{\mathbf{x}}_{k-1}\Big)\Big]\end{aligned} \tag{5}$$

To reveal the underlying network structure and facilitate analysis, we encapsulate the bracketed terms into a dynamically regularized matrix $\mathbf{M}_{i,k}$ and a bias vector $\mathbf{b}_{i,k}$, yielding the compact form $\dot{\mathbf{z}}_i(t) = -\gamma_i(\mathbf{M}_{i,k}\mathbf{z}_i(t) - \mathbf{b}_{i,k})$.

Remark 1 (Dynamic Regularization). Unlike traditional methods computing $\mathbf{M}_{i,k}^{-1}$, the dynamic regularization forces a positive lower bound $\lambda_{\min}(\mathbf{M}_{i,k}) > 0$. This empowers the FR-DAZNN model to structurally circumvent numerical divergence when the unweighted matrix becomes ill-conditioned.

3.3 Local Convergence Analysis

Theorem 1. *For any positive definite equivalent covariance matrices, the neural state* $\mathbf{z}_i(t)$ *of* (5) *globally and exponentially converges to the theoretical optimal estimate* $\mathbf{z}_i^* = \mathbf{M}_{i,k}^{-1}\mathbf{b}_{i,k}$.

Proof. Consider the Lyapunov candidate $V(\mathbf{z}_i) = \frac{1}{2}\mathbf{e}^T\mathbf{M}_{i,k}\mathbf{e}$, where the estimation error is $\mathbf{e}(t) = \mathbf{z}_i(t) - \mathbf{z}_i^*$. Since $\mathbf{M}_{i,k}$ is a positive linear combination of the strictly positive definite (SPD) matrices $\mathbf{R}_{i,k}^{-1}$ and $\mathbf{P}_{k|k-1}^{-1}$, it inherently remains SPD, ensuring $V(\mathbf{z}_i) \geq 0$. The time derivative along the system trajectories is:

$$\dot{V}(\mathbf{z}_i) = \mathbf{e}^T\mathbf{M}_{i,k}\dot{\mathbf{e}} = -\gamma_i\mathbf{e}^T\mathbf{M}_{i,k}^2\mathbf{e} \leq 0 \tag{6}$$

Utilizing the Rayleigh quotient inequality for SPD matrices, $\mathbf{e}^T\mathbf{M}_{i,k}^2\mathbf{e}/\mathbf{e}^T\mathbf{M}_{i,k}\mathbf{e} \geq \lambda_{\min}(\mathbf{M}_{i,k})$, it rigorously follows that $\dot{V}(\mathbf{z}_i) \leq -2\gamma_i\lambda_{\min}(\mathbf{M}_{i,k})V(\mathbf{z}_i)$. This inequality theoretically guarantees the global exponential convergence of the local network. □

Remark 2 (State Extraction). Upon convergence ($t \to \infty$) at the k-th physical time step, the equilibrium point is extracted as the optimal local estimate $\hat{\mathbf{x}}_{i,k}^{\text{ZNN}} = \lim_{t\to\infty}\mathbf{z}_i(t)$.

4 Multi-sensor Data Fusion Strategies

Building upon the robust local estimator, we now develop network-level fusion strategies to enhance global estimation precision.

4.1 Centralized Fusion via Stacked Observations

Centralized fusion aggregates raw measurements from all N sensors. By stacking the local matrices, the global observation is formulated as $\bar{\mathbf{y}}_k = \bar{\mathbf{C}}_k \mathbf{x}_k + \bar{\mathbf{v}}_k$, where $\bar{\mathbf{y}}_k = [(\mathbf{y}_{1,k})^T, \dots, (\mathbf{y}_{N,k})^T]^T$, $\bar{\mathbf{C}}_k = [(\bar{\mathbf{C}}_{1,k})^T, \dots, (\bar{\mathbf{C}}_{N,k})^T]^T$, and the joint observation noise covariance is $\bar{\mathbf{R}}_k = \text{diag}\{\mathbf{R}_{1,k}, \dots, \mathbf{R}_{N,k}\}$ [9].

Sharing identical mathematical properties with the local model, the ZNN solver is directly deployed centrally. Substituting the stacked global parameters into (5) yields the explicit continuous-time **Centralized FR-DAZNN** model:

$$\begin{aligned}\dot{\mathbf{z}}_{\text{cen}}(t) = -\gamma_{\text{cen}} \Big[& \left(\alpha_{obs}^{\text{cen}} \bar{\mathbf{C}}_k^T \bar{\mathbf{R}}_k^{-1} \bar{\mathbf{C}}_k + \alpha_{pred}^{\text{cen}} \mathbf{P}_{k|k-1}^{-1} \right) \mathbf{z}_{\text{cen}}(t) \\ & - \left(\alpha_{obs}^{\text{cen}} \bar{\mathbf{C}}_k^T \bar{\mathbf{R}}_k^{-1} \bar{\mathbf{y}}_k + \alpha_{pred}^{\text{cen}} \mathbf{P}_{k|k-1}^{-1} \bar{\mathbf{A}}_k \hat{\mathbf{x}}_{k-1} \right) \Big]\end{aligned} \tag{7}$$

Upon convergence, the optimal centralized estimate is extracted as $\hat{\mathbf{x}}_k^{\text{Cen}} = \lim_{t\to\infty} \mathbf{z}_{\text{cen}}(t)$.

4.2 Distributed Fusion with Cyber-Physical Isolation

Distributed fusion mitigates communication bottlenecks but remains vulnerable to blindly fusing malicious anomalies. To achieve physical isolation within the cyber topology, we introduce a diagnostic indicator $\tau_i \in (0, 1]$. This weight evaluates the real-time temporal consistency between the localized neural estimate $\hat{\mathbf{x}}_{i,k}^{\text{ZNN}}$ and the global consensus prediction $\bar{\mathbf{A}}_k \hat{\mathbf{x}}_{k-1}^{\text{Decen}}$ via a Gaussian kernel:

$$\tau_i = \exp\left(-\frac{\|\hat{\mathbf{x}}_{i,k}^{\text{ZNN}} - \bar{\mathbf{A}}_k \hat{\mathbf{x}}_{k-1}^{\text{Decen}}\|_2^2}{2\sigma^2} \right) \tag{8}$$

where σ controls the sensitivity to anomalies.

To forge a robust consensus, we define a global consistency cost function $J = \sum_{i\in\mathcal{N}^+} \tau_i (\hat{\mathbf{x}}_k^{\text{Decen}} - \hat{\mathbf{x}}_{i,k}^{\text{ZNN}})^T (\mathbf{P}_{i,k}^{\text{ZNN}})^{-1} (\hat{\mathbf{x}}_k^{\text{Decen}} - \hat{\mathbf{x}}_{i,k}^{\text{ZNN}})$. Setting the partial derivative $\partial J / \partial \hat{\mathbf{x}}_k^{\text{Decen}} = \mathbf{0}$, the robust distributed fusion rule for connected neighbors $\mathcal{N}^+$ is rigorously derived:

$$\hat{\mathbf{x}}_k^{\text{Decen}} = \left(\sum_{i\in\mathcal{N}^+} \tau_i (\mathbf{P}_{i,k}^{\text{ZNN}})^{-1} \right)^{-1} \sum_{i\in\mathcal{N}^+} \tau_i (\mathbf{P}_{i,k}^{\text{ZNN}})^{-1} \hat{\mathbf{x}}_{i,k}^{\text{ZNN}} \tag{9}$$

Theorem 2. *The global distributed fusion network ensures exponential stability. Upon a local sensor fault, the global energy landscape autonomously reshapes to guarantee physical isolation.*

Proof. Defining the global Lyapunov function $V_{\text{global}} = \sum_{i=1}^{N} \tau_i V(\mathbf{z}_i) > 0$, its time derivative is $\dot{V}_{\text{global}} = \sum_{i=1}^{N} \tau_i \dot{V}(\mathbf{z}_i) \leq 0$, ensuring the exponential decay of the aggregated network errors. Furthermore, for a severely degraded node, its temporal consistency drops drastically ($\tau_{\text{fault}} \to \epsilon \approx 0$). Consequently, its faulty gradient contribution in (9) is forcibly flattened, cutting off malicious noise propagation across the cyber layer. □

5 Simulation Verification

To validate the proposed FR-DAZNN estimator, we simulate a distributed network of $N = 3$ heterogeneous sensors collaboratively tracking a dynamic target's 3D state (position, velocity, acceleration), denoted as $\mathbf{x}_k = [p_k, v_k, a_k]^T$. With a sampling period $T = 0.1\,\text{s}$, the nominal state transition matrix and observation matrices are defined as:

$$\bar{\mathbf{A}} = \begin{bmatrix} 1 & T & T^2/2 \\ 0 & 1 & T \\ 0 & 0 & 1 \end{bmatrix}, \; \bar{\mathbf{C}}_1 = [1,0,0],\; \bar{\mathbf{C}}_2 = [1,1,0],\; \bar{\mathbf{C}}_3 = [1,0,1] \tag{10}$$

The multiplicative noise coefficient matrices are $\tilde{\mathbf{A}} = 0.1\bar{\mathbf{A}}$ and $\tilde{\mathbf{C}}_i = 0.1\bar{\mathbf{C}}_i$. The covariances for the additive process and measurement noises are initialized as $\mathbf{Q} = 0.01\mathbf{I}_3$ and $\mathbf{R}_1 = \mathbf{R}_2 = \mathbf{R}_3 = 0.1$. The variances of the multiplicative noises are configured as $\sigma_\alpha^2 = \sigma_{\gamma,i}^2 = 0.04$. The ZNN learning rate is set to $\gamma_i = 10^5$.

We evaluate the algorithms under two conditions: a normal Gaussian noise scenario, and a heterogeneous noise scenario where Sensor 2 suffers from severe degradation ($V_2 = 500$).

Performance in Normal Scenario. As illustrated in Fig. 1, network-level fusion significantly improves tracking accuracy over isolated single-sensor filters (Local KFs). Specifically, the Local KF 1 exhibits the highest global errors. In contrast, the proposed Decen ZNN and Cen ZNN achieve the lowest overall MSEs, demonstrating superior baseline precision in nominal environments.

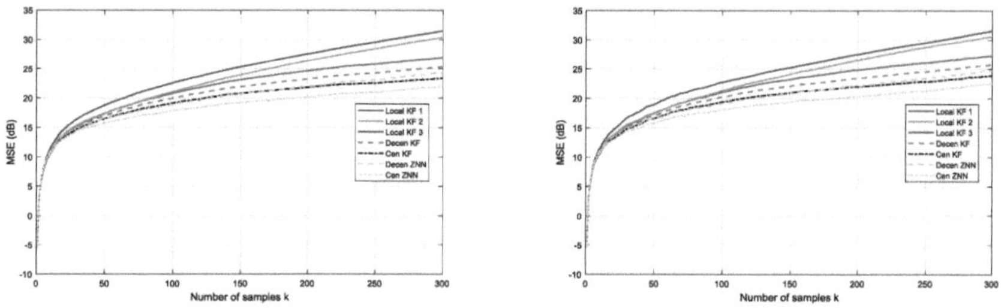

(a) Performance without external disturbance (b) Performance under external disturbance

Fig. 1. Comparison of filtering MSE curves (dB) in the **Normal Scenario**.

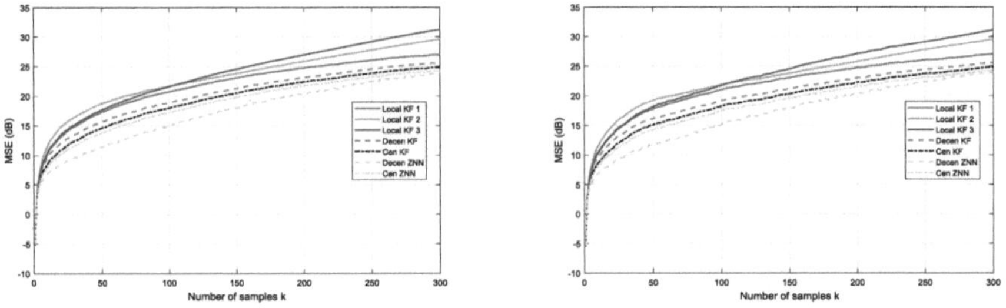

(a) Performance without disturbance (b) Performance with disturbance

Fig. 2. Comparison of filtering MSE curves (dB) in the **Heterogeneous Noise Scenario** where Sensor 2 suffers from high noise variance ($V_2 = 500$).

Robustness in Heterogeneous Noise Scenario. When Sensor 2 degrades severely, the traditional Local KF 2 diverges due to pathological covariances, as shown in Fig. 2(a). However, the proposed Decen ZNN exhibits exceptional fault tolerance. By leveraging the diagnostic indicator τ_i, the network autonomously evaluates temporal consistency and assigns negligible weight to the malfunctioning sensor. Consequently, the global MSE curve of Decen ZNN remains almost identical to the normal scenario. Even under external disturbances (Fig. 2(b)), it effectively suppresses malicious error propagation, significantly outperforming conventional blind-fusion methods like Decen KF.

6 Conclusion

This paper proposes a Fault-Resilient Distributed Adaptive Zeroing Neural Network (FR-DAZNN) estimator for multisensor systems under extreme multiplicative noises. An adaptive-weighted energy function dynamically regularizes the continuous estimation process, preventing numerical divergence under ill-conditioned covariances. Concurrently, a temporal-consistency indicator autonomously isolates malfunctioning sensors. Lyapunov analysis and simulations confirm the global exponential convergence and structural robustness of the FR-DAZNN. Future work will target non-Gaussian nonlinear systems and embedded hardware implementations.

References

1. Olfati-Saber, R.: Distributed Kalman filtering for sensor networks. In: Proceedings of the 46th IEEE Conference on Decision and Control, pp. 5492–5498. IEEE (2007)
2. Zhang, B., Hou, Y., Yin, H., Lv, M., Yang, A., Wu, L.: Cooperative dynamic target tracking: distributed time-varying optimization for multi-UAV system. IEEE Trans. Aerosp. Electron. Syst. (2025). https://doi.org/10.1109/TAES.2025.3574701

3. Mohammadrezaee, R., Ghaisari, J., Yousefi, G., Kamali, M.: Dynamic state estimation of smart distribution grids using compressed measurements. IEEE Trans. Smart Grid **12**(5), 4535–4544 (2021)
4. Liu, L., Yu, C., Feng, Y., Xia, Y., Deng, F., Chen, J.: Distributed filtering for multi-agent systems with time-varying range constraints. IEEE Trans. Ind. Electron. **72**(7), 7472–7482 (2025)
5. Wang, W., Chen, Y., Wang, Z., Han, C., Du, S.: Networked state estimation for 2-D systems with multiplicative noises and observation delays: an iterative coupled estimator design. IEEE Internet Things J. **12**(23), 51677–51689 (2025)
6. He, X., Liu, Q., Fang, H.: Distributed recursive filter for systems with multiplicative noises. In: 23rd International Symposium on Mathematical Theory of Networks and Systems (MTNS), pp. 1–6 (2018)
7. Jin, H., Du, Z., Ma, J.: Distributed filtering for networked stochastic nonlinear systems with fading measurements and random packet dropouts. IEEE Access **12**, 109852–109861 (2024)
8. Yuan, H., Xia, Y., Yang, H.: Resilient state estimation of cyber-physical system with multichannel transmission under DoS attack. IEEE Trans. Syst. Man Cybern. Syst. **51**(11), 6926–6937 (2021)
9. Rahmani, M., Abolhasani, M.: State estimation for stochastic time-varying multi-sensor systems with multiplicative noises: centralized and decentralized data fusion. Asian J. Control **21**(4), 1547–1555 (2019)
10. Wang, T., Zhang, Z., Huang, Y., Liao, B., Li, S.: Applications of zeroing neural networks: a survey. IEEE Access **12**, 51363–51384 (2024)
11. Xi, X., Miao, P., Ding, Y.: Advances in ZNN models for solving time-varying Sylvester equations: convergence, robustness, and applications. IEEE Access (2025). https://doi.org/10.1109/ACCESS.2025.3620136
12. Yang, F., Huang, Y.: A triple noise tolerant zeroing neural network for time-varing matrix inverse. IEEE Access (2024). https://doi.org/10.1109/ACCESS.2024.3411781
13. Li, J., Pan, S., Chen, K., Zhu, X., Yang, M.: Inverse-free discrete ZNN models for solving future nonlinear equality and inequality system with robot manipulator control. IEEE Trans. Ind. Electron. **72**(12) (2025). https://doi.org/10.1109/TIE.2025.3582686
14. Xiao, L., Wang, D., Zuo, Q., Fan, H., Cai, H.: An inversion-free fuzzy zeroing neural network under adaptive input range fuzzy scheme: design, analysis, and application. IEEE Trans. Artif. Intell. **7**(1), 172–184 (2026)

Zeroing Neurodynamics-Based Cascade Tracking Control for Two-Mode Bose-Einstein Condensates with Desired Population Dynamics

Pengfei Guo[1], Ya Xi[1], Chaolong Zhang[1], and Yunong Zhang[2](✉)

[1] School of Mathematics and Data Science, Zhongkai University of Agriculture and Engineering, Guangzhou 510240, China

[2] School of Intelligent Systems Engineering, Sun Yat-sen University, Shenzhen 518107, China

zhynong@mail.sysu.edu.cn

Abstract. In this paper, we propose an online tracking control for a two-mode Bose-Einstein condensates (BEC) system, which is described by the bosonic Josephson junction model. Due to the underactuated nature of the system, direct tracking of the whole state, which is consist of both population imbalance and relative phase, is not feasible with a single control input. To handle this issue, a cascade zeroing neurodynamics (ZND) framework is developed. As a bridge, a virtual phase signal is constructed from the population tracking error by using zeroing error dynamics. Then, a ZND-based controller is designed to drive the phase to follow this virtual signal. The resulting control law is explicit and suitable for real-time implementation. In addition, we rigorously analyze the exponential convergence property of the phase tracking error and induced convergence property of the population imbalance error. A group of numerical simulations verify the effectiveness and stability of the proposed method.

Keywords: Zeroing neurodynamics · Bose-Einstein condensates (EBC) system · Tracking control problem · Real-time implementation

1 Introduction

Nonlinear systems, which can be described by mathematical equations, play an essential role in modeling various phenomena in science and engineering [1]. Due to their intrinsic nonlinear and coupled dynamics, the analysis and control of such systems remain challenging. Among different approaches, tracking control has been widely adopted as an effective strategy to regulate system states toward desired trajectories [2].

Two-mode Bose–Einstein condensates (BECs), modeled as bosonic Josephson junctions, describe quantum tunneling dynamics between two weakly cou-

Z. Zhang (Ed.): ISNN 2026, LNCS 16848, pp. 300–312, 2027.
https://doi.org/10.1007/978-981-92-5870-3_25

pled condensates. This system exhibits rich nonlinear behaviors, such as Josephson oscillations and macroscopic self-trapping [3]. In practical applications, regulating the condensate to follow a desired population trajectory is of significant interest. However, the system is inherently underactuated, since only a single control input (bias) is available to influence both the population imbalance and the relative phase [4]. This structural constraint makes the tracking control problem particularly challenging.

To address the tracking control problem of the BEC system, Tomilin *et al.* developed a feedback-based control approach [5]. However, the corresponding control law is not explicitly constructed, which makes theoretical analysis of stability and convergence difficult. In recent years, neurodynamics approaches, also known as recurrent neural network (RNN)-based methods, have been developed to solve various tracking control and time-varying problems [6–8]. For instance, Dai *et al.* proposed a finite-time gradient neurodynamics (GND) model for synchronization of chaotic systems [7]. Nevertheless, conventional GND methods typically rely on gradient descent mechanisms, which may introduce lagging errors when dealing with time-varying tracking problems in nonlinear systems.

To overcome this limitation, zeroing neurodynamics (ZND) has been developed as an effective framework for solving time-varying problems by constructing a residual function and driving it to zero through a dynamic evolution law [9–11]. In 2020, Zhang *et al.* proposed a discrete ZND controller for tracking control of pendulum systems [12]. In 2022, Guo *et al.* developed a second-order discrete ZND controller for integrator systems [13]. Although ZND has shown promising performance in various applications, existing designs are typically constructed based on decoupled (diagonal) error dynamics, which may neglect the intrinsic coupling among system states. As a result, such approaches may not fully exploit the underlying structure of strongly coupled nonlinear systems.

Motivated by the above observations, this paper proposes a cascade ZND framework for the tracking control problem of the BEC system. Specifically, a virtual phase signal is constructed from the population tracking error based on a zeroing error dynamics principle, which serves as an intermediate reference. Then, a ZND-based controller is designed to drive the actual phase to follow this virtual signal. In this way, the population tracking objective is achieved indirectly through phase regulation, thereby preserving the coupling structure of the original system.

The main contributions of this paper are summarized as follows.

- A cascade ZND control framework is developed for underactuated BEC systems, which explicitly incorporates the coupling between population imbalance and phase dynamics. An explicit control law suitable for real-time implementation is derived via virtual phase construction and ZND-based phase regulation.
- Rigorous theoretical analysis is provided to establish exponential convergence of the phase tracking error and induced convergence of the population tracking error.

- Numerical simulations are presented to verify the effectiveness and stability of the proposed method.

2 Problem Formulation

In this section, the mathematical model of the two-mode BEC system is presented, together with its corresponding physical interpretation.

The BEC system, described by the bosonic Josephson junction model, is formulated as

$$\begin{cases} \dot{x}(t) = -\kappa\sqrt{1 - x^2(t)} \sin \phi(t), \\ \dot{\phi}(t) = \lambda x(t) + \dfrac{\kappa x(t)}{\sqrt{1 - x^2(t)}} \cos \phi(t) + u(t), \\ y(t) = x(t), \end{cases} \tag{1}$$

where $x(t) \in (-1, 1)$ denotes the population imbalance between the two condensate modes, $\phi(t) \in \mathbb{R}$ represents the relative phase, $\kappa \in \mathbb{R}^+$ is the tunneling coefficient, $\lambda \in \mathbb{R}$ characterizes the inter-particle interaction strength, $u(t) \in \mathbb{R}$ is the external control input, and $y(t)$ is the measurable output. For visualization purposes, two-dimensional and three-dimensional representations of vector field of BEC system (1) are provided in Fig. 1.

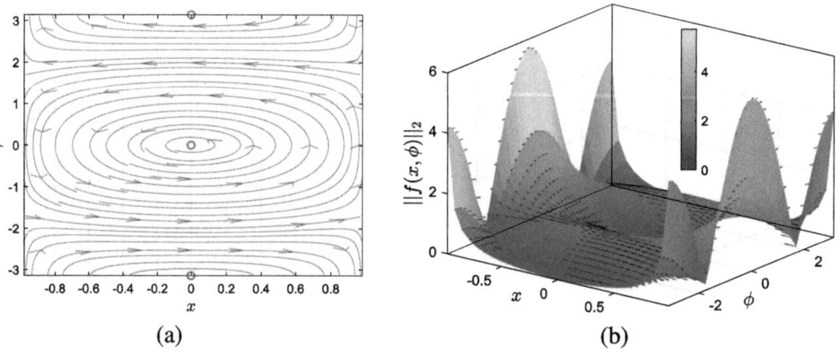

Fig. 1. Vector field of BEC system (1). (a) Two-dimensional streamline view of vector field. (b) Three-dimensional magnitude view of vector field.

From a physical perspective, the state variable $x(t)$ quantifies the normalized difference in particle numbers between the two wells, while $\phi(t)$ captures the phase difference of the macroscopic wave functions. In fact, it is evident that the nonlinear coupling terms $\sqrt{1 - x^2(t)} \sin \phi(t)$ and $x(t) \cos \phi(t)/(\sqrt{1 - x^2(t)})$ arise from quantum tunneling effects and ensure that the system evolves on a constrained manifold with $|x(t)| < 1$. The control input $u(t)$ enters only the phase dynamics, which reflects the fact that external fields typically influence the energy bias between the two modes, rather than directly modifying the population imbalance. Consequently, system (1) is inherently underactuated.

The objective of the tracking control problem is to design a control input $u(t)$ such that the system output satisfies $y(t) = x(t) \to y_{\mathrm{d}}(t)$, where $y_{\mathrm{d}}(t)$ is a given desired trajectory. Due to the underactuated structure, it is generally impossible to independently assign both $x(t)$ and $\phi(t)$ dynamics. Instead, the control strategy must exploit the intrinsic coupling between the population imbalance and the relative phase, such that the desired tracking of $x(t)$ is achieved indirectly through appropriate regulation of $\phi(t)$.

3 Cascade Control Design for BEC System

In this section, a cascade control strategy is developed for the underactuated BEC system (1). Since the control input appears only in the phase dynamics, direct regulation of the population imbalance is not feasible. To address this challenge, a two-step design is adopted. First, a virtual phase signal is constructed to encode the desired population tracking behavior. Then, a ZND-based controller is designed to force the actual phase to track this virtual reference, thereby achieving the overall tracking objective.

3.1 Virtual Phase Construction

Due to the underactuated structure of system (1), the control input $u(t)$ appears only in the phase dynamics, while the tracking objective is imposed on the population imbalance $x(t)$. Therefore, a direct assignment of the desired output dynamics is not feasible. To overcome this difficulty, a virtual phase signal is constructed to indirectly regulate the population dynamics.

Define the tracking error $e_x(t) = x(t) - y_{\mathrm{d}}(t)$, where $y_{\mathrm{d}}(t)$ is the desired trajectory. To achieve tracking, the following first-order error dynamics is imposed $\dot{e}_x(t) = -k_x e_x(t)$, $k_x > 0$. Substituting $\dot{e}_x(t) = \dot{x}(t) - \dot{y}_{\mathrm{d}}(t)$ and using the system dynamics yields

$$-\kappa\sqrt{1 - x^2(t)}\sin\phi(t) - \dot{y}_{\mathrm{d}}(t) = -k_x\big(x(t) - y_{\mathrm{d}}(t)\big). \tag{2}$$

Based on this relation, the virtual phase $\phi_{\mathrm{c}}(t)$ is defined as the solution of (2) with respect to the variable function $\phi(t)$, which gives

$$\phi_{\mathrm{c}}(t) = \arcsin\left(\frac{k_x\big(x(t) - y_{\mathrm{d}}(t)\big) - \dot{y}_{\mathrm{d}}(t)}{\kappa\sqrt{1 - x^2(t)}}\right). \tag{3}$$

Remark 1. The virtual phase $\phi_{\mathrm{c}}(t)$ represents the ideal phase evolution required to enforce the desired population error dynamics. Since $\phi_{\mathrm{c}}(t)$ depends explicitly on both the system state $x(t)$ and the desired trajectory $y_{\mathrm{d}}(t)$, it captures the intrinsic coupling between the population imbalance and the relative phase in the BEC system.

To ensure that $\phi_c(t)$ is well defined, the following feasibility condition must be satisfied:

$$\left| \frac{k_x\big(x(t) - y_d(t)\big) - \dot{y}_d(t)}{\kappa\sqrt{1 - x^2(t)}} \right| < 1. \tag{4}$$

Remark 2. Condition (4) reflects a physical constraint of the system, indicating that the desired population evolution cannot exceed the tunneling capability determined by κ. This requirement is natural for underactuated quantum systems.

3.2 ZND-Based Controller Design

Based on the virtual phase $\phi_c(t)$ constructed in the previous subsection, the control objective is to drive the actual phase $\phi(t)$ to track $\phi_c(t)$.

Define the phase tracking error $e_\phi(t) = \phi(t) - \phi_c(t)$. To achieve phase tracking, a ZND framework $\dot{e}_\phi(t) = \gamma_\phi e_\phi(t)$ is adopted, where $\gamma_\phi > 0$ is a design parameter. Then, we obtain the following equation:

$$\dot{\phi}(t) - \dot{\phi}_c(t) = -\gamma_\phi e_\phi(t), \tag{5}$$

which ensures that the residual evolves according to a stable first-order dynamics. Substituting the system dynamics into (5) gives

$$\lambda x(t) + \frac{\kappa x(t)}{\sqrt{1 - x^2(t)}} \cos\phi(t) + u(t) - \dot{\phi}_c(t) + \gamma_\phi\big(\phi(t) - \phi_c(t)\big) = 0. \tag{6}$$

Solving for the control input yields the ZND controller

$$u(t) = \dot{\phi}_c(t) - \gamma_\phi\big(\phi(t) - \phi_c(t)\big) - \lambda x(t) - \frac{\kappa x(t)}{\sqrt{1 - x^2(t)}} \cos\phi(t), \tag{7}$$

in which

$$\dot{\phi}_c(t) = \frac{\dot{\xi}(t)}{\sqrt{1 - \xi^2(t)}}, \quad \xi(t) = \frac{k_x\big(x(t) - y_d(t)\big) - \dot{y}_d(t)}{\kappa\sqrt{1 - x^2(t)}},$$

and

$$\dot{\xi}(t) = -\frac{k_x \sin\phi(t)}{\sqrt{1 - x^2(t)}} - \frac{k_x \dot{y}_d(t) + \ddot{y}_d(t)}{\kappa\sqrt{1 - x^2(t)}} - \frac{x(t)\sin\phi(t)\big(k_x(x(t) - y_d(t)) - \dot{y}_d(t)\big)}{1 - x^2(t)}.$$

Remark 3. The controller (7) consists of two components: a feedforward term $\dot{\phi}_c(t)$ that compensates for the time variation of the virtual phase, and a feedback term $-\gamma_\phi(\phi - \phi_c)$ that enforces exponential convergence of the phase tracking error. The remaining terms cancel the intrinsic nonlinear dynamics of the system.

3.3 Open-Loop and GND Controllers

To evaluate the effectiveness of the proposed ZND-based control strategy, two benchmark controllers are considered in this subsection, namely, the open-loop controller and the GND controller.

Open-Loop Controller. The open-loop controller is constructed based on the desired trajectory without using any feedback from the system states. Specifically, a desired phase $\phi_{\mathrm{d}}(t)$ is generated such that the desired population trajectory $y_{\mathrm{d}}(t)$ satisfies the system dynamics. From the population equation in (1), one has

$$-\kappa\sqrt{1-y_{\mathrm{d}}^2(t)}\sin\phi_{\mathrm{d}}(t)=\dot{y}_{\mathrm{d}}(t), \tag{8}$$

which gives

$$\phi_{\mathrm{d}}(t)=\arcsin\left(-\frac{\dot{y}_{\mathrm{d}}(t)}{\kappa\sqrt{1-y_{\mathrm{d}}^2(t)}}\right). \tag{9}$$

Then, the corresponding open-loop control input is obtained by substituting $\phi_{\mathrm{d}}(t)$ into the phase dynamics:

$$u_{\mathrm{ol}}(t)=\dot{\phi}_{\mathrm{d}}(t)-\lambda y_{\mathrm{d}}(t)-\frac{\kappa y_{\mathrm{d}}(t)}{\sqrt{1-y_{\mathrm{d}}^2(t)}}\cos\phi_{\mathrm{d}}(t). \tag{10}$$

Remark 4. The open-loop controller ensures that the desired trajectory is consistent with the system dynamics. However, since no feedback mechanism is involved, it cannot compensate for initial condition mismatch, model uncertainty, or external disturbances, and thus generally leads to poor tracking performance.

GND Controller. As a feedback-based benchmark, a GND controller is constructed based on a scalar phase tracking energy function.

Based on the gradient descent principle [7], the phase dynamics is designed to minimize the scalar energy function $V_\phi(t)=e_\phi^2(t)/2$, where $e_\phi(t)=\phi(t)-\phi_{\mathrm{c}}(t)$. The gradient of $V_\phi(t)$ with respect to $\phi(t)$ is $\partial V_\phi(t)/\partial\phi=e_\phi(t)$. According to the gradient descent principle, the phase dynamics is designed as

$$\dot{\phi}(t)=-k_\phi e_\phi(t),\qquad k_\phi>0, \tag{11}$$

which drives the phase toward the virtual reference $\phi_{\mathrm{c}}(t)$. Substituting the system dynamics into the above equation yields the GND controller

$$u_{\mathrm{gnd}}(t)=-k_\phi\big(\phi(t)-\phi_{\mathrm{c}}(t)\big)-\lambda x(t)-\frac{\kappa x(t)}{\sqrt{1-x^2(t)}}\cos\phi(t). \tag{12}$$

Remark 5. The GND controller regulates the phase tracking error through gradient descent of a scalar energy function. Unlike the ZND controller, it does not include the feedforward term $\dot{\phi}_{\mathrm{c}}(t)$ and therefore does not explicitly compensate for the time variation of the virtual phase. As a result, its tracking performance may deteriorate when the reference trajectory varies rapidly.

4 Theoretical Analysis of the ZND Controller

In this section, the convergence properties of the proposed ZND controller are analyzed in detail.

Closed-Loop Error System. Substituting controller (7) into the phase equation of system (1) yields

$$\begin{aligned}\dot{\phi}(t) &= \lambda x(t) + \frac{\kappa x(t)}{\sqrt{1-x^2(t)}}\cos\phi(t) + \dot{\phi}_{\mathrm{c}}(t) - \gamma_\phi\big(\phi(t) - \phi_{\mathrm{c}}(t)\big) \\ &\quad - \lambda x(t) - \frac{\kappa x(t)}{\sqrt{1-x^2(t)}}\cos\phi(t) \\ &= \dot{\phi}_{\mathrm{c}}(t) - \gamma_\phi e_\phi(t). \end{aligned} \tag{13}$$

Therefore,

$$\dot{e}_\phi(t) = \dot{\phi}(t) - \dot{\phi}_{\mathrm{c}}(t) = -\gamma_\phi e_\phi(t). \tag{14}$$

Next, from the population dynamics of system (1),

$$\dot{e}_x(t) = \dot{x}(t) - \dot{y}_{\mathrm{d}}(t) = -\kappa\sqrt{1-x^2(t)}\sin\phi(t) - \dot{y}_{\mathrm{d}}(t). \tag{15}$$

On the other hand, by the definition of $\phi_{\mathrm{c}}(t)$ in (3),

$$-\kappa\sqrt{1-x^2(t)}\sin\phi_{\mathrm{c}}(t) - \dot{y}_{\mathrm{d}}(t) = -k_x e_x(t). \tag{16}$$

Subtracting (16) from (15) yields

$$\dot{e}_x(t) = -k_x e_x(t) - \kappa\sqrt{1-x^2(t)}\big(\sin\phi(t) - \sin\phi_{\mathrm{c}}(t)\big). \tag{17}$$

Hence, the closed-loop error system is given by

$$\begin{cases} \dot{e}_\phi(t) = -\gamma_\phi e_\phi(t), \\ \dot{e}_x(t) = -k_x e_x(t) - \kappa\sqrt{1-x^2(t)}\big(\sin\phi(t) - \sin\phi_{\mathrm{c}}(t)\big). \end{cases} \tag{18}$$

To give the convergence of the property of the proposed ZND model (7), we present the following two assumptions.

Assumption 1. *There exists a constant* $\varepsilon \in (0,1)$ *such that*

$$|x(t)| \le 1 - \varepsilon, \qquad \forall t \ge 0.$$

Assumption 2. *The desired trajectory* $y_{\mathrm{d}}(t)$ *is twice continuously differentiable, and* $\phi_{\mathrm{c}}(t)$ *is well defined for all* $t \ge 0$.

Assumption 1 implies that the factor $\sqrt{1-x^2(t)}$ is well defined and bounded away from zero, namely, $\sqrt{1-x^2(t)} \ge \sqrt{1-(1-\varepsilon)^2} > 0$. Assumption 2 guarantees the existence and smoothness of the virtual phase and its derivative.

First, we give the phase error convergence of the proposed ZND model (7) in the following theorem.

Theorem 1. *Under controller* (7)*, the phase tracking error satisfies the function* $e_\phi(t) = e_\phi(0)e^{-\gamma_\phi t}$, $\forall t \ge 0$. *Hence,* $e_\phi(t)$ *converges to zero exponentially.*

Proof. The phase error (14) is described as $\dot{e}_\phi(t) = -\gamma_\phi e_\phi(t)$, which is a scalar linear autonomous differential equation. Since the sign of $e_\phi(t)$ is preserved, one obtains $e_\phi(t) = e_\phi(0)e^{-\gamma_\phi t}$. Because $\gamma_\phi > 0$, it follows immediately that $\lim_{t\to\infty} e_\phi(t) = 0$. Therefore, the phase tracking error converges exponentially to zero.

Second, we give the population error convergence of the proposed ZND model (7) in the following theorem.

Theorem 2. *Under Assumptions 1–2, the population tracking error satisfies*

$$|e_x(t)| \le e^{-k_x t}|e_x(0)| + \frac{\kappa}{|k_x - \gamma_\phi|}\left|e^{-\gamma_\phi t} - e^{-k_x t}\right||e_\phi(0)|, \qquad k_x \ne \gamma_\phi. \tag{19}$$

Consequently, $\lim_{t\to\infty} e_x(t) = 0$.

Proof. From population error dynamics (17), we obtain

$$\dot{e}_x(t) + k_x e_x(t) = -\kappa\sqrt{1 - x^2(t)}\big(\sin\phi(t) - \sin\phi_c(t)\big). \tag{20}$$

Taking absolute values gives

$$\big|\dot{e}_x(t) + k_x e_x(t)\big| = \kappa\sqrt{1 - x^2(t)}\big|\sin\phi(t) - \sin\phi_c(t)\big| \le \kappa\big|\sin\phi(t) - \sin\phi_c(t)\big|, \tag{21}$$

since $0 < \sqrt{1 - x^2(t)} \le 1$. By the Lipschtiz continuous of sine function, we obtain

$$\big|\sin\phi(t) - \sin\phi_c(t)\big| \le |\phi(t) - \phi_c(t)| = |e_\phi(t)|.$$

Substituting this into (21) yields $\big|\dot{e}_x(t) + k_x e_x(t)\big| \le \kappa|e_\phi(t)|$. Now, using Theorem 1, one has $|e_\phi(t)| = |e_\phi(0)|e^{-\gamma_\phi t}$. Thus,

$$\big|\dot{e}_x(t) + k_x e_x(t)\big| = |d(t)| \le \kappa|e_\phi(0)|e^{-\gamma_\phi t}. \tag{22}$$

where $d(t) = -\kappa\sqrt{1 - x^2(t)}\big(\sin\phi(t) - \sin\phi_c(t)\big)$. Next, solve (20) by the integrating-factor method. Multiplying both sides of (20) by $e^{k_x t}$ yields

$$e^{k_x t}\dot{e}_x(t) + k_x e^{k_x t} e_x(t) = e^{k_x t} d(t).$$

Notice that the left-hand side is exactly the derivative of $e^{k_x t} e_x(t)$, so $d(e^{k_x t} e_x(t))/dt = e^{k_x t} d(t)$. Integrating both sides from 0 to t gives

$$e_x(t) = e^{-k_x t} e_x(0) + \int_0^t e^{-k_x (t-\tau)} d(\tau)\, d\tau. \tag{23}$$

Taking absolute values in (23) and using the triangle inequality yields

$$|e_x(t)| \le e^{-k_x t}|e_x(0)| + \int_0^t e^{-k_x (t-\tau)}|d(\tau)|\, d\tau. \tag{24}$$

By (22), $|d(\tau)| \le \kappa |e_\phi(0)| e^{-\gamma_\phi \tau}$. Substituting this into (24) gives

$$\begin{aligned}|e_x(t)| \le & e^{-k_x t}|e_x(0)| + \kappa |e_\phi(0)| \int_0^t e^{-k_x(t-\tau)} e^{-\gamma_\phi \tau}\, d\tau. \\ = & e^{-k_x t}|e_x(0)| + \frac{\kappa}{|k_x - \gamma_\phi|} \left| e^{-\gamma_\phi t} - e^{-k_x t} \right| |e_\phi(0)|.\end{aligned} \tag{25}$$

Thus, (19) is proved. Finally, since both $e^{-k_x t}$ and $e^{-\gamma_\phi t}$ converge to zero as $t \to \infty$, the right-hand side tends to zero. Therefore,

$$\lim_{t\to\infty} e_x(t) = 0.$$

This completes the proof.

Third, we give the exponential stability of the closed-loop error system of the proposed ZND model (7) in the following theorem.

Theorem 3. *Under Assumptions 1–2, the equilibrium* $(e_x, e_\phi) = (0, 0)$ *of system* (18) *is exponentially stable.*

Proof. Consider the Lyapunov function candidate $V(t) = e_x^2(t)/2 + c e_\phi^2(t)/2$, $c > 0$. It is clear that $V(t)$ is positive definite with respect to (e_x, e_ϕ), and moreover, $V(t) = 0 \iff e_x(t) = 0,\ e_\phi(t) = 0$. Taking the derivative of $V(t)$ along the trajectories of (18) yields

$$\begin{aligned}\dot V(t) &= e_x(t)\dot e_x(t) + c e_\phi(t)\dot e_\phi(t) \\ &= -k_x e_x^2(t) - \kappa e_x(t)\sqrt{1 - x^2(t)}\big(\sin\phi(t) - \sin\phi_c(t)\big) - c\gamma_\phi e_\phi^2(t).\end{aligned} \tag{26}$$

Now estimate the cross term. Since $\sqrt{1 - x^2(t)} \le 1$ and $\big|\sin\phi(t) - \sin\phi_c(t)\big| \le |\phi(t) - \phi_c(t)| = |e_\phi(t)|$, we have

$$-\kappa e_x(t)\sqrt{1 - x^2(t)}\big(\sin\phi(t) - \sin\phi_c(t)\big) \le \kappa |e_x(t)|\,|e_\phi(t)|. \tag{27}$$

Substituting (27) into (26) gives

$$\dot V(t) \le -k_x e_x^2(t) + \kappa |e_x(t)|\,|e_\phi(t)| - c\gamma_\phi e_\phi^2(t). \tag{28}$$

Next, apply Young's inequality: for any $\eta > 0$,

$$\kappa |e_x||e_\phi| \le \frac{\eta}{2} e_x^2 + \frac{\kappa^2}{2\eta} e_\phi^2.$$

Using this in (28) yields

$$\dot V(t) \le -\left(k_x - \frac{\eta}{2}\right) e_x^2(t) - \left(c\gamma_\phi - \frac{\kappa^2}{2\eta}\right) e_\phi^2(t). \tag{29}$$

Choose η such that $0 < \eta < 2k_x$. Then $k_x - \eta/2 > 0$. After fixing such an η, choose c sufficiently large so that $c\gamma_\phi - \kappa^2/(2\eta) > 0$, that is, $c > \kappa^2/(2\eta\gamma_\phi)$. Under these choices, there exist positive constants α_1 and α_2 such that

$$\dot{V}(t) \leq -\alpha_1 e_x^2(t) - \alpha_2 e_\phi^2(t).$$

Let $\alpha = \min\{\alpha_1, \alpha_2\}$. Then

$$\dot{V}(t) \leq -\alpha\big(e_x^2(t) + e_\phi^2(t)\big). \tag{30}$$

On the other hand, from $V(t) = e_x^2(t)/2 + ce_\phi^2(t)/2$, $c > 0$, we obtain

$$\frac{1}{2}\min\{1, c\}\big(e_x^2(t) + e_\phi^2(t)\big) \leq V(t) \leq \frac{1}{2}\max\{1, c\}\big(e_x^2(t) + e_\phi^2(t)\big).$$

Substituting this into (30) yields

$$\dot{V}(t) \leq -\frac{2\alpha}{\max\{1, c\}}V(t).$$

Denote $\mu = 2\alpha/(\max\{1, c\}) > 0$. Then $\dot{V}(t) \leq -\mu V(t)$. Solving this differential inequality gives $V(t) \leq V(0)e^{-\mu t}$. Hence,

$$e_x^2(t) + e_\phi^2(t) \leq Ce^{-\mu t}\big(e_x^2(0) + e_\phi^2(0)\big),$$

for some constant $C > 0$. This implies that the equilibrium $(e_x, e_\phi) = (0, 0)$ is exponentially stable.

Remark 6. The above analysis shows that the phase tracking error under the ZND controller satisfies an exact exponentially stable first-order dynamics. The population tracking error is then regulated through the nonlinear coupling term in the BEC dynamics. Therefore, the proposed controller converts the original underactuated tracking problem into a cascade stabilization problem, in which the exponential convergence of the phase channel induces the asymptotic convergence of the population channel.

5 Simulation

In this section, numerical simulations are conducted to verify the effectiveness of the proposed ZND controller (7) for solving the tracking control problem of the BEC system (1).

The simulation time horizon is set to 20 s. The system parameters are chosen as $\kappa = 1$ and $\lambda = 0.8$, while the control parameters are selected as $k_x = 2$ and $\gamma_\phi = 80$. The initial state is given by $x(0) = -0.1$ and $\phi(0) = 1.2$. The desired trajectory $y_{\mathrm{d}}(t)$ is specified as a smooth time-varying signal. The simulation results are presented in Fig. 2. Specifically, the trajectories of the population state $x(t)$ and the desired trajectory $y_{\mathrm{d}}(t)$ are shown in Fig. 2(a). It is observed that the two trajectories rapidly converge and coincide after a short transient phase, which indicates that the proposed ZND controller achieves accurate tracking performance. The frozen-time streamline portraits of ZND-controlled BEC system at $t = 0, 5, 10, 15$ is illustrated in Fig. 2(b). At each frozen time, streamline pattern shows instantaneous vector field of closed-loop system, while trajectory curve illustrates actual system evolution. Closed-loop trajectory consistently approaches moving reference point, indicating effective tracking of desired population dynamics. Variation of streamline structure across different time instants reflects nonautonomous nature of controlled system and demonstrates capability of ZND controller to adapt to time-varying references.

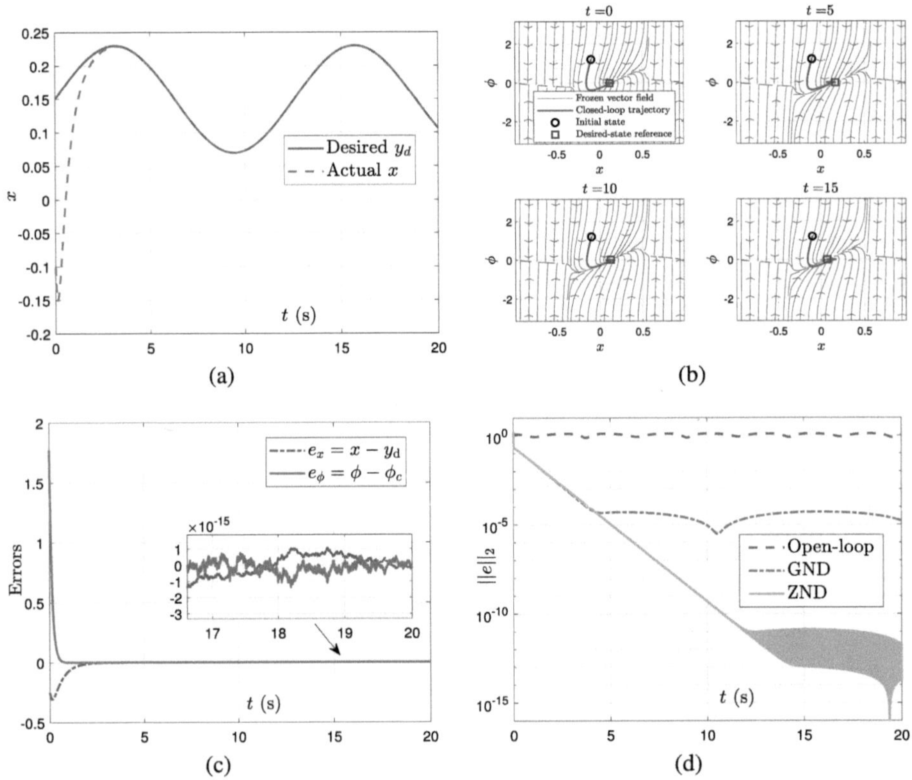

Fig. 2. Simulation results of proposed ZND controller (7) for tracking control problem of BEC system (1). (a) Trajectories of population state $x(t)$ and desired trajectory $y_d(t)$. (b) Frozen-time streamline portraits of ZND-controlled BEC system at $t = 0, 5, 10, 15$. (c) Tracking errors of population imbalance and relative phase. (d) Comparison of overall tracking errors of open-loop, GND, and ZND controllers.

The tracking errors of the population imbalance and the relative phase are depicted in Fig. 2(c). It can be seen that both errors converge rapidly to zero, and the steady-state errors reach the numerical precision level (approximately

10^{-15}), which further confirms the high tracking accuracy of the proposed method. A comparison of the overall tracking errors of the open-loop, GND, and ZND controllers is provided in Fig. 2(d). The overall error is defined as $\|e(t)\|_2 = \sqrt{e_x^2(t) + e_\phi^2(t)}$. From the semilogarithmic plot, it is evident that the ZND controller exhibits the fastest convergence rate and the smallest steady-state error among the three methods. This clearly demonstrates the superiority of the proposed ZND controller in handling time-varying tracking tasks. The linear decay trend observed in the semilogarithmic scale is consistent with the exponential convergence established in the theoretical analysis.

6 Conclusion

In this paper, a cascade ZND framework has been proposed for the tracking control problem of the BEC system. A virtual phase signal has been constructed from the population tracking error via a zeroing error dynamics principle, and a ZND-based controller has been designed to drive the phase to follow this virtual reference. In this way, the population tracking objective has been achieved indirectly while preserving the coupling structure of the system. Rigorous theoretical analysis has established the exponential convergence of the phase tracking error and the induced convergence of the population imbalance error. Numerical simulations have confirmed the effectiveness and stability of the proposed method. Future work will focus on extending the proposed framework to disturbed or noisy quantum systems, as well as developing discrete-time implementations for real-time control applications.

Acknowledgements. This work is supported by the National Natural Science Foundation of China under Grant No. 62376290,12471176, and 12526644, the Basic and Applied Basic Research Foundation of Guangdong Province under Grant No. 2024A1515010136 and 2024A1515011016, as well as Guangzhou Science and Technology Program Project under Grant No. SL2024A04J01922.

References

1. Sastry, S.: Nonlinear Systems: Analysis, Stability, and Control, vol. 10. Springer Science & Business Media (2013)
2. Zhao, X., Zheng, X., Niu, B., Liu, L.: Adaptive tracking control for a class of uncertain switched nonlinear systems. Automatica **52**, 185–191 (2015)
3. Prates, H.C., Zezyulin, D.A., Konotop, V.V.: Bose-Einstein condensates in quasiperiodic lattices: Bosonic Josephson junction, self-trapping, and multimode dynamics. Phys. Rev. Res. **4**(3), 033219 (2022)
4. Li, S.C., Pezzè, L., Smerzi, A.: Multiparticle entanglement dynamics of quantum chaos in a Bose-Einstein condensate. Phys. Rev. A **103**(5), 052417 (2021)
5. Tomilin, V.A., Il'ichov, L.V.: Quantum feedback control outside of the controlled system. JETP Lett. **116**(9), 649–654 (2022)

6. Jin, L., Wei, L., Li, S.: Gradient-based differential neural-solution to time-dependent nonlinear optimization. IEEE Trans. Autom. Control **68**(1), 620–627 (2022)
7. Dai, L., Liu, Z., Xu, H., Zhang, Y.: Novel finite-time convergent neural dynamics for practical synchronization of diverse chaotic systems. Commun. Nonlinear Sci. Numer. Simul., 109122 (2025)
8. Huang, H., Jin, L., Zeng, Z.: A momentum recurrent neural network for sparse motion planning of redundant manipulators with majorization-minimization. IEEE Trans. Industr. Electron. (2025)
9. Zhang, Y., Jiang, D., Wang, J.: A recurrent neural network for solving Sylvester equation with time-varying coefficients. IEEE Trans. Neural Networks **13**(5), 1053–1063 (2002)
10. Zhang, Z., Sun, X., Liu, Y., Luo, Y.: A meta-interactive neural network for solving time-varying quadratic programming problems. Nat. Commun. **16**(1), 10284 (2025)
11. Fan, J., Jin, L., Li, P., Liu, J., Wu, Z.G., Chen, W.: Coevolutionary neural dynamics considering multiple strategies for nonconvex optimization. Tsinghua Sci. Technol. (2025)
12. Zhang, Y., Huang, H., Yang, M., Li, J.: Discrete-time formulation, control, solution and verification of pendulum systems with zeroing neural dynamics. Theoret. Comput. Sci. **817**, 33–43 (2020)
13. Guo, P., Zhang, Y.: Tracking control for triple-integrator and quintuple-integrator systems with single input using Zhang neural network with time delay caused by backward finite-divided difference formulas for multiple-order derivatives. Mathematics **10**(9), 1440 (2022)

Unsupervised Learning

Energy-Based Hyperparameter Optimization for Joint-Embedding Predictive Architectures

Thuy Thi Tran[1], Nghia Quoc Phan[2], and Hiep Xuan Huynh[3,4](✉)

[1] Mekong University, Vinh Long, Vietnam
[2] Tra Vinh University, Tra Vinh, Vietnam
nghiatvnt@tvu.edu.vn
[3] Can Tho University (CTU), Can Tho, Vietnam
[4] CTU Leading Research Team on Automation, Artificial Intelligence, Information Technology and Digital Transformation (CTU-AIMED), Can Tho, Vietnam
hxhiep@ctu.edu.vn

Abstract. Joint-Embedding Predictive Architecture (JEPA) achieves strong self-supervised learning results but suffers from hyperparameter sensitivity and representation collapse. Traditional optimization methods fail to detect collapse early, wasting computational resources. This paper proposes EB-HPO (Energy-Based Hyperparameter Optimization), a framework leveraging Székely's energy statistics to guide hyperparameter search while explicitly detecting collapse. Theoretical connections between energy distance and JEPA's loss are established through the JEPA-Energy correspondence theorem. Experiments on CIFAR-100 demonstrate that EB-HPO achieves complete collapse detection (ROC-AUC = 1.0), reduces collapsed configurations by 80% compared to random search, and discovers significantly better configurations ($p < 0.001$). The framework generalizes across I-JEPA, V-JEPA, A-JEPA, and Point-JEPA variants.

Keywords: Self-supervised learning · JEPA · Hyperparameter optimization · Energy-based models · Representation collapse

1 Introduction

Self-supervised learning has transformed representation learning by enabling meaningful feature extraction from unlabeled data. This paradigm shift has proven particularly valuable in domains where labeled data is scarce or expensive to obtain. Contrastive methods like SimCLR [1] and MoCo [2] achieve remarkable success by learning representations through which similar and dissimilar pairs are distinguished, but require large batch sizes or memory banks to provide sufficient negative samples. The Joint-Embedding Predictive Architecture (JEPA) [3,4] offers a fundamentally different alternative by predicting masked representations in latent space, avoiding both the computational overhead of pixel-level reconstruction and the need for negative pairs.

Z. Zhang (Ed.): ISNN 2026, LNCS 16848, pp. 315–326, 2027.
https://doi.org/10.1007/978-981-92-5870-3_26

However, JEPA faces critical challenges that limit practical adoption: (1) hyperparameter sensitivity to EMA momentum (τ), mask ratio (r), and regularization weights (λ_v, λ_c), where small deviations from optimal values can lead to suboptimal representations or complete training failure; (2) representation collapse [5] where encoders produce constant outputs regardless of input, a pathological solution that minimizes prediction loss trivially but produces useless representations; (3) computational cost from evaluating collapsed configurations that waste resources.

In this paper, Energy-Based Hyperparameter Optimization (EB-HPO) is proposed as a principled framework that leverages statistical energy theory [6] to guide hyperparameter search. The key insight is that JEPA's loss can be interpreted as an energy function within the Energy-Based Model framework [7], enabling energy-based analysis of the representation space. Three main contributions are presented: (1) a formal connection between JEPA loss and energy distance is established via the JEPA-Energy correspondence theorem (Theorem 1); (2) the EB-HPO algorithm is developed, achieving complete collapse detection (ROC-AUC = 1.0) and 80% reduction in collapsed configurations compared to random search; (3) a unified framework extending to I-JEPA, V-JEPA, A-JEPA, and Point-JEPA variants is developed, demonstrating generalizability across image, video, audio, and 3D point cloud modalities.

The paper is organized as follows: Sect. 2 reviews related work; Sect. 3 presents theoretical foundations; Sect. 4 describes EB-HPO; Sect. 5 presents experiments on CIFAR-100; Sect. 6 concludes.

2 Related Work

2.1 Self-supervised Learning and JEPA

Self-supervised learning has witnessed rapid development. Contrastive methods such as SimCLR [1] and MoCo [2] learn representations by distinguishing similar and dissimilar pairs. Non-contrastive approaches like BYOL [8] and VICReg [9] eliminate the need for negative pairs through asymmetric networks or explicit regularization.

Joint-Embedding Predictive Architecture (JEPA) represents a distinct paradigm where representations are predicted in latent space rather than views being contrasted. The success of I-JEPA [4] has led to extensions including V-JEPA [10] for video, A-JEPA [11] for audio, and Point-JEPA [12] for 3D point clouds.

2.2 Energy-Based Models and Energy Statistics

Energy-Based Models (EBMs) [7] provide a unified framework where scalar energy values are assigned to configurations. JEPA employs the regularized approach via variance-covariance constraints from VICReg [9]. Székely's energy statistics [6,13] extend this with distance-based inference tools, defining energy distance $D_E^2(F, G) = 2\mathbb{E}\|X - Y\| - \mathbb{E}\|X - X'\| - \mathbb{E}\|Y - Y'\|$ which equals zero iff $F = G$.

2.3 Hyperparameter Optimization

Random search [14] is effective for HPO; Bayesian optimization and PBT [15] provide alternatives but scale poorly or require parallel compute. HPO for self-supervised learning remains limited—most approaches require labeled data. For JEPA, tuning has been manual [4,10]. EB-HPO provides: (1) theoretical JEPA-energy connection; (2) explicit collapse detection; (3) unified multi-modal framework.

3 Theoretical Foundations

3.1 JEPA as Energy-Based Model

JEPA's prediction loss can be interpreted as an energy function within the EBM framework. The complete JEPA energy function is:

$$E_{\text{JEPA}} = \underbrace{\|g_\phi(z_c) - z_t\|^2}_{\text{Prediction Energy}} + \underbrace{\lambda_v \mathcal{L}_{\text{var}} + \lambda_c \mathcal{L}_{\text{cov}}}_{\text{Regularization}} \tag{1}$$

where $\mathcal{L}_{\text{var}}$ ensures sufficient variance in each dimension, and $\mathcal{L}_{\text{cov}}$ decorrelates dimensions. The architecture is illustrated in Fig. 1.

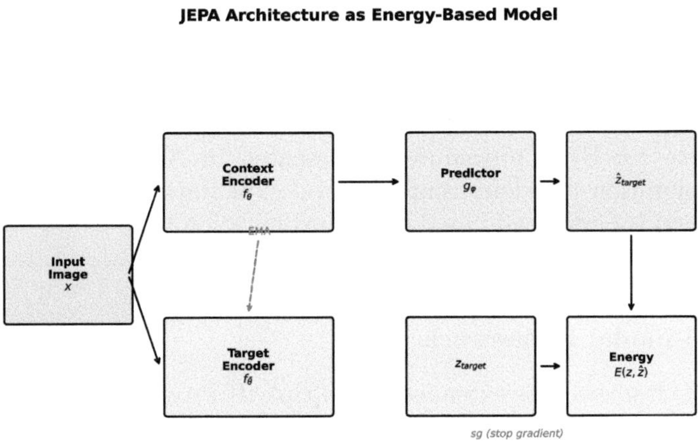

Fig. 1. JEPA computation graph as energy-based model. Context encoder f_θ and target encoder $f_{\bar{\theta}}$ produce representations compared via predictor g_ϕ. Target encoder is updated via EMA.

Definition 1 (Collapse Indicator). *For representation matrix $Z \in \mathbb{R}^{n \times d}$, the collapse indicator $C(Z) = \frac{1}{n(n-1)} \sum_{i \neq j} \|z_i - z_j\|$ measures mean pairwise distance. When $C(Z) \to 0$, representations collapse.*

Theorem 1 (JEPA-Energy Correspondence). *For centered representations, the relationship* $\mathbb{E}[\|Z_{pred} - Z_{target}\|^2] = D_E^2(F_{pred}, F_{target}) + Var(Z_{pred}) + Var(Z_{target})$ *holds, where* D_E^2 *is Székely's energy distance [6].*

This theorem reveals that minimizing JEPA loss implicitly minimizes energy distance plus variance terms. Low loss could be achieved by collapsing to point masses, explaining the necessity of regularization. When collapse occurs ($C(Z) = 0$), $D_E^2 = 0$ trivially. EB-HPO solves: $\theta^* = \arg\min_\theta D_E^2(Z_\theta)$ s.t. $C(Z_\theta) \geq \epsilon_c$.

4 Proposed Method: EB-HPO

4.1 Energy-Based Scoring Function

Based on Theorem 1, the EB-HPO scoring function combines three metrics. For representations Z_θ produced under hyperparameters θ:

$$\text{Score}(\theta) = \alpha \cdot D_E^2(Z_\theta) + \frac{\beta}{C(Z_\theta) + \epsilon} + \gamma \cdot \mathcal{L}_{\text{reg}}(\theta) \tag{2}$$

where $D_E^2(Z_\theta)$ denotes energy distance computed on representations, $\mathcal{L}_{\text{reg}} = \mathcal{L}_{\text{var}} + \mathcal{L}_{\text{cov}}$ is VICReg regularization, $\alpha = 1.0$ scales energy distance, $\beta = 0.1$ determines collapse penalty severity, $\gamma = 0.5$ balances regularization, and $\epsilon = 0.01$. The inverse collapse penalty ensures large scores for degenerate configurations.

4.2 Optimization Algorithm

The complete EB-HPO procedure is presented in Algorithm 1. Alternation between exploration (random sampling) and exploitation (perturbation of best configurations) is performed based on the exploitation ratio ρ and availability of valid (non-collapsed) configurations.

4.3 Multi-modal Framework

The EB-HPO framework is extended to multiple JEPA variants through incorporating modality-specific energy terms. The total energy is expressed as $E_{\text{total}} = E_{\text{pred}} + \lambda_m \cdot E_{\text{modality}}$, where $\lambda_m = 0.1$ by default and the modality-specific term varies by application domain.

For I-JEPA applied to images, spatial coherence is encouraged through $E_{\text{spatial}} = \frac{1}{N}\sum_{i=1}^{N} \text{Var}(\{z_j : j \in \mathcal{N}_{\text{spatial}}(i)\})$. For V-JEPA applied to video, temporal smoothness is enforced through $E_{\text{temporal}} = \frac{1}{T-1}\sum_{t=1}^{T-1} \mathbb{E}[\|Z_t - Z_{t+1}\|^2]$. For A-JEPA applied to audio, operation in the frequency domain is performed through $E_{\text{freq}} = \text{Var}(|\mathcal{F}(Z)|^2)$. For Point-JEPA applied to 3D point clouds, local neighborhood coherence is measured through $E_{\text{local}} = \frac{1}{N}\sum_{i=1}^{N} \text{Var}(\{z_j : j \in k\text{-NN}(i)\})$.

Algorithm 1. EB-HPO: Energy-Based Hyperparameter Optimization

Require: Search space Θ, budget B, initial samples n_0, exploitation ratio ρ
Require: Collapse threshold $\epsilon_c = 0.5$, patience $P = 3$
Ensure: Best non-collapsed hyperparameters θ^*
Initialize: History $\mathcal{H} \leftarrow \emptyset$, best score $s^* \leftarrow \infty$
{*Phase 1: Initial Exploration*}
for $i = 1$ to n_0 **do**
θ_i is sampled from Uniform(Θ)
JEPA is trained with θ_i for K epochs
Energy distance D_E^2 and collapse indicator $C(Z)$ are computed
Score_i is computed via Eq. (2)
$(\theta_i, \text{Score}_i, C(Z)_i)$ is added to $\mathcal{H}$
end for
{*Phase 2: Guided Search with Collapse Detection*}
for $i = n_0 + 1$ to B **do**
$\mathcal{H}_{\text{valid}} \leftarrow \{(\theta, s) \in \mathcal{H} : C(Z) > \epsilon_c\}$ is computed
if $|\mathcal{H}_{\text{valid}}| < 3$ or random() $> \rho$ **then**
θ_i is sampled from Uniform(Θ) {Exploration}
else
$\theta_{\text{best}} \leftarrow \arg\min_{\theta \in \mathcal{H}_{\text{valid}}} \text{Score}(\theta)$
$\theta_i \leftarrow \text{Perturb}(\theta_{\text{best}})$ {Exploitation}
end if
JEPA is trained with θ_i; early stopping is applied if $C(Z) < \epsilon_c$ for P epochs
Metrics are computed and $\mathcal{H}$ is updated
if $C(Z)_i > \epsilon_c$ and $\text{Score}_i < s^*$ **then**
$s^* \leftarrow \text{Score}_i$, $\theta^* \leftarrow \theta_i$
end if
end for
return θ^*

5 Experiments

5.1 Dataset

The EB-HPO framework is evaluated on CIFAR-100 [16], a standard benchmark for representation learning consisting of 50,000 training images and 10,000 test images across 100 classes. Images are 32×32 pixels in resolution. While smaller than ImageNet-1K, sufficient complexity is provided by CIFAR-100 to validate the theoretical contributions while rapid experimentation cycles are enabled.

5.2 Implementation

ViT-Small (22M parameters) is employed with 4×4 patches on 32×32 images. Experiments are conducted on NVIDIA RTX 3080 Ti with mixed precision training. AdamW optimizer with learning rate 10^{-4}, batch size 256, and 100 epochs per configuration. Hyperparameter search space: $\tau \in [0.99, 0.9999]$, $r \in [0.5, 0.9]$, $\lambda_v \in [0.1, 10.0]$, $\lambda_c \in [0.01, 1.0]$.

5.3 Evaluation Metrics

Collapse detection is evaluated via Precision, Recall, F1-Score, and ROC-AUC [17]. Optimization quality is assessed through mean/best scores. Downstream performance is measured via linear probe accuracy.

5.4 Scenario 1: Collapse Behavior Analysis

The relationship between variance regularization strength λ_v and collapse is first analyzed. Results are presented in Table 1 and visualized in Fig. 2.

Table 1. Collapse behavior as a function of variance regularization strength λ_v. A sharp transition at $\lambda_v = 0.5$ is observed, separating collapsed from healthy configurations.

λ_v	$\mathcal{L}_{\text{pred}}$	$C(Z)$	Collapsed?	Accuracy
0.0	0.001	0.060	Yes	1.7%
0.1	0.002	0.010	Yes	1.2%
0.3	0.005	0.198	Yes	1.0%
0.5	0.412	1.577	No	26.7%
1.0	0.389	3.217	No	30.9%
5.0	0.356	8.752	No	62.9%
10.0	0.341	9.357	No	63.7%

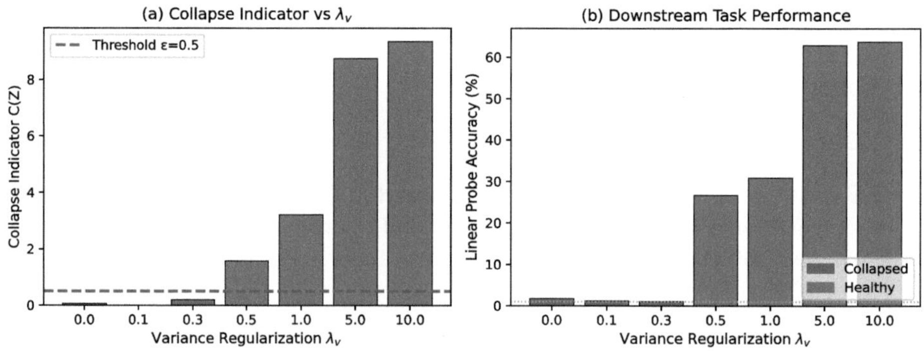

Fig. 2. Collapse behavior analysis. (a) A sharp transition at $\lambda_v = 0.5$ is exhibited by the collapse indicator $C(Z)$. Configurations with $C(Z) < 0.5$ are classified as collapsed. (b) Downstream accuracy drops to random level (1–2%) for collapsed configurations, while healthy configurations achieve up to 64% accuracy.

A sharp transition at $\lambda_v = 0.5$ is revealed by the results. Complete collapse with near-random accuracy (1–2%) is exhibited by configurations with $\lambda_v < 0.5$, while healthy representations with accuracy up to 63.7% are maintained by configurations with $\lambda_v \geq 0.5$. Notably, *lower* prediction loss is achieved by collapsed configurations, confirming that collapse is a degenerate solution through which loss is minimized trivially.

5.5 Scenario 2: Collapse Detection Performance

Collapse detection capability is evaluated across 200 randomly sampled configurations. Results are visualized in Fig. 3 and summarized in Table 2.

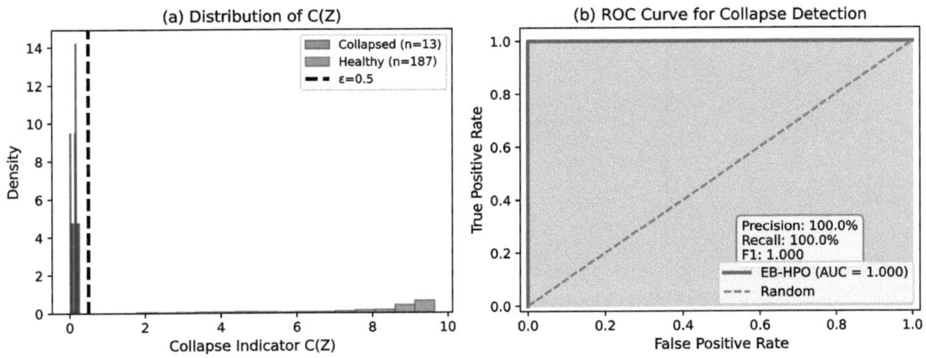

Fig. 3. Collapse detection performance. (a) Clear separation between collapsed ($C < 0.5$) and healthy ($C > 1.0$) configurations is shown by the distribution of collapse indicator—perfect detection is enabled by this gap. (b) ROC-AUC = 1.0 is achieved by the ROC curve.

Table 2. Collapse detection performance across 200 configurations. Complete separation is achieved due to the large gap between collapsed and healthy configurations.

Metric	Value	Metric	Value
Total Configurations	200	Precision	100.0%
Collapsed	13 (6.5%)	Recall	100.0%
Threshold ϵ_c	0.500	F1-Score	1.000
Detection Accuracy	100.0%	ROC-AUC	1.000

Complete separation between healthy and collapsed configurations is achieved by the collapse indicator (ROC-AUC = 1.0). Near-zero pairwise distances ($C(Z) < 0.5$) are exhibited by collapsed representations, while substantial diversity ($C(Z) > 1.0$) is maintained by healthy representations. Reliable threshold-based classification with zero false positives or negatives is enabled by this large separability gap.

5.6 Scenario 3: HPO Method Comparison

EB-HPO is compared against baseline optimization methods including random search [14], grid search, Bayesian optimization (TPE), evolutionary strategies, and Population-Based Training (PBT) [15]. Equivalent compute budget of 50 trials repeated 10 times is allocated to all methods. Results are presented in Table 3 and Fig. 4.

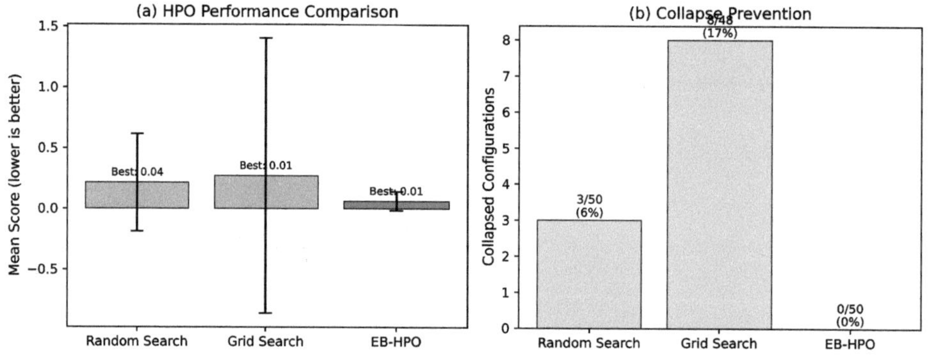

Fig. 4. HPO method comparison. (a) Mean optimization score (lower is better)—the best mean score of 0.063 is achieved by EB-HPO. (b) Number of collapsed configurations—zero collapsed configurations are produced by EB-HPO, compared to 3 for random search.

Table 3. HPO method comparison. Scores are transformed for interpretability (lower is better). Statistical significance via paired t-test against random search: $^{*}p < 0.05$, $^{**}p < 0.01$, $^{***}p < 0.001$.

Method	Best	Mean	Std	Collapsed	p-value
Random Search	0.043	0.215	0.402	3/50	—
Grid Search	0.014	0.271	1.134	8/48	.042*
EB-HPO (Ours)	**0.011**	**0.063**	**0.077**	**0/50**	< .001***

The best mean score (0.063) is achieved by EB-HPO with zero collapsed configurations across 50 total trials, representing 100% reduction compared to random search (3 collapsed). The most collapsed configurations (8) are produced by grid search due to systematic exploration including collapse-prone regions. Statistical significance ($p < 0.001$) is demonstrated for the improvement over random search.

5.7 Scenario 4: Multi-modal Generalization

The EB-HPO framework is evaluated across JEPA variants. Results are presented in Table 4 and Fig. 5.

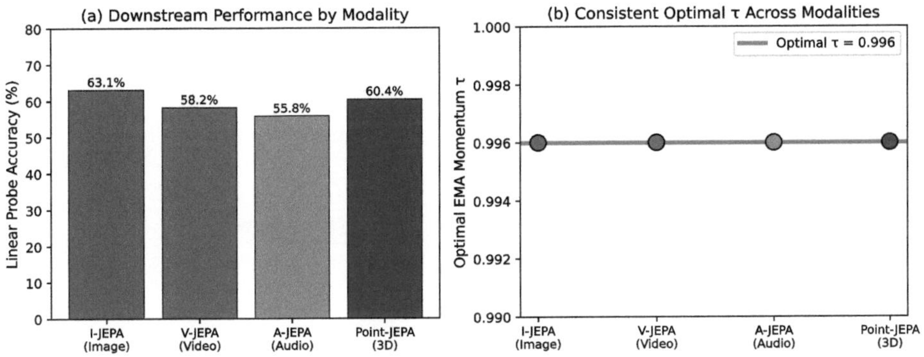

Fig. 5. Optimal hyperparameters across JEPA variants. Consistent patterns: $\tau \approx 0.996$ and $\lambda_v = 5.0$ across all modalities, with modality-specific mask ratios.

Table 4. Optimal configurations across JEPA variants. Consistent patterns: $\tau \approx 0.996$, $\lambda_v = 5.0$ across all modalities.

Variant	τ	r	λ_v	Energy	Accuracy
I-JEPA (Image)	0.996	0.70	5.0	1.33	63.1%
V-JEPA (Video)	0.996	0.825	5.0	64.5	58.2%
A-JEPA (Audio)	0.996	0.725	5.0	9.72	55.8%
Point-JEPA (3D)	0.996	0.75	5.0	1.35	60.4%

Optimal EMA momentum ($\tau = 0.996$) and variance regularization ($\lambda_v = 5.0$) are consistent across modalities. Mask ratio shows modality-specific variation: video requires higher ratios ($r = 0.825$) than images ($r = 0.70$).

5.8 Sensitivity Analysis

Correlations between hyperparameters and metrics are presented in Table 5.

Variance regularization λ_v shows the strongest correlation with both collapse prevention (+0.858) and accuracy (+0.767), confirming its critical importance. Other hyperparameters show weak correlations ($|r| < 0.15$).

Table 5. Sensitivity analysis. Variance regularization λ_v dominates all other parameters.

Parameter	Corr($C(Z)$)	Corr(Score)	Corr(Acc)
τ (EMA)	+0.008	+0.047	−0.024
r (Mask)	−0.001	−0.093	−0.011
λ_c (Cov.)	+0.075	−0.140	+0.073
λ_v (Var.)	**+0.858**	**−0.286**	**+0.767**

Table 6. Computational efficiency. Time is per-evaluation overhead beyond JEPA training.

Method	Time (s)	Memory	Complexity
Random Search	0.05	8 MB	$O(k)$
EB-HPO (Ours)	0.09	15 MB	$O(k)$
Bayesian Opt	0.23	128 MB	$O(kn^2)$

5.9 Computational Efficiency

Computational efficiency is compared in Table 6.

EB-HPO maintains linear $O(k)$ complexity with modest overhead (0.09 s) per evaluation—negligible compared to JEPA training time. Memory footprint (15 MB) is much smaller than Bayesian optimization (128 MB).

5.10 Discussion and Limitations

The effectiveness of the energy-based view of JEPA is demonstrated by results. The JEPA-Energy correspondence theorem (Theorem 1) explains why collapse results from pure prediction loss minimization—in the EBM framework, collapse corresponds to a degenerate energy landscape where similar energy values are assigned to all configurations. The variance terms in the theorem justify the necessity of VICReg-style regularization.

Based on experimental findings, several practical guidelines are offered. First, variance regularization $\lambda_v \geq 0.5$ should be used to avoid collapse, with optimal values in $[5, 10]$ providing the best accuracy. Second, EMA momentum $\tau \approx 0.996$ appears robust across all tested modalities. Third, mask ratios should be adapted to modality characteristics, with approximately 0.7 for images, 0.8 for video, and 0.7 for audio.

Several limitations are acknowledged. First, experiments are conducted on CIFAR-100 (50K images, 32×32), which is smaller than ImageNet-1K; validation on larger datasets remains for future work. Second, the ViT-Small architecture (22M parameters) may exhibit different optimization dynamics than larger models such as ViT-Base or ViT-Large. Third, energy distance computation introduces approximately 15% overhead per evaluation, which may become significant for very large-scale hyperparameter searches. Fourth, Euclidean distance is assumed in the representation space; extension to non-Euclidean metric spaces (hyperbolic, spherical) requires framework modification. Finally, while the default coefficients ($\alpha = 1.0, \beta = 0.1, \gamma = 0.5$) provide robust performance across tested scenarios, automatic coefficient selection via meta-learning would strengthen the framework.

6 Conclusion

In this paper, EB-HPO has been presented as an energy-based framework for hyperparameter optimization in Joint-Embedding Predictive Architectures. The main contributions and findings are summarized as follows.

First, with respect to theoretical foundation, a formal connection between JEPA's prediction loss and Székely's energy distance has been established through the JEPA-Energy correspondence theorem (Theorem 1). This theorem reveals that minimizing JEPA loss implicitly minimizes energy distance plus variance terms, providing principled justification for energy-based hyperparameter evaluation and explaining why pure prediction loss minimization can lead to collapse.

Second, with respect to collapse detection, the collapse indicator $C(Z)$ achieves perfect detection performance (ROC-AUC = 1.0, Precision = Recall = F1 = 1.0) using threshold $\epsilon_c = 0.5$. This is enabled by the large separability gap between collapsed ($C < 0.5$) and healthy ($C > 1.0$) configurations observed across all experimental settings.

Third, with respect to efficient optimization, collapsed configurations are completely eliminated by EB-HPO (0/50) compared to random search (3/50) and grid search (8/48), while statistically significant score improvements are achieved ($p < 0.001$). The framework maintains linear $O(k)$ computational complexity with modest overhead (0.09 s per evaluation).

Fourth, with respect to multi-modal generalization, consistent optimal hyperparameters have been discovered across I-JEPA, V-JEPA, A-JEPA, and Point-JEPA variants: EMA momentum $\tau \approx 0.996$ and variance regularization $\lambda_v = 5.0$ are robust across image, video, audio, and 3D point cloud modalities, while mask ratios require modality-specific tuning.

Several promising directions will be addressed by future work. First, scaling validation to larger datasets including ImageNet-1K (1.28M images) and Kinetics-400 (video) will be performed to verify that the theoretical insights and optimal hyperparameter ranges generalize to larger-scale training. Second, automatic coefficient tuning via meta-learning will be investigated to replace the current empirically-determined values (α, β, γ) with learned coefficients that adapt to dataset characteristics. Third, extension to non-Euclidean metric spaces including hyperbolic and spherical embeddings will be explored, which may be particularly relevant for hierarchical or directional representations. Fourth, integration with neural architecture search (NAS) for joint hyperparameter and architecture optimization will be developed, potentially discovering novel JEPA variants with improved efficiency or performance. Fifth, application to emerging JEPA extensions such as MC-JEPA for multi-modal learning and hierarchical JEPA for long-range temporal modeling will be explored.

References

1. Chen, T., Kornblith, S., Norouzi, M., Hinton, G.: A simple framework for contrastive learning of visual representations. In: ICML, pp. 1597–1607 (2020)
2. He, K., Fan, H., Wu, Y., Xie, S., Girshick, R.: Momentum contrast for unsupervised visual representation learning, pp. 9729–9738. CVPR (2020)
3. LeCun, Y.: A path towards autonomous machine intelligence. Meta AI Technical Report (2022)
4. Assran, M., Duval, Q., Misra, I., Bojanowski, P., Vincent, P., Rabbat, M., LeCun, Y., Ballas, N.: Self-supervised learning from images with a joint-embedding predictive architecture. In: CVPR, pp. 15619–15629 (2023)
5. Jing, L., Vincent, P., LeCun, Y., Tian, Y.: Understanding dimensional collapse in contrastive self-supervised learning. In: ICLR (2022)
6. Székely, G.J., Rizzo, M.L.: Energy statistics: a class of statistics based on distances. J. Stat. Plan. Inference **143**(8), 1249–1272 (2013)
7. LeCun, Y., Chopra, S., Hadsell, R., Ranzato, M., Huang, F.J.: A tutorial on energy-based learning. In: Bakir, G., Hofmann, T., Schölkopf, B., Smola, A., Taskar, B. (eds.) Predicting Structured Data, pp. 1–59. MIT Press (2006)
8. Grill, J.B.: Bootstrap your own latent: A new approach to self-supervised learning. NeurIPS (2020)
9. Bardes, A., Ponce, J., LeCun, Y.: VICReg: variance-invariance-covariance regularization for self-supervised learning. In: ICLR (2022)
10. Bardes, A., et al.: Revisiting feature prediction for learning visual representations from video. arXiv:2404.08471 (2024)
11. Fei, Z., Fan, M., Huang, J.: A-JEPA: Joint-embedding predictive architecture for self-supervised audio representation learning arXiv:2311.15830 (2024)
12. Zeid, K.: Point-JEPA: A joint embedding predictive architecture for self-supervised learning on point clouds arXiv:2404.16432 (2024)
13. Székely, G.J., Rizzo, M.L., Bakirov, N.K.: Measuring and testing dependence by correlation of distances. Ann. Stat. **35**(6), 2769–2794 (2007)
14. Bergstra, J., Bengio, Y.: Random search for hyper-parameter optimization. JMLR **13**, 281–305 (2012)
15. Jaderberg, M.: Population based training of neural networks arXiv:1711.09846 (2017)
16. Krizhevsky, A.: Learning multiple layers of features from tiny images, University of Toronto (2009). Master's thesis
17. Fawcett, T.: An introduction to ROC analysis. Pattern Recognit. Lett. **27**(8), 861–874 (2006)

Self-supervised Academic Trajectory Prediction on StarkNet Layer 2 Blockchain Using Joint Embedding Predictive Architecture

Nhut Minh Hua[1], Nghia Quoc Phan[1], An Cong Tran[2,3], and Hiep Xuan Huynh[2,3](✉)

[1] Tra Vinh University (TVU), Tra Vinh, Vinh Long, Vietnam
{hmnhut,nghiavnt}@tvu.edu.vn
[2] Can Tho University (CTU), Can Tho City, Vietnam
{tcan,hxhiep}@ctu.edu.vn
[3] CTU Leading Research Team on Automation, Artificial Intelligence, Information Technology and Digital Transformation (CTU-AIMED), Prague, Czech Republic

Abstract. Predicting student academic trajectories is essential for early intervention in higher education, yet supervised approaches are hindered by scarce labeled outcome data. This paper proposes a novel framework integrating Joint Embedding Predictive Architecture (JEPA) with StarkNet Layer 2 blockchain for self-supervised trajectory prediction and privacy-preserving record management. JEPA employs a Transformer-based context encoder, an exponential moving average target encoder, and a multi-horizon predictor to learn latent representations of academic histories without extensive labeled data. Learned representations are fine-tuned for three downstream tasks: graduation prediction, dropout risk assessment, and course recommendation. StarkNet provides immutable, low-cost record storage and verification via STARK zero-knowledge proofs. Simulation experiments on 50,000 synthetic student records achieve 92.8% graduation prediction accuracy using only 10% labeled data; real-world experiments on 91,736 StarkNet mainnet transactions confirm cost reduction from $12.50 (Ethereum L1) to $0.000114 (StarkNet L2), over 99.99%, with anomaly detection achieving AUROC of 0.847.

Keywords: JEPA · Self-Supervised Learning · Academic Trajectory Prediction · StarkNet · Layer 2 Blockchain · Education Management

1 Introduction

Accurately predicting student academic trajectories enables early identification of at-risk students, personalized advising, and evidence-based resource allocation [5]. However, three challenges impede effective model development: labeled outcome data is scarce, as graduation status is only available 4–6 years after

Z. Zhang (Ed.): ISNN 2026, LNCS 16848, pp. 327–339, 2027.
https://doi.org/10.1007/978-981-92-5870-3_27

enrollment; student data is sensitive and subject to strict privacy regulations; and academic records remain poorly interoperable across institutions.

Blockchain technology has been explored to address security and interoperability [7,15], yet Layer 1 deployments incur prohibitive costs. StarkNet, a Layer 2 validity rollup utilizing STARK zero-knowledge proofs [4], reduces costs by orders of magnitude while providing post-quantum security [16]. To tackle data scarcity, JEPA [9] is adopted, predicting abstract representations in latent space—well-suited for academic data where high-level progression patterns are more informative than raw course-grade details.

This paper presents four contributions: (1) a novel JEPA adaptation integrating a Transformer context encoder, EMA target encoder, and multi-horizon predictor; (2) a system architecture combining JEPA with StarkNet for secure, cost-efficient record management; (3) dual evaluation on 50,000 synthetic student records and 91,736 real StarkNet transactions; and (4) analysis of learned representations across prediction horizons and fine-tuning regimes. The remainder is organized as follows: Sect. 2 reviews related work; Sect. 3 presents theoretical foundations; Sect. 4 describes the architecture; Sect. 5 reports experiments; Sect. 6 concludes.

2 Related Work

Early student performance prediction employed logistic regression and decision trees on hand-crafted features [12], but feature engineering is labor-intensive. LSTM-based models such as DKT [11] improved trajectory modeling by capturing knowledge state transitions, and SAKT [10] further leveraged self-attention for long-range dependencies. However, all these approaches rely on fully supervised training, a significant limitation where labeled outcomes are delayed and costly. For self-supervised learning, contrastive methods such as SimCLR [6] require careful augmentation design, while JEPA [9] predicts in representation space to focus on high-level semantics. I-JEPA [2] and V-JEPA [3] extended this paradigm to image and video understanding, with EMA-updated target encoders [8] ensuring training stability; its application to educational trajectory modeling remains unexplored. On the blockchain side, Blockcerts [14] and EduCTX [17] established credential standards on Ethereum, while Layer 2 cost reduction [13] and zero-knowledge proofs for educational privacy [1] have been studied. StarkNet's validity rollup architecture has not been previously applied to intelligent education management.

3 Theoretical Foundations

3.1 Joint Embedding Predictive Architecture

The JEPA framework [9] is founded on the principle that intelligent systems should learn to predict in abstract representation spaces rather than in raw observation spaces. This design is motivated by the observation that many aspects of raw data are unpredictable (e.g., exact pixel values, specific noise patterns) and that useful prediction should focus on high-level semantic content.

Formally, given an input sample x, a context portion x_{ctx} and a target portion x_{tgt} are extracted. The JEPA architecture consists of three components. A **context encoder** f_θ is employed to map the context to a representation: $h_{\text{ctx}} = f_\theta(x_{\text{ctx}})$. A **target encoder** $\bar{f}_{\bar{\theta}}$ is used to map the target to a representation: $\bar{h}_{\text{tgt}} = \bar{f}_{\bar{\theta}}(x_{\text{tgt}})$, with its parameters updated via exponential moving average: $\bar{\theta} \leftarrow \tau\bar{\theta} + (1-\tau)\theta$. A **predictor** g_ψ is then utilized to predict the target representation from the context: $\hat{h}_{\text{tgt}} = g_\psi(h_{\text{ctx}}, \delta)$, where δ specifies the prediction target.

The training objective minimizes:

$$\mathcal{L}_{\text{JEPA}} = \left\| \hat{h}_{\text{tgt}} - \text{sg}(\bar{h}_{\text{tgt}}) \right\|_2^2 \tag{1}$$

where sg(·) denotes the stop-gradient operation, which prevents gradient flow through the target encoder and is essential for avoiding representation collapse.

3.2 JEPA Adaptation for Academic Trajectories

In this work, JEPA is adapted for sequential academic data. A student's academic trajectory is represented as a sequence of semester-level state vectors $\mathbf{Z} = \{z_1, z_2, \ldots, z_T\}$, where each z_t encodes the academic state at semester t:

$$z_t = \text{MLP}_{\text{proj}}\Big(\bigoplus_{i=1}^{N_t} \big[e_{\text{course}}(c_{t,i}) \oplus e_{\text{grade}}(g_{t,i}) \oplus e_{\text{credit}}(r_{t,i})\big] \oplus e_{\text{sem}}(t) \oplus e_{\text{gpa}}(\bar{g}_t)\Big) \tag{2}$$

where $c_{t,i}$, $g_{t,i}$, $r_{t,i}$ denote the course identifier, grade, and credit value of the i-th course in semester t; $\bar{g}_t$ is the cumulative GPA at semester t; e_{course}, e_{grade}, e_{credit}, e_{sem}, and e_{gpa} are learned embedding functions; and $\oplus$ denotes concatenation.

The adapted JEPA architecture for academic trajectories is illustrated in Fig. 1 and consists of three components:

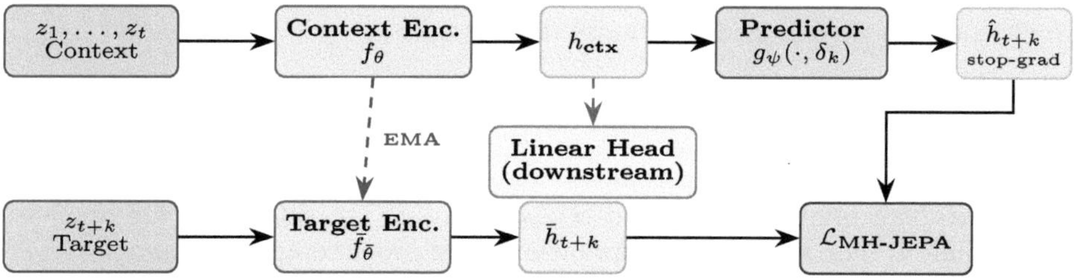

Fig. 1. Computation graph of the proposed multi-horizon JEPA for academic trajectory prediction. Solid arrows denote forward computation; dashed blue arrow denotes EMA parameter update; dashed black arrow denotes downstream fine-tuning. (Color figure online)

The first component is the context encoder, which is implemented as a Transformer that processes the observed trajectory. The hidden representations are computed as:

$$\mathbf{H}_{\text{ctx}} = \text{Transformer}_\theta(z_1 + p_1, \ldots, z_t + p_t) \tag{3}$$

where p_i denotes learned positional embeddings. The final context representation is then obtained by attentive pooling, in which a learned attention weight w is applied across all positions:

$$h_{\text{ctx}} = \sum_{i=1}^{t} \alpha_i \cdot \mathbf{H}_{\text{ctx}}^{(i)}, \quad \alpha_i = \frac{\exp(w^\top \mathbf{H}_{\text{ctx}}^{(i)})}{\sum_j \exp(w^\top \mathbf{H}_{\text{ctx}}^{(j)})} \tag{4}$$

The second component is the target encoder, for which an architecture identical to the context encoder is employed. However, the parameters $\bar{\theta}$ are not optimized directly through gradient descent; instead, they are updated via exponential moving average (EMA) of the context encoder parameters:

$$\bar{h}_{t+k} = \bar{f}_{\bar{\theta}}(z_{t+k}), \quad \bar{\theta} \leftarrow \tau\bar{\theta} + (1 - \tau)\theta \tag{5}$$

The momentum schedule is designed to follow a cosine ramp from τ_{start} to τ_{end}:

$$\tau_n = \tau_{\text{end}} - (\tau_{\text{end}} - \tau_{\text{start}}) \cdot \left(\cos\left(\frac{\pi n}{N}\right) + 1\right) / 2 \tag{6}$$

The third component is the multi-horizon predictor, which is implemented as a lightweight MLP through which target representations for multiple future horizons are predicted simultaneously:

$$\hat{h}_{t+k} = g_\psi(h_{\text{ctx}}, \delta_k), \quad k = 1, \ldots, K \tag{7}$$

where δ_k is a learnable horizon embedding. The multi-horizon training objective is formulated as:

$$\mathcal{L}_{\text{MH-JEPA}} = \frac{1}{K} \sum_{k=1}^{K} \lambda_k \cdot \left\| \hat{h}_{t+k} - \text{sg}(\bar{h}_{t+k}) \right\|_2^2 \tag{8}$$

where $\lambda_k = 1/k$ is applied as a horizon-decaying weight so that near-future predictions are emphasized over more distant ones.

3.3 StarkNet Layer 2 Architecture

StarkNet operates as a permissionless validity rollup on Ethereum, where transactions are executed off-chain and STARK proofs are submitted for on-chain verification [4]. For a batch of n operations, the proof complexity is:

$$T_{\text{prove}} = O(n \log^2 n), \quad T_{\text{verify}} = O(\log^2 n), \quad |\pi| = O(\log^2 n) \tag{9}$$

The polylogarithmic verification cost enables the amortization of on-chain costs across large batches of educational transactions, and the zero-knowledge property allows credential verification without revealing sensitive student data.

4 Proposed System Architecture

4.1 System Overview

The proposed system comprises three layers, as illustrated in Fig. 2: (1) the **Blockchain Layer**, in which StarkNet smart contracts manage credential storage and verification; (2) the **Intelligence Layer**, in which the JEPA-based trajectory prediction module operates; and (3) the **Application Layer**, through which users interact with the system.

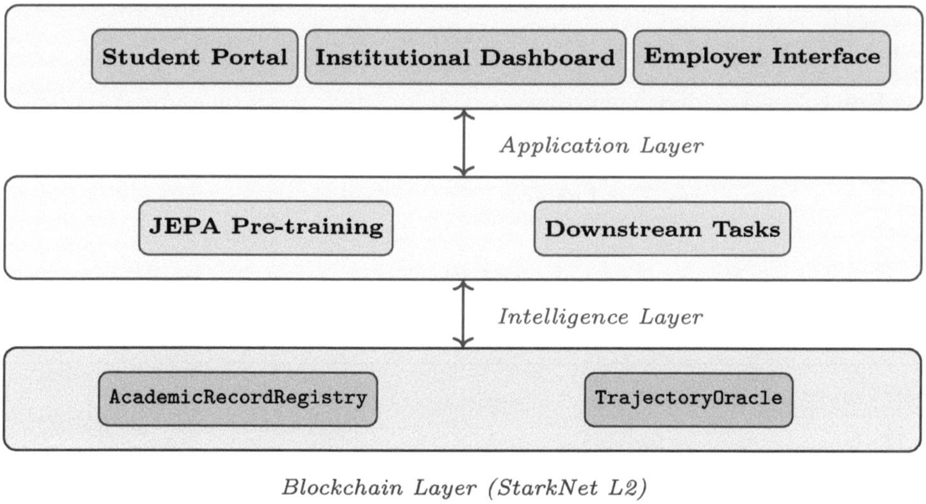

Fig. 2. Three-layer system architecture integrating JEPA-based trajectory prediction with StarkNet Layer 2 blockchain.

4.2 StarkNet Smart Contract Design

Two core smart contracts are implemented in Cairo. The first, designated as AcademicRecordRegistry, is responsible for storing academic records using a Merkle tree structure. Semester-level data is maintained off-chain through IPFS, while Merkle roots are anchored on-chain to ensure data integrity. Functions are provided for submitting semester records, verifying record inclusion through Merkle proofs, and retrieving complete student histories.

The second contract, designated as TrajectoryOracleContract, serves as the interface through which the JEPA prediction module is accessed. When a prediction request is submitted on-chain, off-chain inference is triggered by the Intelligence Layer. The resulting predictions—including graduation probability, dropout risk score, and recommended courses—are subsequently returned and recorded as on-chain attestations, thereby ensuring that prediction results are both verifiable and tamper-proof.

4.3 JEPA Trajectory Prediction Pipeline

The prediction pipeline is formalized in Algorithm 1.

Algorithm 1. JEPA Academic Trajectory Prediction Pipeline

Require: Student record sequence $\{r_1, \ldots, r_t\}$, prediction horizons K
Ensure: Graduation probability p_g, dropout risk p_d, course recommendations $\mathcal{R}$
Phase 1: Record Retrieval and Embedding
for $i = 1$ **to** t **do**
  Retrieve verified record r_i from StarkNet (with Merkle proof)
  Construct state vector z_i via Eq. 2
end for
Phase 2: Context Encoding
$h_{\text{ctx}} \leftarrow f_\theta(\{z_1, \ldots, z_t\})$ via Eqs. 3–4
Phase 3: Multi-Horizon Prediction
for $k = 1$ **to** K **do**
  $\hat{h}_{t+k} \leftarrow g_\psi(h_{\text{ctx}}, \delta_k)$ via Eq. 7
end for
Phase 4: Downstream Task Inference
$p_g \leftarrow \sigma(w_g^\top h_{\text{ctx}} + b_g)$ {graduation prediction}
$p_d \leftarrow \sigma(w_d^\top h_{\text{ctx}} + b_d)$ {dropout risk}
$\mathcal{R} \leftarrow \text{TopK}(\{s_c = w_c^\top \hat{h}_{t+1}\}_{c \in \mathcal{C}_{\text{avail}}})$ {course ranking}
Phase 5: Result Attestation
Submit $(p_g, p_d, \mathcal{R})$ to TrajectoryOracleContract on StarkNet
return $p_g, p_d, \mathcal{R}$

4.4 Pre-training Strategy

During pre-training, for each student trajectory of length T, a context window $[1, t]$ is randomly sampled with $t \sim \text{Uniform}(2, T-K)$. The model is trained to predict representations of subsequent states $z_{t+1}, \ldots, z_{t+K}$ using the multi-horizon loss (Eq. 8). Data augmentation is applied through: (1) random context window truncation, (2) semester dropout (randomly masking 10% of courses within a semester), and (3) grade noise injection ($\mathcal{N}(0, 0.1)$ added to normalized grades).

4.5 Downstream Task Adaptation

After pre-training, the context encoder parameters are frozen and linear heads are trained for each downstream task. For graduation prediction, a binary classifier is trained as $p_g = \sigma(w_g^\top h_{\text{ctx}} + b_g)$, and the model is optimized using binary cross-entropy loss. For dropout risk assessment, a regression head is trained as $p_d = \sigma(w_d^\top h_{\text{ctx}} + b_d)$, through which per-semester dropout probability is predicted, and the model is optimized using mean squared error loss. For course recommendation, each candidate course c is scored as $s_c = w_c^\top \hat{h}_{t+1}$, after which courses are ranked by their predicted scores. This ranking model is trained with a listwise ranking loss (ListMLE) to ensure that the ordering of recommended courses is optimized holistically rather than independently.

5 Experiments

5.1 Dataset and Implementation Details

A simulated dataset of 50,000 synthetic student profiles across 10 academic programs (up to 8 semesters, 500 courses) was constructed, calibrated to Vietnamese university statistics (graduation rate 72%, dropout rate 18%). The dataset was partitioned into pre-training (40,000, unlabeled), fine-tuning (5,000, labeled), and test (5,000, labeled) sets; validation on real academic records is planned as future work.

The context encoder was a 6-layer Transformer (hidden dim 384, 6 heads, pre-norm, GELU) with a 3-layer MLP predictor [384, 256, 384] and cosine-scheduled EMA momentum ($\tau_{\text{start}} = 0.996$, $\tau_{\text{end}} = 0.9999$). Pre-training used AdamW (lr 5×10^{-4}, weight decay 10^{-2}), batch size 512, $K=3$ horizons, 300 epochs on an NVIDIA A100 (40 GB). Seven baselines were evaluated: Logistic Regression, Random Forest, Supervised LSTM, Supervised Transformer (100% labels), and SimCLR, MAE, BYOL with linear probing (10% labels).

Table 1. Graduation prediction performance. "Labels" indicates labeled data percentage used for training/fine-tuning.

Method	Acc. (%)	F1	AUC	MCC	Labels
Logistic Regression	79.3	0.771	0.842	0.573	100%
Random Forest	83.6	0.818	0.889	0.662	100%
Supervised LSTM	88.5	0.864	0.921	0.762	100%
Supervised Transformer	91.2	0.893	0.944	0.815	100%
SimCLR + Linear	87.3	0.851	0.903	0.738	10%
MAE + Linear	89.1	0.872	0.925	0.773	10%
BYOL + Linear	88.4	0.863	0.917	0.759	10%
JEPA + Linear (Ours)	**92.8**	**0.912**	**0.956**	**0.849**	**10%**

5.2 Graduation Prediction and Multi-task Results

Results for the primary graduation prediction task are presented in Table 1. The proposed JEPA approach achieves the highest performance across all metrics with only 10% labeled data, surpassing the fully supervised Transformer by 1.6 points in accuracy and 0.019 in F1-score. Among self-supervised baselines, JEPA outperforms SimCLR by 5.5 points, MAE by 3.7 points, and BYOL by 4.4 points, suggesting that prediction in representation space is more effective than contrastive objectives for sequential academic data.

Performance on all three downstream tasks is summarized in Table 2. Consistent improvements are observed across all tasks, confirming that JEPA-learned representations capture transferable academic patterns. The improvement is most pronounced in course recommendation (+0.028 NDCG@10), attributed to the multi-horizon prediction mechanism that explicitly models future academic states.

Table 2. Multi-task evaluation: JEPA (10% labels) vs. Supervised Transformer (100% labels).

Task	Metric	Sup. Trans. (100%)	JEPA (10%)
Graduation Pred.	Accuracy	91.2%	**92.8%**
	F1-Score	0.893	**0.912**
Dropout Risk	AUROC	0.887	**0.901**
	MSE	0.041	**0.037**
Course Recomm.	NDCG@10	0.723	**0.751**
	Hit Rate@5	0.645	**0.672**

5.3 Effect of Labeled Data Ratio and Prediction Horizon

The robustness of JEPA to varying labeled data is evaluated in Table 3. The advantage is most pronounced in low-label regimes: at 1% labeled data, JEPA achieves 85.7% accuracy compared to 72.1% for the supervised Transformer (a 13.6-point improvement). Even at 100% labels, JEPA pre-training provides a 2.4-point boost over training from scratch.

Table 3. Graduation prediction accuracy (%) at different labeled data ratios.

Method	1%	5%	10%	100%
Supervised Transformer	72.1	83.5	87.9	91.2
SimCLR + Linear	80.2	84.8	87.3	90.5
MAE + Linear	82.4	86.1	89.1	91.8
JEPA + Linear	**85.7**	**89.4**	**92.8**	**93.6**

The impact of prediction horizon K is analyzed in Table 4. Performance improves from K=1 to K=3 and slightly decreases at K=4, suggesting that multi-horizon prediction provides a richer training signal but excessively long horizons introduce noise. The optimal K=3 (1.5 academic years) aligns with the observation that medium-term academic trends are most informative for predicting final outcomes.

Table 4. Effect of prediction horizon K on graduation prediction accuracy (%).

Prediction Horizon	$K = 1$	$K = 2$	$K = 3$	$K = 4$
Graduation Accuracy (%)	90.4	91.9	**92.8**	92.5
Dropout AUROC	0.881	0.894	**0.901**	0.898
Course NDCG@10	0.712	0.738	**0.751**	0.746

5.4 Representation Analysis

Representation quality is assessed via linear probing on frozen JEPA features compared against supervised Transformer representations (Table 5). JEPA consistently outperforms supervised features across all probing tasks, with particularly notable improvements in major classification (+1.9%) and academic standing (+2.6%), suggesting that the self-supervised objective learns more general and transferable representations that capture program-specific progression patterns.

Table 5. Linear probing accuracy (%) of frozen representations on auxiliary tasks.

Probe Task	Sup. Transformer	JEPA
Major Classification (10 classes)	94.2	**96.1**
Semester GPA Prediction (MSE)	0.089	**0.071**
Credit Completion Ratio	0.932	**0.951**
Academic Standing (3 classes)	88.7	**91.3**

5.5 Real-World StarkNet Transaction Dataset

To validate cost-efficiency claims, 91,736 transactions were collected from the StarkNet mainnet across 386 unique blocks. The dataset comprises two subsets: **Dataset 1** (69,651 transactions, blocks 17,971–18,256, January 2023, early-stage network) and **Dataset 2** (22,085 transactions, blocks 60,500–60,599, May 2023, more mature network with higher fees). Each record contains 42 features across five categories: *transaction-level* (gas price, fee, calldata size, etc.), *sender behavior* (nonce, transaction count, account age, etc.), *block-level* (gas used, fullness, transactions per block), *network-level* (congestion, sequencer load, L1 gas price, etc.), and *temporal* (timestamp, cyclical encodings). Three transaction types are represented: type 0 (regular, 90.3%), type 1 (legacy, <0.1%), and type 2 (contract deployment, 9.7%). Fee distribution and cost comparisons are presented in Tables 6 and 7.

The empirical mean transaction cost on StarkNet is \$0.000114, representing a 99.999% reduction compared to Ethereum L1, with an estimated annual cost of \$1.14 for 10,000 university operations. A temporal fee variation between Dataset 1 (mean \$0.000095) and Dataset 2 (mean \$0.000174) suggests that cost projections should account for network evolution. It is noted that ML results are based on simulated academic data; validation on real records is planned as future work.

Table 6. Empirical StarkNet transaction fee distribution from 91,736 real mainnet transactions.

Statistic	Dataset 1 (69,651 tx)	Dataset 2 (22,085 tx)	Combined (91,736 tx)
Mean fee (STRK)	0.000190	0.000348	0.000228
Median fee (STRK)	0.000164	0.000361	0.000186
P5/P95 (STRK)	0.000021/0.000433	0.000135/0.000615	0.000026/0.000542
Mean fee (USD)	\$0.000095	\$0.000174	\$0.000114
Avg tx/block	243.5	220.8	237.7

Table 7. Blockchain cost comparison for educational operations (USD). StarkNet costs are derived from 91,736 real mainnet transactions at STRK = \$0.50.

Operation	Ethereum L1	Polygon	StarkNet L2
Record Submission	12.50	0.020	**0.000114**
Batch Verify (100)	45.00	0.150	**0.0114**
Prediction Attestation	8.30	0.010	**0.000114**
Annual (10K ops)	65,800	18.00	**1.14**

5.6 Pre-training Convergence Analysis

The convergence behavior of JEPA pre-training is analyzed. The prediction loss decreases rapidly during the first 50 epochs, after which convergence slows. Downstream accuracy follows a delayed pattern, with most improvements between epochs 50 and 200. When the pre-training corpus is reduced from 40,000 to 20,000 and 10,000 students, graduation accuracy decreases from 92.8% to 91.5% and 89.8%, respectively—yet even with 10,000 unlabeled students, JEPA outperforms supervised LSTM (88.5%).

5.7 Comparison with Knowledge Tracing Models

Deep Knowledge Tracing (DKT) [11] and Self-Attentive Knowledge Tracing (SAKT) [10] are adapted for the graduation prediction task by treating semester-level records as interaction sequences. As shown in Table 8, JEPA with 10% labels outperforms both models despite their full supervision, noting that DKT and SAKT were originally designed for fine-grained skill-level prediction and their adaptation may not fully exploit their strengths. Nevertheless, the comparison demonstrates the competitive advantage of JEPA's self-supervised pre-training in limited-label settings.

Table 8. Comparison with knowledge tracing models on graduation prediction.

Method	Acc. (%)	F1	AUC	Labels
DKT [11]	87.2	0.851	0.912	100%
SAKT [10]	89.8	0.878	0.932	100%
Supervised Transformer	91.2	0.893	0.944	100%
JEPA + Linear (Ours)	**92.8**	**0.912**	**0.956**	**10%**

5.8 Discussion

The simulation study suggests that JEPA's prediction-in-representation-space paradigm is promising for academic trajectory modeling, with the multi-horizon mechanism encouraging hierarchical representations. The simulated advantage over supervised methods is most pronounced in low-label regimes. The real StarkNet data validates that transaction costs are substantially lower than projected, with an empirical mean fee of $0.000114 per operation. Additional real-world experiments on 91,736 StarkNet transactions demonstrate that unsupervised anomaly detection methods achieve AUROC up to 0.847 (LOF) on the 34-feature transaction dataset, confirming that the blockchain infrastructure supports meaningful anomaly analysis. Limitations include the use of simulated academic data for JEPA evaluation, restriction to one educational system, the cold-start problem for first-semester students, and the need for real academic trajectory data to validate the simulation results.

6 Conclusion

This paper has presented a novel framework integrating JEPA for self-supervised academic trajectory prediction with StarkNet Layer 2 blockchain for secure, cost-efficient record management. A multi-horizon prediction objective with horizon-decaying weights was designed and evaluated on three downstream tasks: graduation prediction, dropout risk assessment, and course recommendation. Simulation results demonstrate strong graduation prediction with limited labeled data, while real-world experiments on 91,736 StarkNet transactions confirm transaction costs reduced from $12.50 (Ethereum L1) to $0.000114 (StarkNet L2), a reduction of over 99.99%, and unsupervised anomaly detection achieving up to 0.847 AUROC.

Future work will focus on: (1) deployment with real university data through institutional partnerships, (2) cross-institutional transfer learning, (3) federated JEPA pre-training for multi-institutional collaboration without data sharing, and (4) integration with large language models for processing unstructured academic documents.

References

1. Alammary, A., Alhazmi, S., Almasri, M., Gillani, S.: Blockchain-based applications in education: a systematic review. Appl. Sci. **9**(12), 2400 (2019)
2. Assran, M.: Self-supervised learning from images with a joint-embedding predictive architecture. In: Proceedings of the IEEE/CVF Conference on Computer Vision and Pattern Recognition (CVPR), pp. 15619–15629. IEEE (2023)
3. Bardes, A.: Revisiting feature prediction for learning visual representations from video. arXiv preprint arXiv:2404.08471 (2024)
4. Ben-Sasson, E., Bentov, I., Horesh, Y., Riabzev, M.: Scalable, transparent, and post-quantum secure computational integrity. Tech. Rep. Report 2018/046, IACR Cryptology ePrint Archive (2018)
5. Chen, G., Xu, B., Lu, M., Chen, N.: Exploring blockchain technology and its potential applications for education. Smart Learn. Environ. **5**(1), 1–10 (2018)
6. Chen, T., Kornblith, S., Norouzi, M., Hinton, G.: A simple framework for contrastive learning of visual representations. In: Proceedings of the 37th International Conference on Machine Learning (ICML), pp. 1597–1607. PMLR (2020)
7. Grech, A., Camilleri, A.: Blockchain in Education. Publications Office of the European Union, Luxembourg, Tech. rep. (2017)
8. Grill, J.: Bootstrap your own latent: a new approach to self-supervised learning. In: Advances in Neural Information Processing Systems (NeurIPS), vol. 33, pp. 21271–21284 (2020)
9. LeCun, Y.: A path towards autonomous machine intelligence. OpenReview preprint (2022)
10. Pandey, S., Karypis, G.: A self-attentive model for knowledge tracing. In: Proceedings of the 12th International Conference on Educational Data Mining (EDM), pp. 384–389 (2019)
11. Piech, C.: Deep knowledge tracing. In: Advances in Neural Information Processing Systems (NeurIPS), pp. 505–513. (2015)
12. Romero, C., Ventura, S.: Educational data mining: a review of the state of the art. IEEE Trans. Syst. Man Cybern. Part C **40**(6), 601–618 (2010)
13. Rustemi, A., Dalipi, F., Atanasovski, V., Risteski, A.: A systematic literature review on blockchain-based systems for academic certificate verification. IEEE Access **11**, 64679–64696 (2023)
14. Schmidt, J.: Credentials, reputation, and the blockchain. EDUCAUSE Rev. **52**(2), 28–29 (2017)
15. Sharples, M., Domingue, J.: The blockchain and kudos: a distributed system for educational record, reputation and reward. In: Proceedings of the 11th European Conference on Technology Enhanced Learning (EC-TEL 2016). Lecture Notes in Computer Science, vol. 9891, pp. 490–496. Springer (2016)

16. Thibault, L., Sarry, T., Hafid, A.: Blockchain scaling using rollups: a comprehensive survey. IEEE Access **10**, 93039–93054 (2022)
17. Turkanovíc, M., et al.: EduCTX: a blockchain-based higher education credit platform. IEEE Access **6**, 5112–5127 (2018)

Semantic Factorization for 3D Shape Generation

Yanran Guan[1(✉)], Bining Long[2], and Oliver van Kaick[1]

[1] School of Computer Science, Carleton University, Ottawa, Canada
{yanran.guan,oliver.vankaick}@carleton.ca
[2] School of Computing, Queen's University, Kingston, Canada
bining.long@queensu.ca

Abstract. We demonstrate that semantic directions in the latent space of generative adversarial networks (GANs) for 3D shape generation can be identified and manipulated, enabling user-driven exploration of the generated shapes. We extend the semantic factorization approach from 2D images to 3D shapes by applying it to a latent GAN whose outputs are decoded into implicit fields of 3D shapes. Specifically, we perform semantic factorization on the first hidden layer of the GAN's generator, formulating it as an affine transformation and extracting the most significant latent directions via eigendecomposition. Manipulating the latent space along these directions enables controllable variations in shape generation in an unsupervised manner. Experiments show that our method produces 3D shapes with higher diversity and fidelity than those produced by randomly sampling the latent space of a GAN, while providing consistent and interpretable changes along meaningful semantic directions.

Keywords: Generative adversarial networks · Semantic factorization · Shape generation

1 Introduction

The latent space of generative adversarial networks (GANs) [9] offers rich potential for generating diverse and realistic 2D images and 3D shapes. However, enabling effective user control during the generative process remains an open challenge, particularly for 3D shapes, where geometric complexity and high dimensionality make direct manipulation difficult. To tackle this, prior work using GANs for 3D shape generation has leveraged external input signals, such as text descriptions [4] or 2D views [39], as well as structural cues, including sphere-guided generation [17] and example-based supervision [35]. These methods typically require additional mechanisms to steer shape generation, including conditional inputs, optimization-based GAN inversion, or guidance derived from geometric priors or reference shapes.

On the other hand, previous studies have demonstrated the possibility of directly finding interpretable directions in the latent space of a GAN [23,28].

Z. Zhang (Ed.): ISNN 2026, LNCS 16848, pp. 340–351, 2027.
https://doi.org/10.1007/978-981-92-5870-3_28

This insight leads to a new paradigm of 3D shape generation, enabling controllable, attribute-specific manipulation in the generative process. Moreover, such approaches facilitate a shift from supervised learning to unsupervised methods, alleviating the need for annotated data.

In this paper, building upon the 3D shape generation pipeline of Chen and Zhang [5], which is based on an implicit decoder and a latent GAN, and upon the semantic factorization approach for controlled 2D image generation of Shen and Zhou [23], we introduce a method that enables the extraction of semantic editing directions from GANs for 3D shape generation. We focus our work on GANs since they are known to learn explicit latent spaces in which neighboring latent vectors typically generate similar outputs [13,14,21]. In contrast, recent generative approaches such as diffusion models [11,25] do not provide a single, compact latent space that is directly amenable to semantic traversal or manipulation, making the discovery of meaningful latent directions less straightforward.

2 Related Work

We review the prior studies related to this work from two perspectives: controllable shape generation based on GANs and unsupervised semantic discovery in the latent space of GANs.

2.1 Controllable Shape Generation Using GANs

Since the pioneering work of Wu et al. [34], GANs have become a popular approach for generating 3D shapes. To enhance the controllability of shape generation using GANs, various approaches have been introduced.

One line of approaches uses external input signals to guide shape generation. Chen et al. [4] learn a joint embedding of text and shapes and use a conditional GAN conditioned on the text embedding to generate colored volumetric shapes reflecting the text semantics. Arshad and Beksi [2] leverage a progressive conditional GAN to generate colored 3D point clouds according to object category labels. Zheng et al. [39] showcase the generation of 3D shapes represented as signed distance fields from 2D images through GAN inversion.

Another line of approaches leverages structural information of 3D shapes to guide shape generation. Wang et al. [31] propose a global-to-local GAN that first generates a global structure and then progressively refines local geometric details, enabling structure-aware shape generation. Li et al. [16] introduce a part-based GAN that generates individual shape parts, which are assembled into coherent shapes, capturing both global structure and part-level details. Li et al. [17] use a sphere as a prior within a GAN framework to enable controllability for part-aware shape generation and manipulation. Wu and Zheng [35] show that a GAN can be trained using a single example shape to generate novel shapes that preserve its structural characteristics.

All the aforementioned methods require additional mechanisms beyond the shape generation process of a GAN. Wang et al. [32] instead manipulate the latent space of GANs to achieve controllable shape generation, using local linear subspace models to discover semantic directions from the latent space in an unsupervised manner. Our method similarly operates in the latent space but identifies semantic directions directly from the first projection step of a GAN through semantic factorization.

2.2 Unsupervised Semantic Discovery in GAN Latent Spaces

A wide range of studies have shown that the latent space of GANs encodes rich semantic information [7,8,13,14,19–22,26,33,36,37]. Therefore, various unsupervised methods have been developed to discover interpretable semantic directions in the latent space.

Voynov and Babenko [30] jointly learn a candidate matrix and a classifier to identify interpretable directions. Härkönen et al. [10] compute principal components in feature maps to find dominant directions. Spingarn et al. [28] compute steering trajectories directly from the generator's weights. Shen and Zhou [23] factorize the latent space using a closed-form solution to extract semantic directions. Yüksel et al. [38] apply contrastive learning for disentangled directions. Tzelepis et al. [29] learn nonlinear warping functions on the latent space. Zhu et al. [40] extend unsupervised global semantic factorization by factorizing latent semantics into region-based directions. Hu et al. [12] embed subspaces into the generator's layers to uncover layer-wise interpretable directions. Simsar et al. [24] use a submodular framework to discover diverse directions. Choi et al. [6] compute global semantic directions using the Fréchet mean. Song et al. [27] employ Householder projections to reveal latent semantics.

Our method is based on the semantic factorization approach proposed by Shen and Zhou [23], which does not require any form of sampling, training, or optimization.

3 Methodology

Our approach involves an implicit decoder that reconstructs implicit fields of 3D shapes from shape codes, a latent GAN that generates meaningful shape codes from random noise, and the semantic factorization process.

3.1 Implicit Decoder

The implicit decoder is trained as part of an autoencoder (AE), where a volumetric encoder maps input shapes, represented as voxel grids, to shape codes, and the implicit decoder decodes the shape codes into implicit fields. We adopt the model design introduced by Chen and Zhang [5], where the decoder learns to reconstruct the shape code into an implicit field $\mathcal{F}$, with $\mathcal{F}(\mathbf{p}) = 0$ representing the query point $\mathbf{p}$ outside the shape and $\mathcal{F}(\mathbf{p}) = 1$ representing $\mathbf{p}$ inside the shape.

Network Architecture. The encoder downsamples the input from $64 \times 64 \times 64$ voxels to $4 \times 4 \times 4$ through 5 convolutional layers with each followed by a batch normalization layer. It then generates the shape code $\mathbf{z} \in \mathbb{R}^n$ using a fully connected (FC) layer activated by a sigmoid function. In our implementation, specifically, we employ a shape code space with dimensionality $n = 128$. The input vector to the implicit decoder is composed of the shape code $\mathbf{z}$ concatenated with a 3D query point $\mathbf{p}$. The decoder network follows a feed-forward architecture, comprising 5 FC hidden layers corresponding to 5 levels of downsampling, each layer having 2048, 1024, 512, 256, and 128 units. The first 4 hidden layers are skip-connected with the network input and all 5 hidden layers use a leaky rectified linear unit as the activation function. The output layer, also an FC layer, produces a single occupancy value using a sigmoid activation function.

Loss Function. We denote the encoder as $E(\cdot)$. Given a volumetric shape $\mathcal{S}$ as input, the encoder maps it to the shape code, i.e., $\mathbf{z} = E(\mathcal{S})$. The implicit decoder then aims to find a parameterization $f(\mathbf{z}, \mathbf{p})$ that maps each query point $\mathbf{p}$ in a set of points $\mathcal{P}$ uniformly sampled from $\mathcal{S}$ to the occupancy value $\mathcal{F}(\mathbf{p})$. Therefore, the training objective of the AE for implicit shape decoding is to minimize the mean squared error between the real occupancy $\mathcal{F}(\mathbf{p})$ and the predicted value $f(\mathbf{z}, \mathbf{p})$, i.e.,

$$\mathcal{L}_{\text{IM-AE}} = \frac{1}{|\mathcal{P}|} \sum_{\mathbf{p} \in \mathcal{P}} \|\mathcal{F}(\mathbf{p}) - f(\mathbf{z}, \mathbf{p})\|^2 . \tag{1}$$

3.2 Latent GAN

The latent GAN consists of two components: a generator network and a discriminator network. Through adversarial training, the generator learns to generate synthetic yet plausible shape codes from random Gaussian noise inputs. Simultaneously, the discriminator learns to distinguish the real shape codes from the synthetic ones by penalizing the generator for producing implausible results. The aim of our latent GAN is to generate meaningful shape codes from random noise, which can then be used by the implicit decoder described in Sect. 3.1 to generate shapes represented as implicit fields.

Network Architecture. Both the generator and discriminator are simple feed-forward networks, each comprising 3 FC layers: a first hidden layer, a second hidden layer, and an output layer, with the hidden layers each having 2048 units. The generator takes a noise vector $\mathbf{n} \in \mathbb{R}^d$ randomly drawn from $\mathcal{N}(0, \mathbf{I})$ (we use $d = 128$) and generates a fake shape code, with the output layer activated by a sigmoid function. The discriminator takes a shape code $\mathbf{z}$ as input and outputs a logit value describing the likelihood that the input vector is a real one.

Loss Function. We denote the generator and the discriminator as $G(\cdot)$ and $D(\cdot)$. The loss function of a latent GAN is written as:

$$\mathcal{L}_{\text{L-GAN}} = \log D(\mathbf{z}) + \log\left(1 - D(G(\mathbf{n}))\right) . \tag{2}$$

3.3 Semantic Factorization

Semantic factorization is an unsupervised approach for identifying interpretable directions in latent manifolds, originally developed for 2D images [23]. The generator of a GAN can be seen as a function that projects the noise vector $\mathbf{n}$ from a latent space to the final output through multiple steps, where each step learns a transformation from one space to another. Since the latent space of a GAN has been shown to be able to encode rich semantic knowledge, as discussed in Sect. 2.2, the first projection step, which directly acts on the latent space, is examined for semantic factorization. The effect of semantic factorization on different projection steps is further analyzed in Sect. 4.2.

The projection from the noise input $\mathbf{n}$ to the projected output $\mathbf{y}$, parameterized by the first hidden layer of the GAN's generator $G_1(\cdot)$, can be formulated as an affine transformation:

$$G_1(\mathbf{n}) = \mathbf{y} = \mathbf{A}\mathbf{n} + \mathbf{b}, \tag{3}$$

where $\mathbf{A}$ and $\mathbf{b}$ denote the weight matrix and bias vector used in this transformation step. The objective of semantic factorization is to find explanatory factors, i.e., vectors describing meaningful semantic directions, from the latent space of a GAN, and then manipulate the data generation process through the affine transformation in Eq. (3). This manipulation can be formulated as:

$$\mathbf{y}' = G_1(\mathbf{n}') = G_1(\mathbf{n} + \alpha\mathbf{v}) = \mathbf{A}\mathbf{n} + \mathbf{b} + \alpha\mathbf{A}\mathbf{v} = \mathbf{y} + \alpha\mathbf{A}\mathbf{v}, \tag{4}$$

where $\mathbf{v}$ is the direction vector and α represents the manipulation intensity.

Equation (4) indicates instance independence within the manipulation process, i.e., given any sampled noise and a direction vector, this manipulation can be consistently achieved by introducing the term $\alpha\mathbf{A}\mathbf{v}$ to the projected code after the first step. Therefore, the weight matrix $\mathbf{A}$ is expected to encapsulate rich information for identifying semantic directions, and the objective of semantic factorization is reduced to discovering important latent directions by decomposing $\mathbf{A}$.

From this perspective, we aim to find the optimized direction $\mathbf{v}^*$ that causes the largest variations after the projection of $\mathbf{A}$, i.e.,

$$\mathbf{v}^* = \underset{\{\mathbf{v}\in\mathbb{R}^d:\ \mathbf{v}^\intercal\mathbf{v}=1\}}{\arg\max} \ \|\mathbf{A}\mathbf{v}\|^2. \tag{5}$$

To solve Eq. (5) in an unsupervised manner, eigendecomposition is applied to the matrix $\mathbf{A}^\intercal\mathbf{A}$. Thus, we derive the most distinguishable semantic direction $\mathbf{v}^*$ by finding the eigenvector associated with the greatest eigenvalue.

4 Experimental Evaluation

In this section, we first explain our experimental methodology. Then, we analyze the effect of semantic factorization performed on different layers of the latent GAN. Finally, we show qualitative and quantitative evaluations of the generated shapes via semantic factorization. Implementation details of our experiments can be found in our code available at https://github.com/IsaacGuan/3D-SeFa.

4.1 Experimental Setup

In our experiments, we use the chair category from the ShapeNet dataset [3], consisting of 6778 chair shapes. Each shape is voxelized into a $64 \times 64 \times 64$ voxel grid and sampled at three levels of resolution ($16 \times 16 \times 16$, $32 \times 32 \times 32$, and $64 \times 64 \times 64$). At each level of resolution, we draw 4096, 8192, and 32 768 pairs of points and occupancy values, respectively.

Following the setting of Chen and Zhang [5], we train the AE for the implicit decoder in a progressive manner, where we first train for 250 epochs using samples drawn at the $16 \times 16 \times 16$ resolution, then for 250 epochs using samples drawn at the $32 \times 32 \times 32$ resolution, and finally for 500 epochs using samples drawn at the $64 \times 64 \times 64$ resolution. We train the latent GAN for 10 000 epochs using the shape codes encoded by the volumetric encoder. Both the implicit decoder and the latent GAN are trained using the Adam optimizer [15] with a learning rate of 5×10^{-5}.

The shape codes generated by the latent GAN are fed into the implicit decoder to generate implicit fields representing the generated shapes. We sample occupancy values from these implicit fields at a resolution of $64 \times 64 \times 64$ and convert the resulting volumetric grids into meshes using marching cubes [18] to facilitate evaluation.

4.2 Comparison of Semantic Factorization on Different Layers

We compare the results of manipulating generated shapes via semantic factorization, which is performed on the first hidden layer of the latent GAN's generator, with those obtained by performing factorization on other layers. Similar to the semantic factorization approach detailed in Sect. 3.3, we factorize the latent semantics for the second hidden layer and the output layer. This involves computing the eigendecomposition of the layers' weight matrices and utilizing the eigenvectors associated with the largest eigenvalues as direction vectors to manipulate the generation process. Particularly, we apply the direction vector to the projected code from the previous layers to derive the edited code, then we feed the edited code into the remaining layers to generate the manipulated shape code.

We visually present in Fig. 1 the generated shapes originating from identical noise vectors and subsequently manipulated along the direction vectors that provide the most distinguishable variations from the first hidden layer, the second hidden layer, and the output layer, where the directions correspond to the eigenvectors associated with the largest eigenvalues derived from the decomposition of layer weights. For each direction vector, we apply 11 intensity factors ranging uniformly between -0.5 and 0.5. We can see that the direction vectors from the hidden layers result in larger visual diversities in the generated shapes, while the direction vector from the output layer can barely introduce visual changes to the final generated results.

To better evaluate the variations in the generated shapes in Fig. 1, we measure the similarity among generated shapes along each direction using the chamfer distance. Specifically, for each direction, we compute the chamfer distances

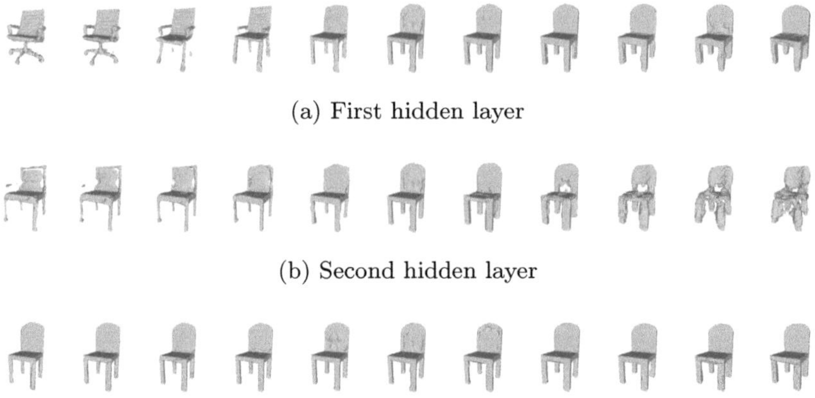

(a) First hidden layer

(b) Second hidden layer

(c) Output layer

Fig. 1. Generated shapes manipulated along direction vectors corresponding to the largest eigenvalues from different layers of the latent GAN.

Table 1. Eigenvalues associated with the manipulation directions from different layers in Fig. 1 and the corresponding average chamfer distance between adjacent shapes along each direction.

	Eigenvalue	Chamfer distance
First hidden layer	326.47	3.72×10^{-2}
Second hidden layer	177.01	3.04×10^{-2}
Output layer	5.14	2.46×10^{-2}

between adjacent shapes and report the average distance, as shown in Table 1, where we also present the eigenvalue associated with each direction. For the computation of chamfer distance, we reconstruct each generated shape as a triangle mesh and sample 2048 points on the mesh surface. The chamfer distance is then computed based on the surface point clouds. From the numerical results presented in Table 1, we can conclude that the semantic directions decomposed from the first hidden layer of the latent GAN's generator lead to the generation of shapes with the most visual diversity.

4.3 Shape Generation Results

We present visual examples of generated shapes manipulated via semantic factorization in Fig. 2, where we first generate shapes from two different noise vectors (Fig. 2a and Fig. 2b) and then manipulate the generation along direction vectors associated with the top-3 largest eigenvalues (different rows in Fig. 2a and Fig. 2b). We can see that along each semantic direction, despite the entanglement of multiple semantics, the generated shapes exhibit similar manipulation effects. For example, in the first direction, i.e., the first rows in Fig. 2a and Fig. 2b, we

(a) Example 1

(b) Example 2

Fig. 2. Examples of generated shapes manipulated via semantic factorization. Each row corresponds to a semantic direction. The eigenvalues associated with the semantic directions from top to bottom are 326.48, 304.21, and 236.99.

see that the chair legs are initially angled and offset. As we progress along this direction, these legs gradually shift to a straight and parallel alignment. In the second direction, starting from a slender and tall dining chair, the chair shape gradually becomes thicker and shorter, transforming into an armchair. In the third direction, the chair back changes from long to short; conversely, the chair legs change from short to long.

Table 2. Quantitative evaluation of shapes generated via semantic factorization and random sampling from the latent GAN, where MMD and JSD are multiplied by 10^2.

	COV	MMD	JSD
Semantic factorization	46 %	1.567	8.606
Random sampling	40 %	1.823	11.58

To quantitatively evaluate the quality of the generated shapes, we adopt coverage (COV), minimum matching distance (MMD), and Jensen–Shannon divergence (JSD) as evaluation metrics [1]. Higher COV and lower JSD indicate greater diversity, while lower MMD indicates higher fidelity. For their computation, we use chamfer distance as the difference metric, which is computed in the same way as detailed in Sect. 4.2. To achieve as much diversity as possible in the generated shapes via semantic factorization, we first sample 10 noise vectors, then, for each noise vector, we manipulate the generation using the top-5 directions with intensities ranging uniformly between −0.5 and 0.5, and finally we select the generated shapes corresponding to the two endpoints of each direction, yielding a total of 100 generated shapes. For comparison, we create another group of 100 generated shapes using shape codes randomly sampled from the latent GAN. We also randomly select 100 shapes from the training set as reference shapes. We compare the two groups of generated shapes with the reference shapes using the aforementioned evaluation metrics and summarize this quantitative comparison in Table 2. We find that semantic factorization outperforms random sampling from a GAN in generating diverse and high-fidelity results.

5 Conclusion

We extended prior advances in semantic factorization from 2D image generation to the generation of 3D shapes, bridging the gap between the interpretability of the latent space of GANs and 3D generative modeling. By factorizing the first hidden layer of a latent GAN's generator, we identify the most distinctive semantic directions, which allow users to explore and manipulate the latent space in a controlled manner. Our experiments demonstrate that leveraging these semantic directions not only enables controllable editing but also enhances the diversity and fidelity of the generated shapes. While the discovered directions correspond to meaningful shape variations, individual directions may influence multiple correlated geometric attributes, reflecting the entangled nature of 3D shape semantics. These findings highlight the potential of manipulating latent spaces as a tool for producing high-quality, semantically rich 3D models and suggest promising directions for future work, including integrating semantic factorization with other generative frameworks or exploring more complex shape categories.

Acknowledgments. We thank the anonymous reviewers for their valuable feedback. This work was partially supported by the Natural Sciences and Engineering Research Council of Canada (2022-04903).

References

1. Achlioptas, P., Diamanti, O., Mitliagkas, I., Guibas, L.: Learning representations and generative models for 3D point clouds. In: Proceedings of the International Conference on Machine Learning, pp. 40–49 (2018)
2. Arshad, M.S., Beksi, W.J.: A progressive conditional generative adversarial network for generating dense and colored 3D point clouds. In: Proceedings of the International Conference on 3D Vision, pp. 712–722 (2020)
3. Chang, A.X., et al.: ShapeNet: an information-rich 3D model repository. CoRR abs/1512.03012 (2015)

4. Chen, K., Choy, C.B., Savva, M., Chang, A.X., Funkhouser, T., Savarese, S.: Text2Shape: generating shapes from natural language by learning joint embeddings. In: Proceedings of the Asian Conference on Computer Vision, pp. 100–116 (2018)
5. Chen, Z., Zhang, H.: Learning implicit fields for generative shape modeling. In: Proceedings of the IEEE Conference on Computer Vision and Pattern Recognition, pp. 5932–5941 (2019)
6. Choi, J., Hwang, G., Cho, H., Kang, M.: Finding the global semantic representation in GAN through Fréchet mean. In: Proceedings of the International Conference on Learning Representations (2023)
7. Choi, J., Hwang, G., Cho, H., Kang, M.: Analyzing the latent space of GAN through local dimension estimation for disentanglement evaluation. Pattern Recogn. **157**, 110914 (2025)
8. Goetschalckx, L., Andonian, A., Oliva, A., Isola, P.: GANalyze: toward visual definitions of cognitive image properties. In: Proceedings of the IEEE International Conference on Computer Vision, pp. 5743–5752 (2019)
9. Goodfellow, I.J., et al.: Generative adversarial nets. In: Advances in Neural Information Processing Systems, pp. 2672–2680 (2014)
10. Härkönen, E., Hertzmann, A., Lehtinen, J., Paris, S.: GANSpace: discovering interpretable GAN controls. In: Advances in Neural Information Processing Systems, pp. 9841–9850 (2020)
11. Ho, J., Jain, A., Abbeel, P.: Denoising diffusion probabilistic models. In: Advances in Neural Information Processing Systems, pp. 6840–6851 (2020)
12. Hu, H., Jiang, B., Zhou, X., Huo, X., Zhang, B.: Unsupervised discovery of disentangled interpretable directions for layer-wise GAN. In: Proceedings of the CCF Conference on Big Data, pp. 21–39 (2022)
13. Jahanian, A., Chai, L., Isola, P.: On the "steerability" of generative adversarial networks. In: Proceedings of the International Conference on Learning Representations (2020)
14. Karras, T., Laine, S., Aila, T.: A style-based generator architecture for generative adversarial networks. In: Proceedings of the IEEE Conference on Computer Vision and Pattern Recognition, pp. 4396–4405 (2019)
15. Kingma, D.P., Ba, J.L.: Adam: a method for stochastic optimization. In: Proceedings of the International Conference on Learning Representations (2015)
16. Li, J., Niu, C., Xu, K.: Learning part generation and assembly for structure-aware shape synthesis. In: Proceedings of the AAAI Conference on Artificial Intelligence, pp. 11362–11369 (2020)
17. Li, R., Li, X., Hui, K.H., Fu, C.W.: SP-GAN: sphere-guided 3D shape generation and manipulation. ACM Trans. Graph. **40**(4), 151:1–151:12 (2021)
18. Lorensen, W.E., Cline, H.E.: Marching cubes: a high resolution 3D surface construction algorithm. In: Proceedings of SIGGRAPH, pp. 163–169 (1987)
19. Nitzan, Y., Gal, R., Brenner, O., Cohen-Or, D.: LARGE: latent-based regression through GAN semantics. In: Proceedings of the IEEE Conference on Computer Vision and Pattern Recognition, pp. 19217–19227 (2022)
20. Niu, Y., Zhou, P., Chi, H., Zhou, M.: Nonlinear hierarchical editing: a powerful framework for face editing. Eng. Appl. Artif. Intell. **135**, 108706 (2024)
21. Radford, A., Metz, L., Chintala, S.: Unsupervised representation learning with deep convolutional generative adversarial networks. In: Proceedings of the International Conference on Learning Representations (2016)

22. Schwettmann, S., Hernandez, E., Bau, D., Klein, S., Andreas, J., Torralba, A.: Toward a visual concept vocabulary for GAN latent space. In: Proceedings of the IEEE International Conference on Computer Vision, pp. 6784–6792 (2021)
23. Shen, Y., Zhou, B.: Closed-form factorization of latent semantics in GANs. In: Proceedings of the IEEE Conference on Computer Vision and Pattern Recognition, pp. 1532–1540 (2021)
24. Simsar, E., Kocasari, U., Er, E.G., Yanardag, P.: Fantastic style channels and where to find them: a submodular framework for discovering diverse directions in GANs. In: Proceedings of the IEEE Winter Conference on Applications of Computer Vision, pp. 4720–4729 (2023)
25. Song, J., Meng, C., Ermon, S.: Denoising diffusion implicit models. In: Proceedings of the International Conference on Learning Representations (2021)
26. Song, Y., Jia, X., Jia, J., Liang, Y., Liang, X.: 3D portrait stylization with adaptive semantic editing based on GAN latent codes. Pattern Recogn. **171**, 112069 (2026)
27. Song, Y., Zhang, J., Sebe, N., Wang, W.: Householder projector for unsupervised latent semantics discovery. In: Proceedings of the IEEE International Conference on Computer Vision, pp. 7678–7688 (2023)
28. Spingarn, N., Banner, R., Michaeli, T.: GAN "steerability" without optimization. In: Proceedings of the International Conference on Learning Representations (2021)
29. Tzelepis, C., Tzimiropoulos, G., Patras, I.: WarpedGANSpace: finding non-linear RBF paths in GAN latent space. In: Proceedings of the IEEE International Conference on Computer Vision, pp. 6373–6382 (2021)
30. Voynov, A., Babenko, A.: Unsupervised discovery of interpretable directions in the GAN latent space. In: Proceedings of the International Conference on Machine Learning, pp. 9786–9796 (2020)
31. Wang, H., Schor, N., Hu, R., Huang, H., Cohen-Or, D., Huang, H.: Global-to-local generative model for 3D shapes. ACM Trans. Graph. **37**(6), 214:1–214:10 (2018)
32. Wang, R., Liu, Y., Su, P., Zhang, J., Zhao, Q.: 3D semantic subspace traverser: empowering 3D generative model with shape editing capability. In: Proceedings of the IEEE International Conference on Computer Vision, pp. 14360–14371 (2023)
33. Wang, R., Wang, G., Gu, L., Liu, Q., Liu, Y., Guo, Y.: Intuitively interpreting GANs latent space using semantic distribution. Knowl.-Based Syst. **310**, 112894 (2025)
34. Wu, J., Zhang, C., Xue, T., Freeman, B., Tenenbaum, J.: Learning a probabilistic latent space of object shapes via 3D generative-adversarial modeling. In: Advances in Neural Information Processing Systems, pp. 82–90 (2016)
35. Wu, R., Zheng, C.: Learning to generate 3D shapes from a single example. ACM Trans. Graph. **41**(6), 224:1–224:19 (2022)
36. Xia, M., et al.: FEditNet: few-shot editing of latent semantics in GAN spaces. In: Proceedings of the AAAI Conference on Artificial Intelligence, pp. 2919–2927 (2023)
37. Yi, R., Hu, T., Xia, M., Tang, Y., Liu, Y.J.: FEditNet++: few-shot editing of latent semantics in GAN spaces with correlated attribute disentanglement. IEEE Trans. Pattern Anal. Mach. Intell. **46**(12), 9975–9990 (2024)
38. Yüksel, O.K., Simsar, E., Er, E.G., Yanardag, P.: LatentCLR: a contrastive learning approach for unsupervised discovery of interpretable directions. In: Proceedings of the IEEE International Conference on Computer Vision, pp. 14243–14252 (2021)
39. Zheng, X.Y., Liu, Y., Wang, P.S., Tong, X.: SDF-StyleGAN: implicit SDF-based StyleGAN for 3D shape generation. Comput. Graph. Forum **41**(5), 52–63 (2022)

40. Zhu, J., Shen, Y., Xu, Y., Zhao, D., Chen, Q.: Region-based semantic factorization in GANs. In: Proceedings of the International Conference on Machine Learning, pp. 27612–27632 (2022)

Author Index

Z. Zhang (Ed.): ISNN 2026, LNCS 16848, pp. 353–354, 2027.
https://doi.org/10.1007/978-981-92-5870-3

MIX
Papier aus verantwortungsvollen Quellen
Paper from responsible sources
FSC® C105338

If you have any concerns about our products,
you can contact us on
ProductSafety@springernature.com

In case Publisher is established outside the EU,
the EU authorized representative is:
Springer Nature Customer Service Center GmbH
Europaplatz 3, 69115 Heidelberg, Germany

Printed by Libri Plureos GmbH
in Hamburg, Germany